Studies in Autonomic, Data-driven and Industrial Computing

Series Editors

The book series Studies in Autonomic, Data-driven and Industrial Computing (SADIC) aims at bringing together valuable and novel scientific contributions that address new theories and their real world applications related to autonomic, data-driven, and industrial computing. The area of research covered in the series includes theory and applications of parallel computing, cyber trust and security, grid computing, optical computing, distributed sensor networks, bioinformatics, fuzzy computing and uncertainty quantification, neurocomputing and deep learning, smart grids, data-driven power engineering, smart home informatics, machine learning, mobile computing, internet of things, privacy preserving computation, big data analytics, cloud computing, blockchain and edge computing, data-driven green computing, symbolic computing, swarm intelligence and evolutionary computing, intelligent systems for industry 4.0, as well as other pertinent methods for autonomic, data-driven, and industrial computing.

The series will publish monographs, edited volumes, textbooks and proceedings of important conferences, symposia and meetings in the field of autonomic, data-driven and industrial computing.

More information about this series at http://www.springer.com/series/16624

Rajeev Mathur · C. P. Gupta · Vaibhav Katewa ·
Dharm Singh Jat · Neha Yadav
Editors

Emerging Trends in Data Driven Computing and Communications

Proceedings of DDCIoT 2021

Springer

Editors
Rajeev Mathur
Department of Electronics
and Communication Engineering
Geetanjali Institute of Technical Studies
Udaipur, India

C. P. Gupta
Department of Computer Science
and Engineering
Rajasthan Technical University
Kota, India

Vaibhav Katewa
Department of Electronics
and Communication Engineering
Indian Institute of Science Bangalore
Bengaluru, India

Dharm Singh Jat
Department of Computer Science
Namibia University of Science
and Technology
Windhoek, Namibia

Neha Yadav
Department of Mathematics
National Institute of Technology
Hamirpur, Himachal Pradesh, India

ISSN 2730-6437 ISSN 2730-6445 (electronic)
Studies in Autonomic, Data-driven and Industrial Computing
ISBN 978-981-16-3914-2 ISBN 978-981-16-3915-9 (eBook)
https://doi.org/10.1007/978-981-16-3915-9

This Springer imprint is published by the registered company Springer Nature Singapore Pte Ltd.
The registered company address is: 152 Beach Road, #21-01/04 Gateway East, Singapore 189721, Singapore

Preface

Data-Driven Computing is the process of computational analysis using available database in any form to derive predictive output. Data-driven computing needs to be explored more extensively with theories and principles other than mainstream computing. A structured database obtained by various processes can be used for further predictions after computational analysis and intelligent manipulation.

The next step is going to be Data-Driven Manufacturing/Industrial Computing, wherein industrial manufacturers will be able to make decisions based on the prediction after data-driven computing. Industries are expected to utilise Internet of Things, Artificial Intelligence, Machine Learning and other technologies to make manufacturing more automated, autonomic, smart and data-driven. Internet of Things (IoT) and the use of sensors in big data and analytics have evolved a new dimension for next generation manufacturing.

This Book on **Emerging Trends in Data Driven Computing and Communications—Proceedings of DDCIoT 2021** addresses design, development and algorithmic aspects of artificial intelligence, machine learning, deep learning, edge computing, communication and networking. A variety of peer-reviewed papers related to wide applications in various fields like industrial IoT, smart systems, healthcare systems, autonomous systems and various other aligned areas are included. The book is a compilation of papers presented in the International Conference on Data Driven Computing & IOT-DDCIoT-2021, Organised by Department of Electronics and Communications and Department of Computer Science Engineering of Geetanjali Institute of Technical Studies, Udaipur, and Rajasthan Technical University, Kota, during 20–21 March 2021. The Conference was sponsored by AICTE and TEQIP-III, India.

The conference was focused to encourage various scholastic revolutionary researchers, scientists and industrial engineers from all around the world and provides them a platform to present the proposed new technologies, share their experiences and discuss emerging trends in Data-Driven and Industrial Computing and IoT-based Smart Systems

In the conference, 130 papers were received; we have selected 30 papers on the basis of blind reviews, quality and originality of the work and research study. All the papers have been carefully reviewed.

This book could be referred to by academicians, researchers, programmers, system infrastructure designers and industrial hardware developers. This book could also be very useful for manufacturers, entrepreneurs and investors.

Udaipur, India — Prof.(Dr.) Rajeev Mathur
Windhoek, Namibia — Prof. Dharm Singh Jat
Kota, India — Dr. C. P. Gupta
Bengaluru, India — Dr. Vaibhav Katewa
Hamirpur, India — Dr. Neha Yadav
May 2021

Contents

1 **Hybrid Deep Learning Model for Real-Time Detection of Distributed Denial of Service Attacks in Software Defined Networks** 1
Auther Makuvaza, Dharm Singh Jat, and Attlee M. Gamundani

2 **CP-ABE Scheme with Decryption Keys of Constant Size Using ECC with Expressive Threshold Access Structure** 15
Rakshit Kothari, Naveen Choudhary, and Kalpana Jain

3 **Facial Expressions Recognization Based on DRLBP and CNN Classification** 37
Surbhi Singh, Akhil Pandey, and Vishal Shrivastav

4 **Post-Stroke Readmission Prediction Model Using Machine Learning Algorithms** 53
Lavanya Kommina, Prasannavenkatesan Theerthagiri, Yogeshwari Payyavula, Prudvi Sai Vemula, and Gurram Deekshith Reddy

5 **Handwriting Recognition Using Deep Learning** 67
Shubh, Muskan Agarwal, Fardin Hassan, Garima Pandey, and Soumalya Ghosh

6 **A Theoretical Approach to Reinforcement Learning Algorithm and Nash Equilibrium to Maximize the Score in Imperfect Information Game** 83
Ruchi Vyas and Alok Singh Gahlot

7 **A Review Paper on Sign Language Recognition Using Machine Learning Techniques** 91
Ritesh Kumar Jain and S. K. Rathi

8 **Deep Learning Methods for the Prediction of Chronic Diseases: A Systematic Review** 99
Gunjan Sahni and Soniya Lalwani

9 Cyberinfrastructure for Advanced Research with High Performance Edge Computing 111
Dharm Singh Jat, Arpit Jain, and Kamwayamunzi Fortune Sepo

10 A Literature Review on Machine Learning Techniques and Strategies Applied to Stock Market Price Prediction 121
Pankaj Rambhau Patil, Deepa Parasar, and Shrikant Charhate

11 A Comparative Analysis of ML Stratagems to Estimate Chronic Kidney Disease Predictions and Progression by Employing Electronic Health Records 137
Shruti Jain, Mayank Patel, and Konika Jain

12 Application of Machine Learning Algorithms in Agriculture: An Analysis 153
Kalpana Jain and Naveen Choudhary

13 Combined DCT-DWT Color Image Digital Watermarking Technique 169
Devender Kumar Dhaked

14 A Performance Analysis of Face and Speech Recognition in the Video and Audio Stream Using Machine Learning Classification Techniques 179
Roshni Wadhwani, Akhil Pandey, and Vishal Shrivastav

15 Comparative Analysis of AFE Cardiac Amplifiers Used in IOT Application 195
Tina Bhati, Mohammad Sabir, Vijendra K. Maurya, and Latif Khan

16 Design and Analysis of Multi-band Fractal Antennas for Wireless Communications 203
Gaurav Kumar Soni, Sonam Gour, and Abha Sharma

17 A Summary on Neurodegenerative Disease Detection 213
Kavita Mundra and Manju Mandot

18 An Ensemble Approach for Classification and Prediction of Diabetes Mellitus Disease 225
Dhyan Chandra Yadav and Saurabh Pal

19 A Novel Approach for Identification of Brain Tumor by Combination of Intelligent Water Drop Algorithm and Convolutional Neural Network 237
Ashish Kumar Dehariya and Pragya Shukla

20 Time Restricted Frame Aggregation for Better QoS in IEEE 802.11ac 247
Mayank Patel, Saurabh Srivastava, Harshita Jain, and Milind D. Jain

21 Self-Powered Intelligent Wastewater Treatment System with IOT-Based Monitoring for Smart Cities 255
Mohammad Ateek Samma, Shubham Vagrecha, Rajeev Mathur, Mr. Latif Khan, Lokesh Sharma, and Sheetal Sharma

22 IoT-based Smart Ultraviolet Disinfecting and Sterilizing Robot for Public Places 267
Ashmit Dabi, Darshan Seth, Divya Soni, Siddhi Chaplot, Khushi Sharma, Latif Khan, and Rajeev Mathur

23 IoT-based Intelligent Intravenous (IV) Stand for Healthcare Sector 275
Mohammad Ateek Samma, Shubham Vagrecha, Mr. Latif Khan, Rajeev Mathur, Lokesh Sharma, and Sheetal Sharma

24 Integrating Lean Green Technique Through Single Combined Strategy in Marble Industry 281
Saurabh Tege and Nilesh Diwakar

25 Industrial Internet of Things: A Survey 291
Uma Sharma, Meenakshi Sharma, Ramendra Singh, Ajay Chauhan, and Pooja Pathak

26 Analysis and Measurement of BER and SNR for Different Block Length in AWGN and Rayleigh Channel 299
Sonam Gour, Gaurav Kumar Soni, and Abha Sharma

27 Software Development Life Cycle Performance Analysis 311
Shilpa Sen, Mayank Patel, and Ajay Kumar Sharma

28 Scrum Framework Based on Agile Methodology in Software Development and Management 321
Upasana Ameta, Mayank Patel, and Ajay Kumar Sharma

29 A Comparative Study and Tool to Early Predict Diabetes Using Various Machine and Deep Learning Based Techniques 333
Shrusti Porwal, Chintal Kumar Patel, Deepa Tilwani, and Shri Krishn Bansal

30 Single-End Current-Based Scheme for Detection and Classification of Distant-End and Adjacent-End Faults in Transmission Lines 345
Asifa, Shoyab Ali, and Gaurav Kapoor

Author Index 353

About the Editors

Dr. Rajeev Mathur is serving as Professor and Dean in Geetanjali Institute of Technical Studies, Udaipur, Rajasthan, India. He received his Ph.D. from Jaipur, Masters in Engineering from NITTTR, Chandigarh, India, and B.E. degree in Electronics from VRCE (now VNIT), Nagpur University, in 1991. He joined industry Punjab Wireless System Ltd. Mohali, Chandigarh, and was working as City Manager for 6 Years. He served as Assistant Professor, Associate Professor, and Head at Engineering Institute at Alwar, Jodhpur. He was Founder Member, Head & Associate Prof. in Faculty of E & T, "Jodhpur National University," Jodhpur and Member of Board of management of the University. He was Director & Principal of Marwar Engineering College & Research Centre, Jodhpur, Rajasthan, India. He is "CMI level 5 certified in Management & Leadership by CMI Royal Chapter, UK." He is Senior Member of professional bodies IEEE, Charted Management Institute-UK, ATMS, and Lions International. He has published more than 50 papers in international journals including SCI/SCOPUS indexed journals. He also published 3 book chapters. He had been Keynote Speaker, Advisor in many conferences & Editor, and Reviewer of many international journals & international conferences. He also have published one patent in his name. His research interests are IoT, data driven computing, advanced antenna techniques, metamaterials, and evaluating nature of electromagnetic waves.

Dr. C. P. Gupta is presently serving at Rajasthan Technical University, Kota, Rajasthan, India, as Associate Professor. He pursued his Ph.D. from University of Kota, Kota in Wireless Sensor Network in 2014, Master of Technology from Indian Institute of Technology, Delhi, India, and Engineering degree from Malviya National Institute of Technology, Jaipur, India, in 1990. His research interests include computer networks, information, security, and wireless sensor networks. He is also an ISO 27000 certified professional. He has published several papers in international & national journals and international conferences. He is appointed as Chief Editor in Seventh International Conference on Recent Trends in Information, Telecommunication and Computing. He is Member of professional society INSTICC.

Dr. Vaibhav Katewa is working as Assistant Professor, Department of Electrical Communication Engineering, and Associate Faculty, Robert Bosch Center for Cyber-Physical Systems, Indian Institute of Science, Bangalore, India. He has also served as Post-Doctoral Scholar in University of California Riverside, USA, in Dec. 2019, as Research Assistant, Department of Electrical Engineering, University of Notre Dame, USA, from May 2010 to Aug. 2015, and as Research Intern in Department of Electrical and Electronic Engineering at University College Cork, Ireland, from May 2006 to July 2006. He also has 3 years of industrial experience as Research Engineer in Ethernet Development Tejas Networks, Bangalore, India, and Senior Research Associate, Intellectual Property Division at Evalueserve, Delhi, India. He pursued his Ph.D. in Electrical Engineering from University of Notre Dame, USA, in 2016, M.S. in Electrical Engineering in May 2012 from University of Notre Dame, USA, and B.Tech in Electrical Engineering from Indian Institute of Technology Kanpur, India, in 2007. He has published more than 35 papers in international journals of repute and two book chapters. He is Reviewer for many conferences and journals and is Guest Editor in Special Issue on "Swarm Intelligence Techniques for Optimum Utilization of Resources in Grid and Utility Computing," International Journal of Grid and Utility Computing, 2019, and Special Issue on "Smart Computing and Optimization," Journal of Information, and Optimization Sciences, 38(6), 2017.

Prof. Dharm Singh Jat is Professor, Department of Computer Science at Namibia University of Science and Technology (NUST), NAMIBIA. He is the author of more than 150 peer-reviewed articles and the author or editor of more than 20 books. He has been the recipient of more than 19 prestigious awards, such as Eminent Scientist Award, Distinguished Academic Achievement, Eminent Engineering Personality, CSI Chapter Patron, CSI Significant Contribution, Best Faculty Researcher, Best Technical Staff, Outstanding University Service Award, and Distinguished ACM Speaker award. Prof. Dharm Singh is Fellow of The Institution of Engineers (I), Fellow of Computer Society of India, Chartered Engineer (I), Senior Member IEEE, and Distinguished ACM Speaker.

Dr. Neha Yadav received her Ph.D. in Mathematics from Motilal Nehru National Institute of Technology, (MNNIT) Allahabad, India, in the year 2013. She completed her post-doctorate from Korea University, Seoul, South Korea. She is a receiver of Brain Korea (BK-21) postdoctoral fellowship given by Government of Republic of Korea. Prior to joining NIT Hamirpur, she taught courses and conducted research at BML Munjal University, Gurugram, Korea University Seoul, S. Korea, and The NorthCap University, Gurugram. Her major research area includes numerical solution of boundary value problems, artificial neural networks, and optimization. She has published more than 15 journal articles and also authored books and editorials.

Chapter 1
Hybrid Deep Learning Model for Real-Time Detection of Distributed Denial of Service Attacks in Software Defined Networks

Auther Makuvaza, Dharm Singh Jat, and Attlee M. Gamundani

Abstract The growth of network devices has brought a lot of problems in managing the networks. The ill-managed networks create different vulnerabilities which attackers can exploit. The attackers take advantage of open-source tools and low-priced Internet to use the networks. Software Defined Networking (SDN) is a good networking architecture that can be managed centrally. The decoupled SDN architecture has the flexibility of programming network devices from the central controller. There is no doubt that SDN addresses the problem of network management; however, SDN comes with a security concern. SDN controller has a vulnerability of a single point of failure. This vulnerability makes the controller vulnerable to different network attacks, including Distributed Denial of Service (DDoS) attacks, among others. To get the best out of SDN, the controller needs security that can protect it from cyber-attacks. The Deep Learning (DL) approach enhanced the selection of the relevant features from the dataset for classification in an unsupervised manner. This paper proposed the hybrid DL model that utilises Long Short-Term Memory (LSTM) and Convolutional Neural Network (CNN) for DDoS attack detection. The proposed hybrid model produced a detection accuracy of 99.72%.

Keywords CNN · Long short-term memory · IDS · Deep learning · DDoS · SDN

1 Introduction

As new technology redefines the digital world, attackers also have their attacking methods by using multiple attacking vectors [1]. Novel technology has changed the human lifestyle and security landscape; this change threatens the security of different

A. Makuvaza (✉) · D. S. Jat · A. M. Gamundani
Namibia University of Science and Technology, Windhoek, Namibia

D. S. Jat
e-mail: dsingh@nust.na

A. M. Gamundani
e-mail: agamundani@nust.na

R. Mathur et al. (eds.), *Emerging Trends in Data Driven Computing and Communications*, Studies in Autonomic, Data-driven and Industrial Computing,
https://doi.org/10.1007/978-981-16-3915-9_1

networks and other Information Technology services. According to [2], the privacy challenges come with novel technologies and can endanger and disturb everyday activities in the networking environment. The legacy network lacks the scalability, which is required by the ever-growing networks; however, SDN has proven to be the future of networking because of its centralised controller which is used to program the whole network. The decoupled SDN architecture has attracted a lot of cyber-attacks [3]. Cyber-attackers target the SDN controller to bring the network down because of the single point failure vulnerability.

With the advancement of technology, DDoS attacks can be launched by different computers within a botnet. Attackers can attack the SDN controller by launching DDoS attacks. According to [1], DDoS attacks are currently the most sophisticated network threat to organisations, and they are difficult to detect and prevent. According to [4], attackers are now using multi-vector attacks to attack the complex network. Attackers can automatically or dynamically change the attacking vectors based on the defence mechanisms they encounter during the attack. According to [4], multi vector attacks are increasing by 13% every year, which brings the debate on which DDoS attack detection mechanism organisations should use. Figure 1 below shows the relationship between artificial intelligence, machine learning and deep learning.

Intrusion Detection Systems (IDS) are used to detect network attacks; however, to detect anomaly-based attacks, the IDS should be trained, tested and evaluated. Cybersecurity is the leading driving force of Artificial intelligence [5]. Deep learning has proven to be one of the best methods cybersecurity officers can use to detect DDoS attacks in SDN. However, DL cannot do the job alone, a flow-based dataset is needed for the SDN environment. CICIDS 2017 dataset is a flow-based dataset, and it has more than 11 different network attacks, with DDoS included.

Deep learning models come with problems of false positives and false negatives. A model can incorrectly predict a positive and negative class [6]. To get the correct accuracy from deep learning models, accuracy should be measured by comparing the model results with the ground truth. According to [1], confusion matrix and standard evaluation parameters can be used to measure the accuracy of a model. This paper will discuss the background of the study and related work, methodology, results analysis and conclusion.

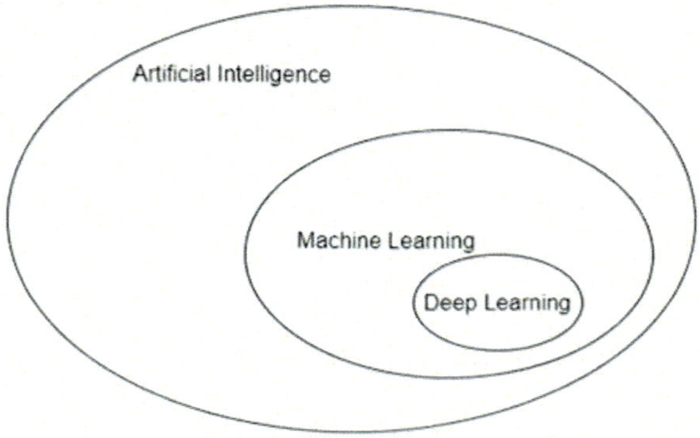

Fig. 1 Relationship between artificial intelligence, machine learning and deep learning

1.1 Background and Related Work

Software-defined networks have changed the game of networking with their flow-based protocol and decoupled architecture [7]. Previously, intrusion detection systems were implemented based on traditional network protocol which is the Internet Protocol (IP). With SDN using OpenFlow protocol, the IP-based intrusion detection systems faced some challenges in detecting DDoS attacks in the SDN environment because of the volume traffic and attack vectors [8]. SDN simplified networking by having a centralised controller where all devices can be programmed and configured from one point [2]. According to [9], recent cybersecurity experts and researchers have designed anomaly intrusion detection systems to detect various network attacks. This direction has seen neural networks being used in this direction of intrusion detection systems.

According to [10], the best way to protect software-defined networks is through intrusion detection systems. Deep learning intrusion systems can provide the last line of defence to software-defined networks. They went on and stated that CNN intrusion systems produce an accuracy of 95% when detecting DDoS attacks in SDN. Reference [11] state that SDN architecture attracts a lot of cyber-attackers because of the vulnerability of the controller. However, an anomaly detection method was used to detect DDoS attacks in SDN. Reference [12] proposed DNN deep learning model for DDoS attack detection in SDN. The proposed DNN intrusion detection system produced an accuracy of 87%.

1.2 Deep Learning

Deep learning has emerged as an effective approach that enables the use of datasets to detect network intrusions [13]. According to [14], deep learning approach has become popular in the field of cybersecurity. Studies have shown that deep learning-based intrusion detection system has surpassed tradition methods [14]. According to [15], deep learning has received a lot of research attention in different fields, including network security. Deep learning has the potential to secure computer networks and information systems [15]. According to [16], traditional machine learning security mechanism faced some difficulties in detecting DDoS attacks due to the high volume of network traffic; however, deep learning has come as a solution to machine learning problems and has shown success in different big data sections.

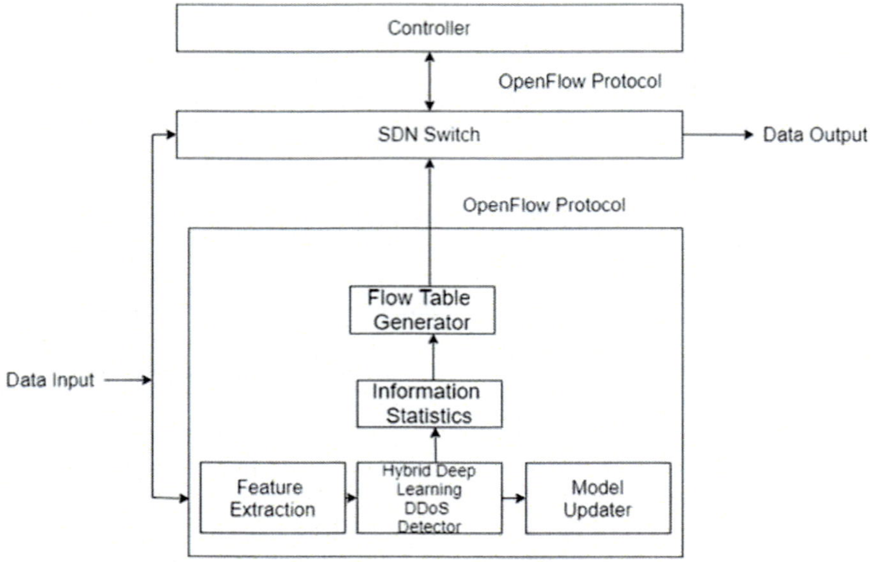

Fig. 2 IDS placement

Figure 2 shows the proposed model placement and traffic flow of the SDN.

Figure 2 shows how the hybrid deep learning IDS was used to detect DDoS attacks in SDN. The communication between the controller and SDN switch was facilitated by OpenFlow protocol. The model has two inputs, one for feature extraction and one for SDN switch. The model has one data output at the SDN switch.

2 Methodology

Feature selection, data distribution, training and testing the classifier and evaluation will be discussed in this section. The proposed model used CICIDS 2017 dataset. The study used literature review [1] to select four best features out of 86 features. Figure 4 shows the hybrid deep learning model block diagram.

Figure 3 shows the proposed model flow diagram. The first step is training, then followed by testing. On the training side, the first step is normalisation. Training of the model follows normalisation. During training, they are forward propagation and weight updates. After training, the model is then tested to detect DDoS attacks.

Figure 3 shows the proposed model flow diagram. The first step is training, then followed by testing. In the training side, the first step is normalisation. Training of the model follows normalisation. During training, they are forward propagation and weight updates. After training, the model is then tested to detect DDoS attacks.

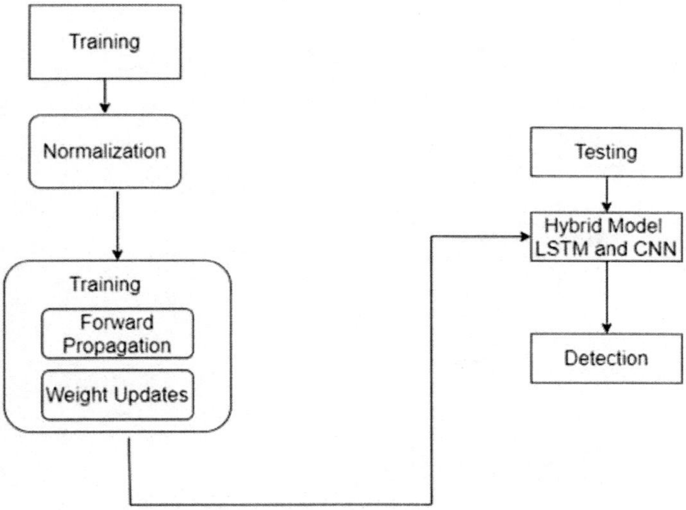

Fig. 3 LSTM and CNN block diagram

For the proposed model to accurately detect DDoS attacks, only relevant features were used to train and test a hybrid DDoS intrusion detection system. According to [1], the following features are the four best features used to build the classifier: [Backward Packet Length Standard Deviation, Flow Duration, Average Packet Size, Forward Inter Arrival Time Standard Deviation].

2.1 Data Distribution

The dataset distribution of the proposed algorithm used an 80:20 ratio, where 80% is the training set and 20% is the testing set.

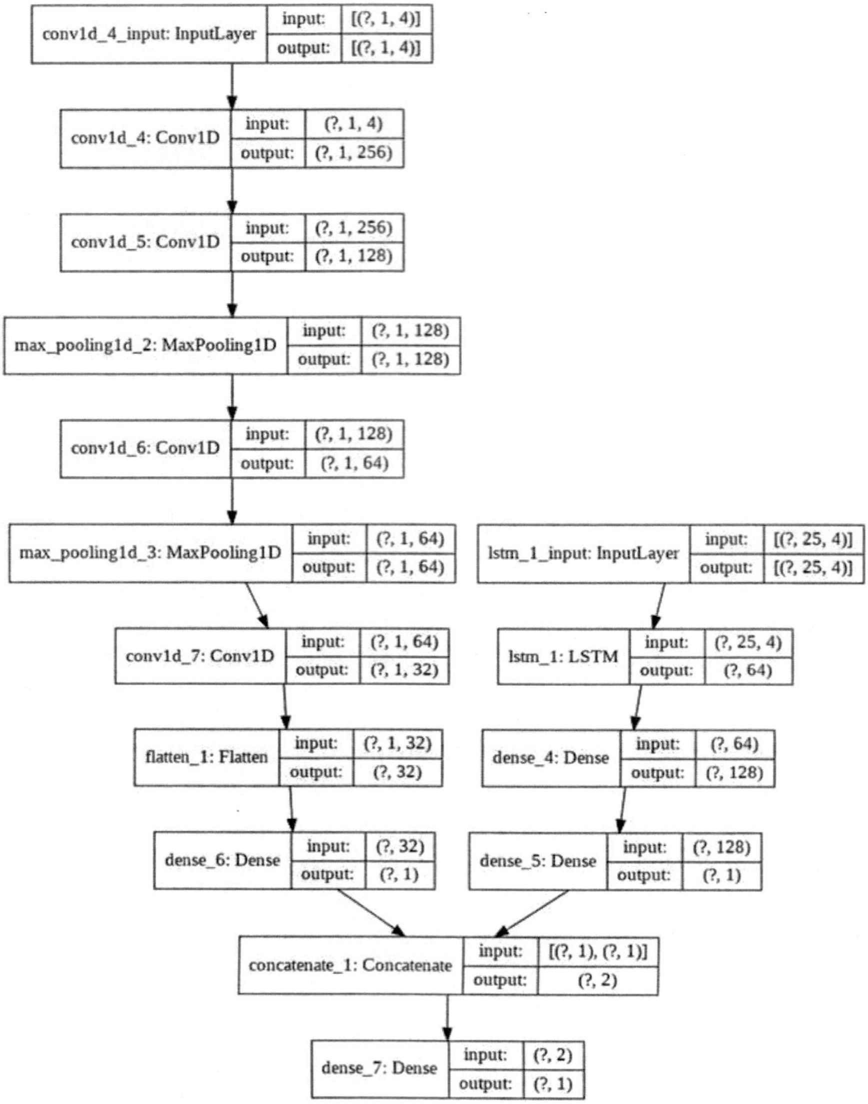

Fig. 4 Proposed model semantic

2.2 *Algorithm*

Input: Four best features

Splitting (Training set 0.8 and Testing data 0.2

Train LTM model

Train CNN model

Merge the predictions from LSTM and CNN

Test the model

Decision using decision tree

Evaluation

Benign traffic

Else

DDoS traffic

2.3 *Proposed Deep Learning Model*

The study proposed a hybrid deep learning model of LSTM and CNN for real-time DDoS attack detection in SDN. Below is the layout of the proposed model (LSTM and CNN).

The hybrid LSTM and CNN model has five layers; the input layer has an input of 64 neurons for CNN and four input neurons for LSTM. CNN has an output of 64 neurons and LSTM has an output of four neurons. The output of CNN, which is 64 neurons, is loaded into the second layer, which produced an output of 32, and LSTM produced an output of 64 neurons. Thirty-two neurons of CNN are then loaded into the next hidden layer, which produced an output of 32 neurons, and the input of LSTM 64 neurons produced 128 neurons. The fourth layer of CNN moved the output of 32 neurons to the next layer, which produced one and 128 neurons of LSTM, then moved to the next hidden layer, which produced the output of one. At this stage, the two models were then joined to form a hybrid model. The input is one and the hybrid produced the result, which is either 0 or 1, where 0 is benign traffic, and 1 is DDoS traffic.

2.4 Evaluation

The study used accuracy, precision, recall and F1-Score as standard evaluation parameters of the hybrid DL distributed denial of service attack detection in SDN.

Accuracy—is used to find the portion of correctly classified values, and it tells how often the classifier is suitable [17].

Precision—The ability of the model to classify positive values correctly [17].

Recall—used to calculate the ability of the model to predict positive values. It shows how often the model predicts the correct positive values [17].

F1-Score—is the average of recall and precision. It is useful when both recall and precision are being used [17].

3 Results

The results section presents and interprets the results of the experiment. Accuracy, Precision, Recall and F1-Score are calculated as:

Accuracy (%)

$$Accuracy = \frac{TP + TN}{TP + TN + FP + FN} * 100$$

Accuracy = 99.72%

Precision (%)

$$Precision = \frac{TP}{TP + FP} * 100$$

$$\boldsymbol{Precision = 99.75\%}$$

Recall (%)

$$Recall = \frac{TP}{TP + FN} * 100$$

$$\boldsymbol{Recall = 99.82\%}$$

F1-Score (%)

$$F1Score = \frac{2 * TP}{2 * TP + FN} * 100$$

$$\boldsymbol{F1Score = 99.9\%}$$

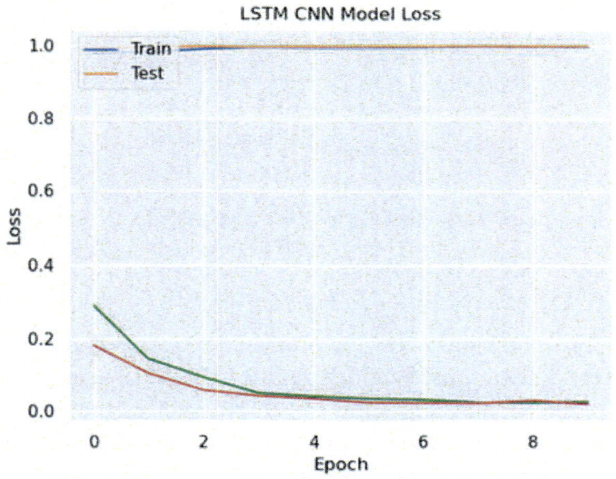

Fig. 5 Proposed model loss

3.1 *Loss*

Figure 5 shows the epoch vs loss graph for the proposed model. The training loss of the proposed model is 0.3, and the test loss of the proposed model is 0.2.

3.2 *Accuracy*

The accuracy of the proposed model is shown in Fig. 6, and it was compared with models which are already in the market.

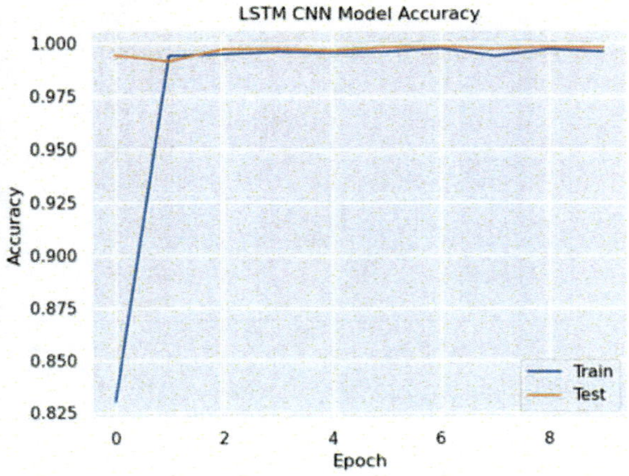

Fig. 6 Accuracy of the proposed model

Table 1 Accuracy comparison table

Model	Accuracy %	Reference
LSTM and CNN (Proposed model)	99.72	–
RNN	68.55	[14]
GRU and RNN	89	[18]
RNN	84	[19]
KNN	98.3	[20]
CNN	97.2	[21]

Figure 6 shows how the proposed model performed in detecting DDoS attacks in SDN. The proposed model produced 99.72% accuracy.

Table 1 and Fig. 7 show the comparison of the proposed model and other existing models. The proposed model (LSTM and CNN) achieved 99.72% accuracy, which is better than existing models.

Figure 8 shows the confusion matrix of the proposed model. The model detects 12,724 instances classified as true positive for benign attacks and 58 instances misclassified as DDoS attacks. The model detects 41 instances of misclassified false positive for DDoS as benign attacks, and the model detects 23,172 instances classified as benign attacks.

Figure 9 shows the value of the evaluation parameters of the proposed model. The proposed model achieved an accuracy of 99.72%, precision 99.73%, recall 99.52% and F1-score 99.9%.

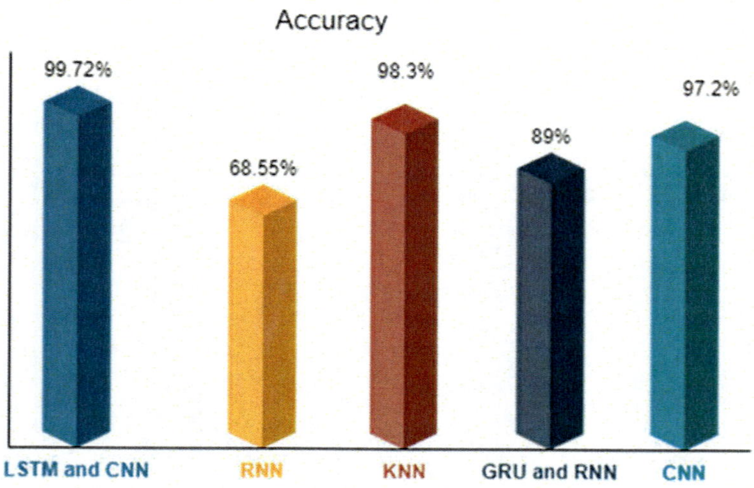

Fig. 7 Comparison graph

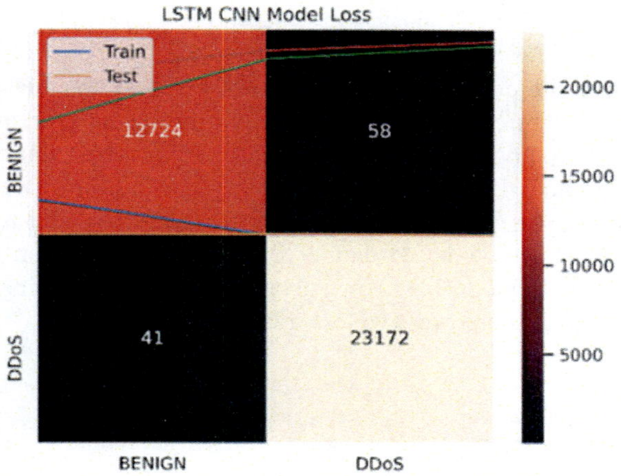

Fig. 8 Confusion matrix of the proposed model

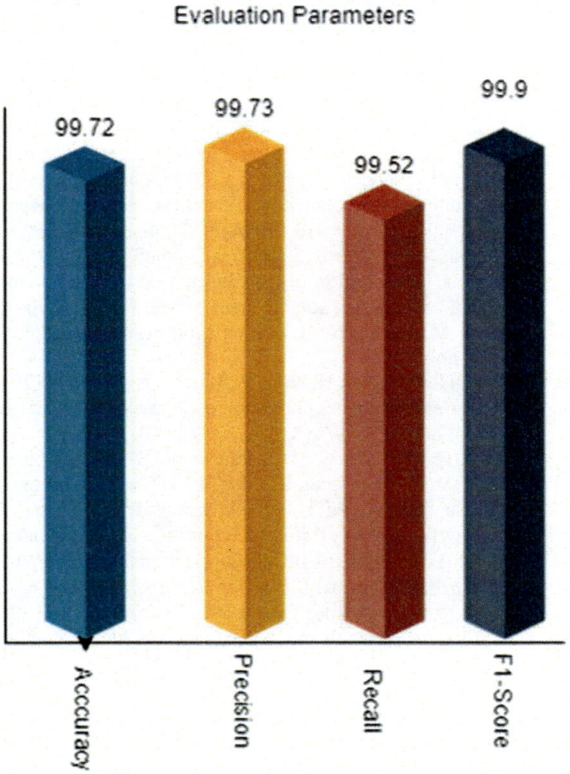

Fig. 9 Evaluation parameters of the proposed model

4 Conclusion

Software-defined networks have shown great potential in changing the networking architecture. Its decoupled architecture has become a significant target of cyber-attacks. To protect SDN architecture from cyber-attacks, the proposed hybrid deep learning DDoS intrusion detection system has shown a strong ability to detect DDoS attacks in SDN. The model is cost-effective and has a high accuracy detection rate. The flow-based dataset CICIDS 2017 has been utilised in this model. This proposed model has shown great potential by producing 99.72% accuracy using less computation power and less training time.

References

1. Haider S et al (2020) A deep CNN ensemble framework for efficient DDoS attack detection in software defined networks 8
2. Haider S, Akhunzada A, Ahmed G, Raza M (2019) Deep learning based ensemble convolutional neural network solution for distributed denial of service detection in SDNs. In: 2019 UK/China Emerging Technologies, pp 1–4. https://doi.org/10.1109/UCET.2019.8881856
3. Ujjan RMA, Pervez Z, Dahal K (2019) Suspicious traffic detection in SDN with collaborative techniques of snort and deep neural networks. In: Proceedings—20th international conference on high performance computing and communications; 16th international conference on smart city; 4th international conference on data science and systems HPCC/SmartCity/DSS 2018, pp 915–920. https://doi.org/10.1109/HPCC/SmartCity/DSS.2018.00152
4. Dimolianis M, Pavlidis A, Kalogeras D, Maglaris V (2019) Mitigation of multi-vector network attacks via orchestration of distributed rule placement. In: 2019 IFIP/IEEE symposium on integrated network and service management IM 2019, vol 731122, no 731122, pp 162–170
5. Learning M, Cookbook R Title: a short review on applications of deep learning for cyber security, no Dl
6. Hwang RH, Peng MC, Huang CW, Lin PC, Nguyen VL (2020) An unsupervised deep learning model for early network traffic anomaly detection. IEEE Access 8(February):30387–30399. https://doi.org/10.1109/ACCESS.2020.2973023
7. Li C et al (2018) Detection and defense of DDoS attack–based on deep learning in OpenFlow-based SDN. Int J Commun Syst 31(5):1–15. https://doi.org/10.1002/dac.3497
8. Santos Da Silva A, Wickboldt JA, Granville LZ, Schaeffer-Filho A (2016) ATLANTIC: a framework for anomaly traffic detection, classification, and mitigation in SDN. In: Proceedings NOMS 2016–2016 IEEE/IFIP network operations and management symposium, no Noms, pp 27–35. https://doi.org/10.1109/NOMS.2016.7502793
9. Garg S, Kumar N, Rodrigues JJPC, Rodrigues JJPC (2019) Hybrid deep-learning-based anomaly detection scheme for suspicious flow detection in SDN: a social multimedia perspective. IEEE Trans Multimed 21(3):566–578. https://doi.org/10.1109/TMM.2019.2893549
10. Čeponis D, Goranin N (2019) Evaluation of deep learning methods efficiency for malicious and benign system calls classification on the AWSCTD. Secur Commun Netw 2019. https://doi.org/10.1155/2019/2317976
11. Hafizah S et al (2018) A review of anomaly detection techniques and distributed denial of service (DDoS) on software defined network (SDN). Technol Appl Sci Res 8(2):2724–2730. https://www.researchgate.net/publication/324830666
12. Dey SK, Rahman MM (2018) Flow based anomaly detection in software defined networking: a deep learning approach with feature selection method. In: 2018 4th international conference

on electrical engineering and information & communication technology, pp 630–635. https://doi.org/10.1109/CEEICT.2018.8628069.
13. Karatas G, Demir O, Koray Sahingoz O (2019) Deep learning in intrusion detection systems. In: 2018 international congress on big data, deep learning and fighting cyber terrorism, pp 113–116. https://doi.org/10.1109/ibigdelft.2018.8625278
14. Yin C, Zhu Y, Fei J, He X (2017) A deep learning approach for intrusion detection using recurrent neural networks. IEEE Access 5:21954–21961. https://doi.org/10.1109/ACCESS.2017.2762418
15. Fadlullah Z, Tang F, Mao B, Kato N et al (2017) State-of-the-art deep learning: evolving machine intelligence toward tomorrow's intelligent network traffic control systems. Surv Tutorials 19(4):2432–2455. https://www.researchgate.net/profile/Imed_Romdhani/post/can_we_use_deep_learning_concepts_in_IoT_security_projects/attachment/5a60c3cbb53d2f0bba4c2b92/AS:584181025361920@1516291018882/download/State-of-the-Art+Deep+Learning+Evolving+Machine+Intelligence+T
16. Diro AA, Chilamkurti N (2018) Distributed attack detection scheme using deep learning approach for Internet of Things. Futur Gener Comput Syst 82:761–768. https://doi.org/10.1016/j.future.2017.08.043
17. Malik J, Akhunzada A, Bibi I, Imran M, Musaddiq A, Kim SW (2020) Hybrid deep learning: an efficient reconnaissance and surveillance detection mechanism in SDN. IEEE Access 8:134695–134706. https://doi.org/10.1109/ACCESS.2020.3009849
18. Tang TA, McLernon D, Mhamdi L, Zaidi SAR, Ghogho M (2019) Intrusion detection in sdn-based networks: deep recurrent neural network approach. IEEE
19. Jakhar K, Hooda N (2019) Big data deep learning framework using keras: a case study of pneumonia prediction, pp 1–5. https://doi.org/10.1109/ccaa.2018.8777571
20. Polat H, Polat O, Cetin A (2020) Detecting DDoS attacks in software-defined networks through feature selection methods and machine learning models. Sustainability 12(3). https://doi.org/10.3390/su12031035
21. Hojjatinia S, Hamzenejadi S, Mohseni H (2019) Android botnet detection using convolutional neural networks, pp 1–6

Chapter 2
CP-ABE Scheme with Decryption Keys of Constant Size Using ECC with Expressive Threshold Access Structure

Check for updates

Rakshit Kothari, Naveen Choudhary, and Kalpana Jain

Abstract With the rapid development of cross-organizational application systems that are geographically distributed, the notion of Virtual Organization (VO) is indispensable. The inevitable existence of global information infrastructure in every field has forced virtual organizations to gain importance as a fitting model for making a large-scale organization of distributed nature. Virtual organization mainly deals with the devolution of responsibilities to other organizations and providing goods and services by having mutual cooperation among organizations. An ancient mechanism is essential to have controlled access to shared resources and proper participation policies. However, this controlled access is challenging in the case of VO because of its distributed nature. Thus, a mechanism which can handle complex access policies is needed. The mechanism is also required to be reasonably scalable and efficient. In this paper, we propose a threshold access structure implemented using Elliptic Curve Cryptography, which is much better in terms of efficiency. Our main contribution is that we propose a threshold access structure that uses ECC and provides constant-size secret keys for the Ciphertext Policy Attribute-based Encryption (CP-ABE) scheme. In essence, we suggest a formal procedure to share secret information using encryption, where secret messages are associated with a policy, and only the users who have this specific attribute set that fulfils the specified policy will be apt to successfully decrypt and gain access to the secret message/information.

Keywords Threshold access structure · Attribute-based encryption · Secret keys of constant size · Ciphertext policy · Distributed system security · Virtual organization · Elliptic curve cryptography

R. Kothari (✉) · N. Choudhary · K. Jain
Department of Computer Science and Engineering, College of Technology and Engineering, MPUAT, Udaipur, India
e-mail: kalpana_jain2@rediffmail.com

R. Mathur et al. (eds.), *Emerging Trends in Data Driven Computing and Communications*, Studies in Autonomic, Data-driven and Industrial Computing,
https://doi.org/10.1007/978-981-16-3915-9_2

1 Introduction

Today, distributed computing systems, deployed over the large geographical area, are used as a model to achieve a single task by many different organizations working together with mutual cooperation. These characteristics can be appropriately sketched by Virtual Organization (VO). Virtual organizations can be thought of as both organizationally and geographically distributed organizations [11] which can aptly satisfy the needs like cooperative resource sharing. Virtual organizations can also be thought of as a model for designing distributed systems for an organization that requires complex resource sharing rules, expressive access structure and tangled inter-process interaction. However, the task of defining and enforcing the access scheme is hard because each organization participating in VO can delegate a large number of representatives, who in turn are at different hierarchical levels. Because of accountability and liability issues, the access of users to shared resources should be controlled with a proper access structure, which is flexible enough to facilitate the required functionality. Therefore, the control over sharing of resources using proper and flexible access structure is an extremely important and challenging task in case of VO.

As an illustration, we present a scenario in which VO is an appropriate model to achieve required goals. The outbreak of Coronavirus in the World led to the establishment of epidemiological and virological surveillance and controlled task forces to combat and find the source of infection. In order to control and combat the spreading of this infection, it is required to closely monitor all the key factors that led to the spreading of this infection. For example, it is mandatory to keep track of records of patients who have COVID-19 symptoms, we may also need to retrieve the immigration records of such patients to exactly locate the origin point of infection. COVID-19 is lethal which spreads through droplets generated when the COVID-19 patient coughs, sneezes or speaks. In order to control and combat the spreading of this infection, it is required to closely monitor all the key factors that led to its spreading.

An appropriate encryption scheme is CP-ABE in which the access policy is associated with the plaintext message by the encrypter and only the users who fulfil the specified policy will be able to decipher the ciphertext. The users are related with attributes based on which organization they belong to and what their designation is in that organization. The policy is also specified by a set of attributes. Hence, CP-ABE is acutely appropriate and suitable for implementing controlled resource sharing in Virtual Organizations because it enables the encrypter to decide and impose access policy while encrypting confidential or sensitive information.

1.1 Related Work

In the literature, ABE [25] is an important encryption scheme given by Sahai and Waters, which can be applied to any role-based system to provide data confidentiality

[10]. Identity-based Encryption [5], in particular, changed the classical way of thinking for public key encryption by letting the public key be some string associated with the identity of a receiver, for example, the email address of the receiver. There exist many identity-based encryption schemes that provide shorter keys for decryption, for example, multi-identity single-key decryption [14, 15, 17] and identity-based encryption with traitor tracing [16].

Sahai-Water's seminal work [25] was followed by the development of several KP-ABE schemes [1, 12, 24, 25] and CP-ABE schemes [3, 7, 10, 18, 21, 22, 25, 26]. CP-ABE permits the user to decide the policy of access during execution of encryption; it is more suitable for access control applications which require implicit authorization. CP-ABE schemes come in two flavours, one with constant-size ciphertext [9, 10, 28], and another with constant-size secret keys [10, 13, 23]. All these schemes are built upon bilinear maps and provide an expressive access structure. Bilinear maps are far less efficient when compared with ECC because they require large-sized security parameter. Thus, ECC is more suitable and a better choice as compared to bilinear maps [2, 20, 29].

Except EMNOS scheme [10], no other scheme facilitates constant-size ciphertext along with constant-size secret keys. EMNOS scheme [10] facilitates (n, n)-threshold structure which can be restrictive for some practical applications. GSWV scheme [13] provides an access structure of AND-gate and facilitates secret keys of constant size. However, both GSWV scheme [13] and EMNOS scheme [10] are based on computationally expensive bilinear maps. ODG scheme [23] uses ECC instead of bilinear maps, thus achieving efficiency, but provides expressive AND gate access structure which is again too restrictive in some practical instances.

The scheme which we propose in this paper is the only one, which achieves threshold access structure and is based on ECC instead of bilinear maps, thus promising in terms of efficiency, while keeping the secret decryption key size constant. Threshold

Table 1 Comparing various CP-ABE techniques

Scheme	LSK	LCT	Access structure
Waters [26]	$(\mathbb{A} + 2)\mathbb{G}$	$(2\mathbb{P} + 1)\mathbb{G} + \mathbb{G}_t$	LSSS
ODG [23]	$2 \times O(\mathbb{G})$	$(n - \lvert\mathbb{P}\rvert + 3)\mathbb{G} + L$	AND gates
Ours	$2 \times O(\mathbb{G})$	$(2n - \lvert\mathbb{P}\rvert - \lvert\mathbb{A}\rvert + 3)\mathbb{G} + L$	Threshold

Note LSSS: scheme on linear secret sharing; LSK: user secret key length; LCT: ciphertext length; L: length of message of plaintext M; $\mathbb{G}$ and $\mathbb{G}_t$: groups of prime order paring; G: base-point of the curve group which is elliptic; O(G): order of base point of the elliptic curve group access structure is flexible and general enough to meet the needs of some practical problems like resource sharing in the case of virtual organizations. A comparison of different attribute-based encryption is presented in Table 1.

1.2 Our Contributions

Our main contribution are summarized below.

1. We propose an efficient mechanism using CP-ABE and Elliptic Curve Cryptography with an expressive (n, k)-threshold access structure. This is the first scheme that provides a provably secure (n, k)-threshold access structure using ECC.
2. Another major contribution of this paper is that it uncovers an error in the KeyGen phase and Decrypt phase of the ODG scheme [23]. We show that the ODG scheme will not work as expected and we also give appropriate changes to make it work correctly.
3. Our proposed CP-ABE technique also achieves secret keys which are of constant size, i.e., user decryption key size is not dependent on the cardinality of universe attribute set $\mathbb{U}$.
4. It is shown that the achieved (n, k)-threshold access structure is secure under the given model.

2 Definitions and Preliminaries

The section provides an outline of the basic preliminary concepts along with the computational hard problems associated with our scheme, which are used in this paper. The representation used throughout this paper is mentioned in Table 2. In the end of this section, we draw a basic CP-ABE technique for a selective game that defines the selective-security against a chosen ciphertext attack.

2.1 Access Structure and Attribute

We represent the Universe attribute set by $\mathbb{U}$ consisting of n number of attributes. Let $\mathbb{U} = \{A_1, A_2, A_3 ... A_n\}$. We represent the ith attribute by A_i. The attribute set corresponding to a user is denoted by $\mathbb{A} \subseteq \mathbb{U}$. In this paper, attribute set is denoted by n-bit string of binary $a_1 a_2 a_3 ... a_n$; also, this binary string is defined by

$$\begin{cases} a_i = 1 \text{ iff } A_i \in \mathbb{A} \\ a_i = 0 \text{ iff } A_i \notin \mathbb{A} \end{cases}$$

Table 2 Notation used

Symbol	Description
ρ	Security parameter
p	Numbers which are prime and large (160 bits)
$E_p(a, b)$	$y^2 = x^3 + ax + b(mod\ p)$ represents Elliptic curve
G	Order of 160-bit number at point of base in $E_p(a, b)$ in $\mathbb{Z}_p$
xG	$G + G + \cdots + G$ (x times), Elliptic curve multiplication by scalar, $G \in E_p(a, b)$
$P + Q$	Point addition for the curve which is elliptic, $P, Q \in E_p(a, b)$
$\mathbb{G}$	Group of curves which are elliptically generated by G
$\mathbb{Z}_p^*$	$\{1, 2, 3, \ldots, p - 1\}$, p is the prime number
p'	Order of the point of base G in $E_p(a, b)$
MSK	Secret key of master
MPK	Public key of master
α, β, γ	Random numbers which serve as master secret key. $\alpha, \beta, \gamma \in \mathbb{Z}_p$
H_1, H_2, H_3, H_4	Collision-resistant for functions which are hash
KDF	Function of Key Derivation
$\mathbb{U}$	Universe attribute set $A_1, A_2, A_3, \ldots, A_n$ for n number of attributes
n	Cardinality of $\mathbb{U}$
$\mathbb{A}$	$\mathbb{A} \subseteq \mathbb{U}$, user attribute set
a_i	ith bit in the bit representation of attribute set $\mathbb{A}$. $a_i = 1$ iff $A_i \in \mathbb{A}$
$\mathbb{P}$	Access Policy, $\mathbb{P} \subseteq \mathbb{U}$
b_i	ith bit in the bit representation of attribute set $\mathbb{P}$. $b_i = 1$ iff $A_i \in \mathbb{P}$
$\|\mathbb{X}\|$	Cardinality of the set $\mathbb{X}$.
$\mathbb{S}^x$	Cartesian Product of $\mathbb{S}$, x times, $\mathbb{S}^x = \mathbb{S} \times \mathbb{S} \times \mathbb{S} \times \cdots \times \mathbb{S}$ (x times)
M	Plaintext message
C	Ciphertext corresponding to M
l_σ	Length of string which is random
l_m	Length of the message which is plaintext M
$\oplus$	Bitwise XOR operator
$\{0, 1\}^*$	String which is binary and of arbitrary length
$\{0, 1\}^l$	String which is binary whose length is represented by l
$f(x, \mathbb{P})$	$\prod_{j=1}^{n} (x + H_4(j))^{1-b_j}$
f_j	Coefficient of x^j in $f(x, \mathbb{P})$
$F(x, \mathbb{A}, \mathbb{P})$	$\prod_{j=1}^{n} (x + H_4(j))^{2-a_j-b_j}$
F_j	Coefficient of x^j in $F(x, \mathbb{A}, \mathbb{P})$

Similarly, we have defined the policy of access by $\mathbb{P} \subseteq \mathbb{U}$ specified in $\mathbb{U}$ with the attributes, thus representing the access policy by n-bit string of binary $b_1b_2b_3...b_n$. We define the binary string for access policy as follows:

$$\begin{cases} b_i = 1 \text{ iff } A_i \in \mathbb{P} \\ b_i = 0 \text{ iff } A_i \notin \mathbb{P} \end{cases}$$

Algorithm 1: Algorithm which checks whether the attribute of user $\mathbb{A}$ fulfils the policy of threshold access structure

```
function CheckThresholdPolicy (A, P, k);
Input  : User attribute A, access policy P, and a non-negative integer k denoting threshold
Output: True if threshold policy is satisfied, otherwise False
matching_attributes = 0;
for i = 1..n do
    if a_i == 1 AND b_i == 1 then
        matching_attributes = matching_attributes + 1
    end
end
if matching_attributes < k then
    return False;
else
    return True;
end
```

2.2 *Assumption*

We assume that policy set $\mathbb{P}$ associated with ciphertext is never an empty set. The policy set contains at least one attribute from the attribute universe $\mathbb{U}$. Formally, we assume that we will always have an i, such that $b_i = 1$. Also by this assumption, we can logically conclude that the attribute set $\mathbb{A}$ corresponding to a user should never be a null set. If this happens, it means that this particular user is not allowed to access any of the encrypted ciphertext. So it makes no sense to have a set of attributes of users as an empty set.

Table 3 Phases with inputs and outputs of different schemes

Phase	Input	Output
Setup phase	$\mathbb{U}, \rho$	MSK, MPK
Encrypt phase	$\mathbb{P}, MPK, M$	$C = \{\mathbb{P}, R_i, S, T, C_{\sigma_m}, C_m\}$
KeyGen phase	$\mathbb{A}, MSK, MPK$	$k_u = (k_1, k_2)$
Decrypt phase	$C, \mathbb{P}, k_u, \mathbb{A}, MPK$	M or ϕ

3 The Proposed CP-ABE Technique with Threshold Access Structure

The intended scheme provides expressive (k, n)-threshold access structure along with constant-size secret keys not depending on number attributes. Figure 1 gives a brief outline of the architecture and working of our protocol. Table 3 shows the phases with inputs and outputs of different schemes.

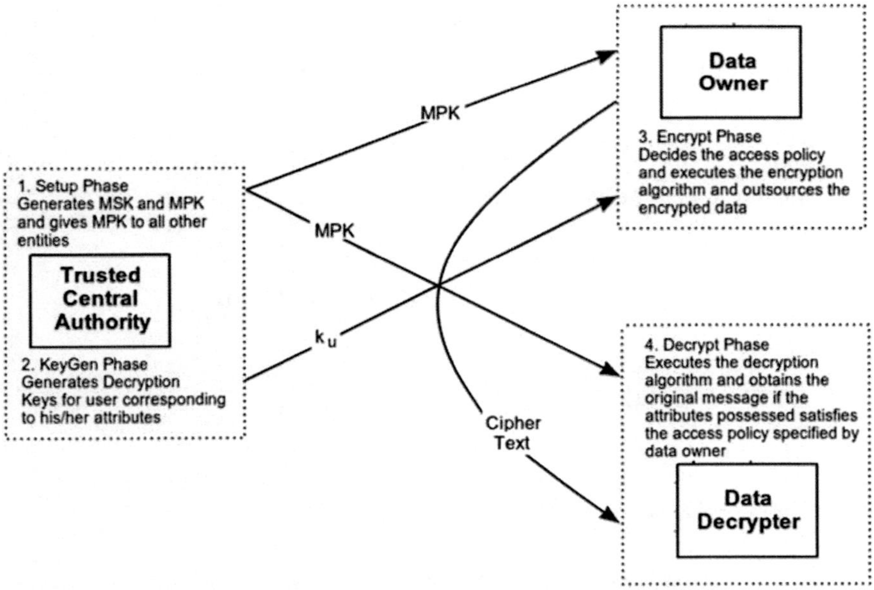

Fig. 1 Architecture and working of proposed scheme

3.1 Setup Phase

Step 1: Selecting a Group based on Curve which is elliptic $\mathbb{G} = \{p, E_p(a, b), G\}$, G denotes the point of base on the curve $E_p(a, b)$ determined over $\mathbb{Z}_p$.
Step 2: Choose the private keys α, β and γ uniformly at random in $\mathbb{Z}_p$
Step 3: Now, compute the vectors X, Y and Z as follows:

$$X_i = \alpha^i G \tag{1}$$

$$Y_j = \beta\alpha^j G \tag{2}$$

$$Z_j = \gamma\alpha^j G \tag{3}$$

for all $j = 0, 1, 2, 3, \ldots, n$ and $i = 0, 1, 2, 3, \ldots, 2n$.
Step 4: Choose four collision-resistant of one-way H_4, H_3, H_2 and H_1 functions of hash, mentioned as follows:

$$\begin{aligned} H_1, H_4 &: \{0, 1\}^* \rightarrow \mathbb{Z}_p \\ H_2 &: \{0, 1\}^* \rightarrow \{0, 1\}^{l_\sigma} \\ H_3 &: \{0, 1\}^* \rightarrow \{0, 1\}^{l_m} \end{aligned}$$

here, l_σ, l_m, M are as described in Table 2.
Step 5: Finally, output MSK and MPK are as follows:

$$\begin{aligned} MSK &= \{\alpha, \beta, \gamma\} \\ MPK &= \{\mathbb{G}, X_i, Y_j, Z_j, H_1, H_2, H_3, H_4\} \end{aligned}$$

for all $j = 0, 1, 2, 3, \ldots, n$, $i = 0, 1, 2, 3, \ldots, 2n$.

3.2 Encrypt Phase

The plaintext message M is to be interrelated with an access policy $\mathbb{P}$, and by using MPK, ciphertext C will be generated. So the input to encryption algorithm is access policy $\mathbb{P} \subseteq \mathbb{U}$ where $|\mathbb{P}| \neq 0$, plaintext message and MPK, and it returns $C = \{\mathbb{P}, R_i, S, T, C_{\sigma_m}, C_m\}$ using the following method:

Step 1: Random number has to be picked $r_m \in \{0, 1\}^{l_\sigma}$ and compute $h_m = H_1(\mathbb{P}, M, r_m)$ and $k_m = KDF(h_m G)$.

Step 2: Let $\mathbb{P} = b_1 b_2 b_3 ... b_n$ be the policy under access string. Calculate the corresponding function of polynomial $f(x, \mathbb{P})$ in $\mathbb{Z}_{p'}[x]$ having degree at most $(n-1)$, where p' is the group of order achieved by base point G, described as follows:

$$f(x, \mathbb{P}) = \prod_{i=1}^{n} (x + H_4(i))^{1-b_i} \tag{4}$$

In $f(x, \mathbb{P})$ polynomial, let f_i be the coefficient of x^i.

Step 3: Computation of the parameter of ciphertext is done as described below.

$$R_i = h_m X_i \text{ for } i = 1, 2, 3, \ldots, 2n - |\mathbb{P}| - |\mathbb{A}| \tag{5}$$

$$S = h_m \sum_{i=0}^{n} f_i Y_i = h_m \beta f(\alpha, \mathbb{P}) G \tag{6}$$

$$T = h_m \sum_{i=0}^{n} f_i Z_i = h_m \gamma f(\alpha, \mathbb{P}) G \tag{7}$$

$$C_{\sigma_m} = H_2(k_m) \oplus r_m \tag{8}$$

$$C_m = H_3(r_m) \oplus M \tag{9}$$

Step 4: Output the ciphertext C as $C = \{\mathbb{P}, R_i, S, T, C_{\sigma_m}, C_m\}$.

3.3 KeyGen Phase

Step 1: Let $\mathbb{A} = a_1 a_2 a_3 ... a_n$ be the attribute string of the user. Compute the corresponding polynomial function $f(x, \mathbb{A})$ in $\mathbb{Z}_{p'}[x]$ which is having a degree of at most $(n-1)$, where order of the group obtained by base point G is p', as described below

$$f(x, \mathbb{A}) = \prod_{i=1}^{n} (x + H_4(i))^{1-a_i} \tag{10}$$

and now compute the value of above polynomial at $x = \alpha$, i.e., $f(\alpha, \mathbb{A})$ which is given by the equation:

$$f(\alpha, \mathbb{A}) = \prod_{i=1}^{n} (\alpha + H_4(i))^{1-a_i} \quad (11)$$

Step 2: Random numbers have to be picked say x_u and z_u. Obtain s_u by the following equation:

$$s_u = \frac{1}{\beta}(f(\alpha, \mathbb{A}) - \gamma x_u)(mod\ p') \quad (12)$$

where the order of the subgroup generated by the curve which is elliptic at base point G is denoted by p'.

Step 3: Now calculate the two user keys k_1 and k_2 as follows:

$$k_1 = x_u + \beta z_u(mod\ p') \quad (13)$$

$$k_2 = s_u - \gamma z_u(mod\ p') \quad (14)$$

Step 4: Output the secret user key $k_u = (k_1, k_2)$.

3.4 Decrypt Phase

The decrypt phase takes the key which is secret to the user $k_u = (k_1, k_2)$ for the user attribute set $\mathbb{A}$ along with ciphertext $C = \{\mathbb{P}, R_i, S, T, C_{\sigma_m}, C_m\}$ for the access policy $\mathbb{P}$, and the threshold value k, and produces the original plaintext message if decryption keys are valid.

The detailed steps to be executed along with the explanation for decryption algorithm are as follows:

Step 1: If the user attributes $\mathbb{A} = a_1a_2a_3...a_n$ do not satisfy the access policy $\mathbb{P} = b_1b_2b_3...b_n$ then output ϕ and abort. If user attributes satisfies the policy, execute the following steps.

Step 2: Compute

$$U = k_2S \quad (15)$$

$$V = k_1T \quad (16)$$

Algorithm 2: Algorithm for Decryption

function Decrypt $(C, \mathbb{P}, k_u, \mathbb{A}, MPK, k)$

if *CheckThresholdPolicy*$(\mathbb{A}, \mathbb{P}, k) == False$ **then**
    return ϕ
    abort
end
$U = k_2 S$
$V = k_1 T$
if $\mathbb{A} == \mathbb{U}$ *AND* $\mathbb{P} == \mathbb{U}$ **then**
    $h_m G = U + V$
else
    $W = \sum_{i=1}^{2n-|\mathbb{P}|-|\mathbb{A}|} F_i R_i$
    $h_m G = \frac{((U+V)-W)}{F_0}$
end
$k'_m = KDF(h_m G)$
$r'_m = H_2(k'_m) \oplus C_{\sigma_m}$
$M' = H_3(r'_m) \oplus C_m$
$h'_m = H_1(\mathbb{P}, M', r'_m)$
if $h_m G == h'_m G$ **then**
    return M'
else
    return ϕ
end

$$
\begin{aligned}
U &= k_2 S = (s_u - \gamma z_u)(h_m \beta f(\alpha, \mathbb{P}))G \\
V &= k_1 T = (x_u + \beta z_u)(h_m \gamma f(\alpha, \mathbb{P}))G \\
U + V &= (s_u - \gamma z_u)(h_m \beta f(\alpha, \mathbb{P}))G + (x_u + \beta z_u)(h_m \gamma f(\alpha, \mathbb{P}))G \\
&= s_u(h_m \beta f(\alpha, \mathbb{P}))G - \gamma z_u(h_m \beta f(\alpha, \mathbb{P}))G \\
&\quad + x_u(h_m \gamma f(\alpha, \mathbb{P}))G + \beta z_u(h_m \gamma f(\alpha, \mathbb{P}))G \\
&= s_u(h_m \beta f(\alpha, \mathbb{P}))G + x_u(h_m \gamma f(\alpha, \mathbb{P}))G \\
&= h_m(s_u \beta + x_u \gamma) f(\alpha, \mathbb{P})G \\
&= h_m f(\alpha, \mathbb{A}) f(\alpha, \mathbb{P})G \\
&= h_m F(\alpha, \mathbb{A}, \mathbb{P})G
\end{aligned}
$$

Step 3: If $\mathbb{A} = \mathbb{U}$ and $\mathbb{P} = \mathbb{U}$, then $F(x, \mathbb{A}, \mathbb{P}) = 1$.

$$U + V = h_m G$$

We don't need to calculate W in this case. So skip Step 4, 5 and go to Step 6.

Step 4: Given $\mathbb{A} = a_1a_2a_3...a_n$ and $\mathbb{P} = b_1b_2b_3...b_n$, compute the corresponding polynomial function $F(x, \mathbb{A}, \mathbb{P})$, of degree at most $(2n-2)$ in $\mathbb{Z}_{p'}[x]$, where p' is the order of group obtained by the point of base G, which is as follows:

$$\begin{aligned} F(x, \mathbb{A}, \mathbb{P}) &= f(x, \mathbb{A}) \times f(x, \mathbb{P}) \\ &= \prod_{i=1}^{n}((x + H_4(i))^{1-a_i} \times (x + H_4(i))^{1-b_i}) \\ &= \prod_{i=1}^{n}(x + H_4(i))^{2-a_i-b_i} \end{aligned}$$

Consider F_i in polynomial function $F(x, \mathbb{A}, \mathbb{P})$ is the coefficient of x^i. Now, compute

$$W = \sum_{i=1}^{2n-|\mathbb{P}|-|\mathbb{A}|} F_i R_i \tag{17}$$

$$\begin{aligned} W &= \sum_{i=1}^{2n-|\mathbb{P}|-|\mathbb{A}|} F_i R_i \\ &= h_m \left(\sum_{i=1}^{2n-|\mathbb{P}|-|\mathbb{A}|} F_i \alpha^i \right) G \\ &= h_m \left(\sum_{i=1}^{2n-|\mathbb{P}|-|\mathbb{A}|} F_i \alpha^i + F_0 - F_0 \right) G \\ &= h_m (F(\alpha, \mathbb{A}, \mathbb{P}) - F_0) G \\ &= h_m F(\alpha, \mathbb{A}, \mathbb{P}) G - h_m F_0 G \end{aligned}$$

Step 5: Now, calculate the temporary key $h_m G$ as follows:

$$\begin{aligned} h_m G &= \frac{((U + V) - W)}{F_0} \\ &= \frac{(h_m F(\alpha, \mathbb{A}, \mathbb{P}) G - (h_m F(\alpha, \mathbb{A}, \mathbb{P}) G - h_m F_0 G))}{F_0} \\ &= h_m G \end{aligned}$$

Step 6: Use the above value of $h_m G$ and compute the following parameters in the below mentioned sequence:

(a) $k'_m = KDF(h_m G)$.
(b) $r'_m = H_2(k'_m) \oplus C_{\sigma_m}$
(c) $M' = H_3(r'_m) \oplus C_m$
(d) $h'_m = H_1(\mathbb{P}, M', r'_m)$

After computing the above parameters, check whether the condition $h_m G = h'_m G$ holds or not.

$$\text{Output} = \begin{cases} M' & \text{if } h_m G = h'_m G \\ \phi & \text{if } h_m G \neq h'_m G \end{cases}$$

4 Analysing the Parameters Based on Security

Here, we provide the verification of security for the given CP-ABE technique for constant-size secret keys, with (n, k)-threshold access structure against some feasible recognized attacks. The pre-eminent aim of security which is selective is to capture the collision resistance of decryption keys and the indistinguishability of messages. We have already defined the computational hard problems in Sect. 2 that we adopt for security reductions.

4.1 Surety Across Attack of Collusion

We examine the case where l attacker, which is having authentic keys for secret interrelated with attributes, coordinates as well as collaborates to provoke the keys of the system which are private (β, γ).

Proposition 1 *Let $a_i x + b_i y = c_i$ for $i = 1, 2, \ldots, l$ be equations with one degree, that is, linear in two variables, x and y, where $b_i = b_j$ also $a_i = a_j$ only when $j = i$. Considering the above linear equations, we have following three cases:*

1. *If value of b_i as well as a_i are familiar, then x and y are two unknowns of linear equations that form a system of l. Then the system has a solution which is unique and is solvable for x and y.*
2. *If any one of b_i or a_i is not known, then the mathematical statement has $l + 2$ unknown variables, namely a_i for $i = 1..l$, x and y. Then the system has many solutions which are infinite and, however, the system is solvable.*
3. *If b_i and a_i are not known, then the mathematical statement has $2l + 2$ unknown variables, namely a_i and b_i for $i = 1..l$, x, y. The system is solvable and we can say that the equation has solutions which are not finite.*

Theorem 1 *The technique is protected across collusion attack by adversaries who desire at acquiring the pair of private keys (β, γ) of the system.*

Proof: Consider a bunch of adversaries who wish to collaborate with each other and derive a new pair of user secret key. Suppose that we have l such adversaries. Let the adversaries be u^i for $i = 1, 2, \ldots, l$ and the corresponding attribute set of adversaries be $\mathbb{A}^i$. Assuming that all the adversaries have their valid secret keys, let their secret keys be $k_{u^i} = (k_1^i, k_2^i)$ where

$$k_1^i = x_u^i + \beta z_u^i (mod\ p') \tag{18}$$

$$k_2^i = s_u^i - \gamma z_u^i (mod\ p') \tag{19}$$

From Eqs. (18) and (19), it is clear that there are $2l + 1$ unknowns with l linear equations. As it is known from Proposition 1 that it endures no unique solution, there are infinite solutions satisfying these l equations. So until the adversaries make a guess for z_u^i and x_u^i to solve for β, and guess of z_u^i and s_u^i to solve for γ, they are unable to recover anything. Hence, l corrupted users, who collaborated, cannot recover the system's private keys (β, γ), and the random numbers x_u^i and z_u^i along with s_u^i are also unknown to the adversary. This in turn implies that as $f(\alpha, \mathbb{A}) = s_u\beta + x_u\gamma\ (mod\ p')$ adversaries after collaborating cannot recover $f(\alpha, \mathbb{A})$ also. Thus, α in turn cannot be derived. So, (α, β, γ) are all unknown to the collaborating adversaries. □

4.2 Security Against Key Recovery Attack

The analysis of the proposed scheme makes an attempt to generate a key which is valid, and the secret corresponding to the set of attributes $\mathbb{A}$ is discussed in this section.

Theorem 2 *This scheme is defended across an adversary who tries to procure an authentic key of user secret $k_u = (k_1, k_2)$ corresponding to $\mathbb{A}$ attribute set.*

Proof: From Theorem 1, it is computationally impractical for $\mathcal{A}$ who is an adversary to enumerate the system private key pair (β, γ). This signifies that it is also computationally absurd for $\mathcal{A}$ adversary who derive the authentic pair of user secret key $k_u = (k_1, k_2)$ correlative to the set $\mathbb{A}$. So even if the adversary makes a guess for random numbers x_u and z_u, he will not be able to calculate s_u that satisfies the condition $f(\alpha, \mathbb{A}) = s_u\beta + x_u\gamma\ (mod\ p')$, because β, γ and $f(\alpha, \mathbb{A})$ are unknown, and until he/she know s_u, he/she is unable to calculate k_1, and similarly, he/she will not be able to calculate k_2 without knowing γ. Therefore, obtaining the valid USK k_u remains a computationally infeasible problem for the adversary $\mathcal{A}$. □

4.3 Adversary Without Secret User Keys Deriving Original Message Against Security

The subdivision elaborates what the adversary could derive without significantly a legitimate key of a secret user.

Proof: A C ciphertext corresponding to $\mathbb{P}$ access policy abides by the following criterion:

$$\begin{aligned}
R_i &= h_m X_i \text{ for } i = 1, 2, 3, \ldots, 2n - |\mathbb{P}| - |\mathbb{A}| \\
S &= h_m \beta f(\alpha, \mathbb{P})G \\
T &= h_m \gamma f(\alpha, \mathbb{P})G \\
C_{\sigma_m} &= H_2(k_m) \oplus r_m \\
C_m &= H_3(r_m) \oplus M
\end{aligned}$$

It is hard to compute $h_m G$ using R_i, S and T due to the hard problem, namely elliptic curve discrete logarithm problem. The following is the explanation. As the adversary has access to the ciphertext, he can compute

$$\begin{aligned}
\sum_{i=1}^{2n-|\mathbb{P}|-|\mathbb{A}|} f_i R_i &= \sum_{i=1}^{2n-|\mathbb{P}|-|\mathbb{A}|} f_i (h_m X_i) \\
&= h_m \left(\sum_{i=1}^{2n-|\mathbb{P}|-|\mathbb{A}|} \alpha^i f_i \right) G \\
&= h_m (f(\alpha, \mathbb{P})G - f_0)G \\
&= h_m f(\alpha, \mathbb{P})G - h_m f_0 G
\end{aligned}$$

Now the adversary has access to $h_m f(\alpha, \mathbb{P})G - h_m f_0 G$, $h_m \beta f(\alpha, \mathbb{P})G$ and $h_m \gamma f(\alpha, \mathbb{P})G$, but to deal with discrete logarithm drawback for elliptic curve there are more obstacles; thus, he/she is unable to derive $h_m G$. □

Consider another scenario, where given $R_i = h_m X_i = h_m \alpha^i G$ for $i = \{1, \ldots, 2n - |\mathbb{P}| - |\mathbb{A}|\}$; we need to calculate $h_m G$. We can bring this problem to problem of $(q-1)$-DHI as mentioned. Let $Q = \alpha h_m G$; we can rephrase the parameter $R_i = h_m X_i = h_m \alpha^i G = \alpha^{i-1} Q$. Now we have modelled this problem as a $(q-1)$-DHI problem as mentioned below.

The adversary has access to $\alpha^{i-1} Q$, and he/she is required to calculate the value of $(1/\alpha)Q$ in order to derive $h_m G$. This means that the adversary should have the aptness to resolve the problem of $(q-1)$-DHI in order to get $h_m G$. Once he gets the value of $h_m G$, he can directly execute Step 6 of the Decrypt phase and get the original message. But as getting the value of $h_m G$ is as hard as solving the $(q-1)$-DHI problem, the adversary cannot successfully decipher C ciphertext.

5 Comparing Performance

We analyse the execution of the proposed CP-ABE technique for threshold access structure and compare it with other existing techniques in this section.

Table 1 gives a comparison of ciphertext length and size of decryption keys for existing CP-ABE schemes. It can be concluded from this table that the EMNOS scheme [10] achieves perpetual size of the ciphertext along with keys which are secret, and it anticipates barely (n, n)-threshold; also, it is not difficult to plan such a scheme, as constructed by the ODG scheme [23]. GSWV scheme [13] provides an efficient technique with expressive AND gate access structure, but it works efficiently only for shorter secret keys. This scheme is the only scheme which does not use bilinear maps and achieves threshold access structure, thus providing promising efficiency in contrast to other CP-ABE techniques that use bilinear maps. ODG scheme [23] also achieves the same secret key size, but it provides only AND gate access structure, which is restrictive in nature. As bilinear maps require larger size security parameters, they are not preferred over ECC because of efficiency requirements. The key size which is secret in our technique is $|k_u| = O(G) \times 2 = 320$ bits, and 163-bit ECC imparts the 80-bit security [20]. The secret key size of the GSWV scheme [13], $|k_u| = |\mathbb{G}_x| + |\mathbb{G}_y| = 160 \times 2 + 512 \times 2 = 1344$ bits achieves security of 80-bit, where $\mathbb{G}_x$, $\mathbb{G}_y$ are bilinear groups of curve which are elliptic as described in [13].

6 Implementation Details

To verify the above-stated claims regarding the performance and to show the correctness of our scheme, we have implemented the proposed technique in C++. For this, we used Crypto++ as the cryptographic library whose source code is available and therefore the Crypto++ library is freely available and contains the execution of various cryptographic schemes and formulas.

The bottleneck will be in calculating the ith coefficient $f(x, \mathbb{P})$. To calculate the coefficient of x^i of these polynomials, we have the following approaches:

1. Dynamic Programming approach
2. Divide and conquer approach
   (a) Karatsuba Method (modified)
   (b) Fast Fourier Transformation.

Figure 2 shows the execution time for various phases with cardinality of attribute universe set $\mathbb{U}$ as 1000. To avoid trivial cases in the decryption algorithm, we have considered that the policy set $\mathbb{P}$ and the user attribute set $\mathbb{A}$ are equal and none of them contains all the attributes from $\mathbb{U}$. This assure that the $\mathbb{A}$ attribute set will always satisfy the $\mathbb{P}$ policy. In Fig. 2, we have used the dynamic programming approach.

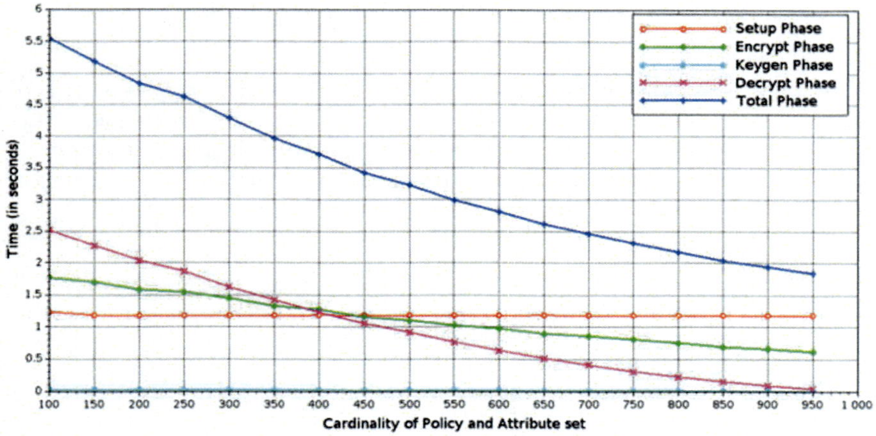

Fig. 2 Comparing different phases with execution time—calculating coefficients of $f(x) = \prod_{k=1}^{n'} (x + c_k)$ using dynamic programming for $|\mathbb{U}| = n = 1000$

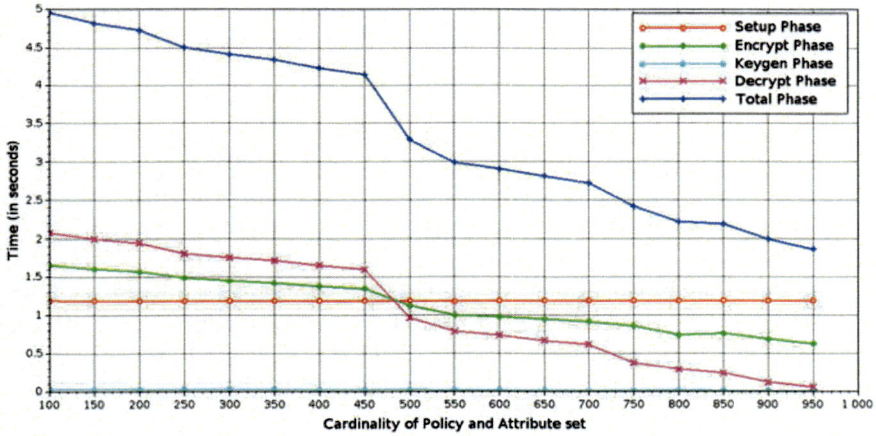

Fig. 3 Comparing different phases with execution time—calculating coefficients of $f(x) = \prod_{k=1}^{n'} (x + c_k)$ using Karatsuba algorithm for $|\mathbb{U}| = n = 1000$

Figure 3 shows the execution time for various phases with same parameters and conditions as in Fig. 2. The only difference is that we have used a modified Karatsuba algorithm instead of dynamic programming for computing the coefficients.

6.1 Description of Dynamic Programming Algorithm

The strategy in all dynamic programming solutions is to store the solution of sub-problems when computed and to reuse these solutions to avoid re-computation. Let the table that we use to store the solutions to sub-problems be denoted by $Table$, which is a 2-dimensional table. Let the polynomial be denoted as $f(x)$ of degree n' defined as $f(x) = \prod_{k=1}^{n'} (x + c_k)$. We have the following interpretation for the (i, j)th cell of this table corresponding to the polynomial $f(x)$.

$$\text{In polynomial } \left(\prod_{k=1}^{i}(x + c_k)\right) \text{ the coefficient of } x^{i-j} = Table(i, j)$$

for $i, j = 0, 1, 2, \ldots, n'$.

The multiplication of the first i terms of $f(x)$ is represented by $\left(\prod_{k=1}^{i} (x + c_k)\right)$. We proceed in bottom-up fashion for filling the table, with base cases and recurrence relation as follows:

$$Table(i, j) = \begin{cases} 0 \text{ if } j \neq 0 \\ 1 \ otherwise \end{cases}$$

And then we incrementally update the table as follows:

$$Table(i, j) = c_i \times Table(i-1, j-1) + Table(i-1, j)$$

for $i = 1..n'$ and $j = 1..i$.

As at each point of calculation we just need the solution to sub-problems from the previous row of table, we can reduce the space complexity from $\mathcal{O}(n'^2)$ to $\mathcal{O}(n')$, however, the complexity of time is still $\mathcal{O}(n'^2)$.

6.2 Divide and Conquer Approach

We will discuss two divide and conquer approaches which are classically used to multiply two polynomials and describe how these approaches are used as building blocks for solving our original problem efficiently.

6.2.1 Karatsuba Algorithm

Here, we discuss how the Karatsuba algorithm [19] for multiplication of two n-bit numbers can be modified to multiply two polynomials as described in [27]. Time complexity for the algorithm can be calculated by using the recurrence relation mentioned as follows:

$$T(n') = 3T(n'/2) + \mathcal{O}(n')$$

Therefore, the complexity of time by the Karatsuba algorithm for multiplication of polynomial is $\mathcal{O}(n'^{log_2(3)})$ where the degree of polynomials is denoted by n'.

6.2.2 Fast Fourier Transformation

Fast Fourier Transformation is a divide and conquer approach which is used to obtain the product with degree n' of two polynomials with time complexity $\mathcal{O}(n'log(n'))$.

Using the above two stated algorithms for polynomial multiplication, we can compute the coefficient of x^i in $f(x) = \prod_{k=1}^{n'} (x + c_k)$. Algorithm 3 describes the algorithm to compute the coefficients. It calls the function $PolynomialMultiply$ which is assumed to take two coefficient vectors corresponding to two polynomials and return the coefficient vector of resultant product polynomial. The complexity of time of $PolynomialMultiply$ is intended to be $\mathcal{O}(n'log(n'))$ if FFT is used, and $\mathcal{O}(n'^{log_2(3)})$ if the Karatsuba algorithm is used.

If we use Karatsuba for $PolynomialMultiply$, the recurrence relation for Algorithm 3 will be

$$T(n') = 2T(n'/2) + \mathcal{O}(n'^{log_2(3)}).$$

So, the overall time complexity is $\mathcal{O}(n'^{log_2(3)})$.

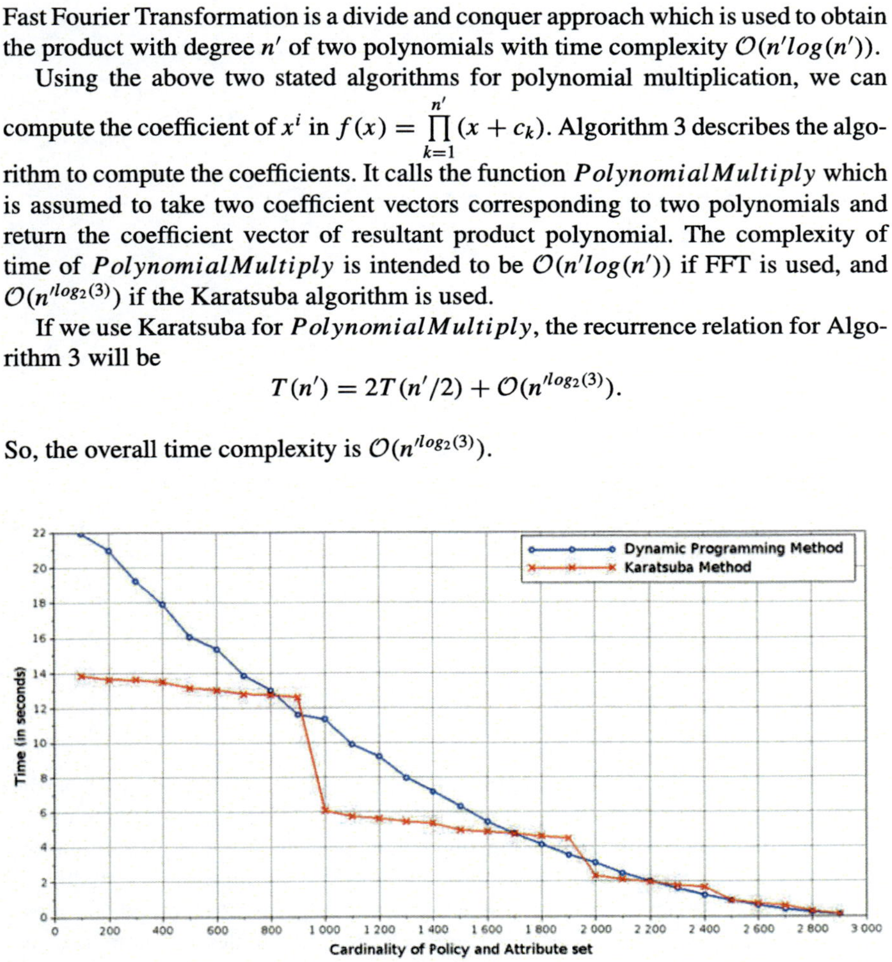

Fig. 4 Comparison of dynamic programming approach and Karatsuba algorithm with execution time for decryption phase for $|\mathbb{U}| = n = 3000$

And similarly if we use FFT for $PolynomialMultiply$, the recurrence relation for Algorithm 3 will be

$$T(n') = 2T(n'/2) + \mathcal{O}(n'log(n')).$$

So the overall complexity is $\mathcal{O}(n'log(n'))$ when FFT is used for $PolynomialMultiply$.

Figure 4 illustrates the comparison of the execution time of the Decryption phase with $|\mathbb{U}| = 3000$ for the Dynamic programming approach and modified Karatsuba algorithm.

Algorithm 3: Algorithm to get coefficient vector for $f(x) = \prod_{k=1}^{n'} (x + c_k)$

---

function GetCoefficients ($c_k; k = 1, 2, 3, \ldots n'$);

Input : a vector of n' elements: c_k corresponding to $f(x) = \prod_{k=1}^{n'} (x + c_k)$

Output: coefficient vector for $f(x)$

if $n' == 1$ **then**
  return $< c_1, 1 >$
else if $n' == 2$ **then**
  $coeff_1 = < c_1, 1 >;$
  $coeff_2 = < c_2, 1 >;$
  return $PolynomialMultiply(coeff1, coeff2)$
else
  $coeff_1 = GetCoefficients(c_k; k = 1, 2, 3, \ldots, n'/2 - 1);$
  $coeff_2 = GetCoefficients(c_k; k = n'/2, n'/2 + 1, n'/2 + 2, \ldots, n);$
  return $PolynomialMultiply(coeff1, coeff2)$

---

7 Conclusion

We have proposed and formalized a novel CP-ABE technique based on ECC with constant-size secret user keys with a threshold access structure. Notably, the scheme we have implemented is not based on bilinear maps, which are computationally expensive when compared to ECC. The proposed scheme for threshold access structure offers secret user keys of constant size and is as meagre as 320 bits and provides equivalent security of 80 bits. Each existing technique in the literature provides an access structure with an expressive AND gate that is very restrictive. Besides, our scheme makes the access structure more flexible by providing a threshold access structure. Therefore, the technique which is proposed is much adaptable for scenarios where devolution of responsibilities and complex access structure is to be achieved. The analysis showed that our scheme is the only scheme that provides a threshold access structure based on CP-ABE in which decryption keys are smaller when compared to other techniques that exist in the literature, and can be stored in lightweight devices. We have also shown that our technique is impregnable under different available known attacks like recovery of key attack, collusion attack along with message recovery attack. It is the foremost CP-ABE technique based on ECC which provides threshold access structure with constant-size secret user keys.

Future research work will include optimally handling the problem of realignment of the attributes corresponding to a user, allowing the user to decrypt data in accordance with his/her current attribute. More importantly, the scheme should take care of revoking the access for data which should now not be accessible by the new attribute set but was previously accessible by the user. Thus, we would like to handle the revocation of user secret keys also. Furthermore, we would like to extend our scheme for providing constant-size ciphertext along with a constant-size secret key to improve the efficiency and to extend the applicability of the scheme. Another major extension can be to optimally extend/introduce a traitor tracing scheme with a flexible access structure.

References

1. Attrapadung N, Libert B, de Panafieu E (2011) Expressive key-policy attribute-based encryption with constant-size ciphertexts. Proc Public Key Cryptogr 6571:90–108
2. Barreto PSLM, Kim HY, Lynn B, Scott M (2002) Efficient algorithms for pairing-based cryptosystems. In: 22nd international conference on cryptology (CRYPTO 2002). Springer, Santa Barbara, USA, pp 354–369
3. Bethencourt J, Sahai A, Waters B (2007) Ciphertext-policy attribute-based encryption. In: IEEE symposium security privacy. IEEE Computer Society, pp 321–334
4. Boneh D, Boyen X (2004) Efficient selective-id secure identity-based encryption without random oracles. In: 34th annual international conference on the theory and applications of cryptographic techniques (EUROCRYPT). Switzerland Springer, pp 223–238
5. Boneh D, Franklin MK (2001) Identity-based encryption from the weil pairing. In: Proceedings CRYPTO, vol. 2139, pp 213–229
6. Chen C, Chen J, Lim HW, Zhang Z, Feng D, Ling S, Wang H (2013) Fully secure attribute-based systems with short ciphertexts/signatures and threshold access structures. In: CT-RSA, Lecture notes in computer science, vol 7779, pp 50–67
7. Cheung L, Newport CC (2007) Provably secure ciphertext policy abe. In: ACM conference on computer and communications security, pp 456–465
8. Delerablée C (2007) Identity-based broadcast encryption with constant size ciphertexts and private keys. ASIACRYPT 4833:200–215
9. Doshi N, Jinwala DC (2014) Fully secure ciphertext policy attribute-based encryption with constant length ciphertext and faster decryption. Secur Commun Netw 7(11):1988–2002
10. Emura K, Miyaji A, Nomura A, Omote K, Soshi M (2009) A ciphertext-policy attribute-based encryption scheme with constant ciphertext length. In: 5th international conference on information security practice and experience (ISPEC 2009). Springer, pp 13–23
11. Foster I, Kesselman C, Tuecke S (2001) The anatomy of the grid: enabling scalable virtual organizations. Int J Supercomput Appl 15(3):200–222

12. Goyal V, Pandey O, Sahai A, Waters B (2006) Attribute-based encryption for fine-grained access control of encrypted data. In: Juels A, Wright RN, Di Vimercati SDC (eds) ACM conference on computer and communication security. ACM, pp 89–98
13. Guo F, Mu Y, Susilo W, Wong DS, Varadharajan V (2014) CP-ABE with constant-size keys for lightweight devices. IEEE Trans Inf Forensics Secur 9(5):763–771
14. Guo F, Mu Y, Chen Z (2007) Identity-based encryption: how to decrypt multiple ciphertexts using a single decryption key. In: Proceedings pairing, vol 4575, pp 392–406
15. Guo F, Mu Y, Chen Z, Xu L (2007) Multi-identity single-key decryption without random oracles. In: Proceedings inscrypt, vol 4990, pp 384–398
16. Guo F, Mu Y, Susilo W (2012) Identity-based traitor tracing with short private key and short ciphertext. In: Proceedings ESORICS, vol 7459, pp 609–626
17. Guo H, Xu C, Li Z, Yao Y, Mu Y (2013) Efficient and dynamic key management for multiple identities in identity-based systems. Inf Sci 221:579–590
18. Herranz J, Laguillaumie F, Ráfols C (2010) Constant size ciphertexts in threshold attribute-based encryption. In: Public key cryptography, vol 6056, pp 19–34
19. Karatsuba A, Ofman Y (1963) Multiplication of multidigit numbers on automata. Sov Phys Dokl 7
20. Lauter K (2004) The advantages of elliptic curve cryptography for wireless security. IEEE Wirel Commun 11(1):62–67
21. Lewko AB, Okamoto T, Sahai A, Takashima K, Waters B (2010) Fully secure functional encryption: attribute-based encryption and (hierarchical) inner product encryption. In: Gilbert H (ed) EUROCRYPT. Lecture notes in computer science, vol 6110, pp 62–91
22. Lewko AB, Waters B (2012) New proof methods for attribute-based encryption: achieving full security through selective techniques. CRYPTO 7417:180–198
23. Odelu V, Das AK, Goswami A (2015) An efficient CP-ABE with constant size secret keys using ECC for lightweight devices. IACR Cryptology ePrint Archive, Report 2015/841. https://eprint.iacr.org/2015/841
24. Ostrovsky R, Sahai A, Waters B (2007) Attribute-based encryption with non-monotonic access structures. In: Proceedings ACM conference on computer and communication security, pp 195–203
25. Sahai A, Waters B (2005) Fuzzy identity-based encryption. In: Cramer R (ed) EUROCRYPT. Lecture notes in computer science, vol 3494. Springer, pp 457–473
26. Waters B (2011) Ciphertext-policy attribute-based encryption: an expressive, efficient, and provably secure realization. Public Key Cryptogr 6571:53–70
27. Weimerskirch A, Paar C (2002) Generalizations of the Karatsuba algorithm for polynomial multiplication
28. Zhang Y, Zheng D, Chen X, Li J, Li H (2014) Computationally efficient ciphertext-policy attribute-based encryption with constant-size ciphertexts. In: 8th international conference on provable security (ProvSec 2014). Springer, Hong Kong, pp 259–273
29. Zheng M, Xiang Y, Zhou H (2015) A strong provably secure IBE scheme without bilinear map. J Comput Syst Sci 81(1):125–131

Chapter 3
Facial Expressions Recognization Based on DRLBP and CNN Classification

Surbhi Singh, Akhil Pandey, and Vishal Shrivastav

Abstract Facial expression appreciation typically can be a distinct one of significant investigation in the field of AI or example appreciation. The facial expression contains rich imperceptible information, which can appreciate human emotions or intentions, which extensively investigate value. Facial expression analysis is a compelling and challenging trouble that affects essential purposes in various fields such as human–computer interface and data-driven animation. Developing useful facial expressions from original face images is a necessary step in realizing face recognition. The actual evaluation is based on local statistical traits; Dominant Rotated Local Binary Patterns (DRLBP) facial expressions. Several machine learning methods have been thoroughly observed in various databases. Researchers typically use the DRLBP function, which is efficient and capable of face recognition. Cohn Canada is the current work database or programming language used in MATLAB. First, divide the face area into small areas, drag the histogram, Dominant Rotated Local Binary Patterns out of the area, and connect it to a single function vector. This feature vector outlines a well-organized face illustration and helps determine the similarity between images. It is still a severe face recognition problem to create powerful and unique features, while increasing interpersonal differences. In this article, the researchers explained how to use Dominant Rotated Local Binary Patterns (DLBP) for feature extraction and Convolutional Neural Networks (CNN) for image classification to enhance the face recognition capabilities system. The post-workout correspondence helps CNN converge faster and achieve better accuracy. Compared to other traditional methods, it has also been significantly improved to evaluate this new method's completion.

S. Singh (✉) · A. Pandey · V. Shrivastav
Arya College of Engineering and IT, Jaipur, Rajasthan, India

Rajasthan Technical University, Kota, India

A. Pandey
e-mail: akhil@aryacollege.in

V. Shrivastav
e-mail: vishalshrivastava.cs@aryacollege.in

R. Mathur et al. (eds.), *Emerging Trends in Data Driven Computing and Communications*, Studies in Autonomic, Data-driven and Industrial Computing,
https://doi.org/10.1007/978-981-16-3915-9_3

Keywords DRLBP · CNN · Face detection · Facial expression · Machine learning (ML)

1 Introduction

In the last few years, many FER algorithms have been proposed in the literature, including face appreciation from frontal or non-frontal facial imagery. Compared to frontal FER, non-frontal FER is more difficult or more suitable for the actual setting. In the various proposed methods, however, only a small part of the algorithms solve this problem. For positive and non-positive FER issues, there has been a common recognition framework in most previous cases. The work can be separated into two main steps; one is features of extraction, or the other is classification structure. To remove facial skin texture, different image functions were used in before papers, such as Dominant Rotated Local Binary Patterns (DRLBP) local phase quantization, direction gradient histogram, and scale-invariant function transformation (SIFT). SIFT has shown encouraging presentation among diverse facial appearance due to its powerful presentation on image scaling, rotation, occlusion, or light differences. Communication between two or more individuals can take place in oral or non-verbal language. Non-verbal communication involves many different aspects as well as near meaning (physical and interpersonal space: distance and location), kinematics (body orientation: posture, body movements, and gestures), appearance (physical attraction and clothing), touch (touch), Paralanguage (voice, intonation, speech speed, pitch, and volume), and facial expressions. These non-verbal communications need to be explained and occupy more than 60% of the position in the communication process (Foley and Gentile 2010). This non-verbal communication plays a significant role in the transmission of emotions, attitudes, and feelings. Emotions stem from the psycho-physiological process of any event or object related to a person's mental state, characteristics, and character of conscious and unconscious consciousness (College et al. 2014). Emotions play an essential role in communication between individuals. Personal feelings can affect relationships with others, such as family, relatives, friends in the family, the workplace, or other environments where you can connect with others. Emotions are expressed through tone, gesture, and posture, with the most common being facial expressions [1, 2].

Automatic face recognition aims to use pattern recognition technology to recognize people in photos or videos. Automatic face recognition has become widespread in applications ranging from social media to advanced authentication systems. Signaling technology is very serious. In today's world of timelessness, recognizing the face automatically with a digital camera is still very difficult as there are significant changes in business relationships, appearance, and character change. Therefore, the biggest challenge in this field and the advanced solutions and applications based on image processing and machine learning technology are the introduction of automatic computer face recognition. Face Acceptance Problems of the FACE research platform is attractive and challenging in computer vision and biometric technology, with theoretical methods and software systems for machine recognition based on digital photos of people. Many academic researchers and industrial designers need to work harder at this. Face recognition can be used for security applications (access to authorized areas, computers, airports, etc.), surveillance equipment in public places (such as playgrounds, stations, stores), forensic applications (verification/management of information, identification of victims), identity in photo/video archives, contact requests with human–computer, best-in-class card solutions (improved ATM security, biometric passports, also called e-passports), and targeted advertising. Signaling technology uses knowledge in many fields of research, such as image processing (IP), pattern recognition (PR), machine learning (ML), visual acuity, psychophysics, and neuroscience. And one of its most successful fields is biometric research. The other two are identified by fingerprints and iris. Although fingerprints and iris recognition are mature technologies that can be implemented in real-world submission, although many systems have been anticipated to do our best, face recognition still has many challenges that need to be addressed with more effective methods. And this is the most prominent challenge for it [3].

Machine learning. In the last two decades, with the emergence of large data sets in various fields of science, technology, business, and everyday life, my second research area focuses on machine learning, aiming to improve the understanding of data and the patterns. And, use it to solve non-trivial and challenging problems, such as diagnosis, prediction of the event, or detection of events, face recognition. My work covers the development of new machine learning algorithms and their interdisciplinary applications in biomedicine, multimedia, video surveillance, and technology [4].

1.1 Facial Expression Recognition System

The most popular application of emotion recognition systems is face recognition systems. Most studies of facial expressions have cited Professor Paul Ekman's work. He proposed the Facial Action Coding System (FACS) encode facial expressions extracted from thousands of photos and thousands of photographed and recorded facial expressions (Ekman 2003) [5–7]. FACS can be extracted from the face and used as a function for facial expression classification. The facial expression recognition system can be realized in a similar manner using biometric design. The facial expression recognition system uses the following more specific steps: The first step is to use sensors to collect physiological signals about humans' facial expressions. For this purpose, a camera can be used to generate a still image of the face. Alternatively, to continuously monitor facial expressions, video or webcams will be more suitable for different conditions and environments. In the face recognition system, the essential part is the process extraction process. The right process extraction process will provide a better recognition system and achieve more accurate results. Algorithms for extracting facial features can be divided into two categories, namely, geometry-based methods that collect function points or point movements by tracing function points from facial images or classifying expressions based on traced features. Appearance method, this method gathers all or part of facial marks, arranges them in a wide range of function vectors, and applies them in the classification process. Compared to geometry-based methods, appearance-based methods have several advantages [7].

Objective The main goal of this work is to develop practical algorithms for face recognition. In the current research work, the main focus is on face recognition's three main problems through the lighting variance in the image, the expression variance, and the moderate attitude change. Since it may not be possible to get more images of a person for training in most practical situations, attempts have been made to design algorithms to minimize the number of images required for face recognition.

2 Related Work

Emotion recognition Understanding emotions has become the essence of successful communication with others. However, sometimes misinterpretation of feelings between two or more individuals occurs, making communication unsuccessful. Sometimes, this event can aggravate the situation and thereby affect the relationship between the communication partners, for example, relationship with family, relatives, or friends. It can affect the relationship between workers and employers, students and teachers, and even doctors and patients in a more formal context. These situations triggered the development of a system that can recognize human emotions to minimize misinterpretation of emotions. Since then, emotion recognition systems have become very popular in recent years and have made significant contributions to many applications that can interpret human emotions based on various physiological signals. It can also be monitored with teachers, therapists. Chao [7] in his paper offers a new way of accepting language based on knowledge and binary maps.

First, the method is based on the LBP operator to exploit face. Second, the facial area is divided into six small regions in shape by using a pseudo-3D model. In this case, the image in part and the world's form use the LBP mapping method for feature origin. The supporting machine and the two classification softmax are two models of emotion organization, namely, the basic sensation model or rotating emotion model. Finally, we conducted a relative experiment on Cohn-Kanade (CK +) facial response and test data composed of ten volunteers. The results show that this method can efficiently confiscate disturbing elements in the image. Also, the results of using emotional models are much better than traditional expressive models. Through research related to human knowledge, we have confirmed that eyes or mouth are more sensitive to emotions [7].

He Jun et al. [8] for recognizing facial features, consider LBP function as an important method of color design. Still, the whole image is usually considered as a mining area, and the values of facial expressions are not considered. Based on the preceding LBP form extraction method and the division of facial expressions, it is suggested how to accept words based on the fusion combination of the key areas of the LBP. Multiple sections: eyes, eyebrows, eyebrows, nose, mouth, and then, we get the essential areas for independent shape appearance, which means that the global shapes are preserved. The shapes are also removed from an entire face. After that, the two functions are combined to form a new function called a fusion of mixed fusion. Then, the combined categories are SVM and NN categories to identify the different languages. This article has conducted experiments in the JAFFE database, or fallout shows that the rate of acceptance of the form factor has improved [9].

According to Swapna Agarwal et al. (2019), to confuse the proper forms of those who express emotions, we need to define success in pure form (as unexpected) and mixed (as random). In this post, we have trained emoticons (XM) that can be used effectively to express emotions. We present an algorithm that uses XM to coordinate the expression of the stimulus according to the facial constitution of the target object. The methods planned can also control the different divisions of this emotional language, and when they are mixed, they can lead to emotionally charged forms. Unlike several existing techniques, our expression-synthesis model only requires a target human form with nothing to do with language. The qualitative and qualitative experiments in the four data groups showed encouraging results. On average, we verified the integrity of the language adopted by our method by 92.4%. We also offer that the way we work may result in better speech details for basic emotional expression and mixed emotional similarity compared to the latest technical work [10].

According to Yiming Wang et al. [10], the main reason for creating a face shape is to combine the front face shape with the non-front image. In official terms, how to evaluate the front shape is a big but challenging problem. Most existing techniques use a single-form model to adapt to the external form, which will result in the loss of speech-related sequence. In this work, we present a new way of seeing facial expressions, which directly studies the relationship between non-frontal and adjacent shapes. Find a supportive door to train to develop a developmental model. Assuming that the relationship between the couple is not related, a cascade method is used that could design and improve the model. Using the front design, the exterior design is streamlined through the design process to solve simple problems for improvement. The proposed method was validated in an unobstructed viewport. The experiments' results show that the visual effects of the restoration of the form factor and the acceptance of the form factor are good [11].

Facial expressions are modified by those kinds of facial expressions that reflect a person's personality, purpose or public conversation. When it comes to appearance changes, all human faces are the primary means of conveying and defining emotional states. Eye contact has become a significant research area because it plays an essential role in computer and human communication. Face recognition has essential applications in social relations and social intelligence. This article presents the various technologies used in conjunction with the form control system [12].

Zhang Haifeng et al. [12] have shown that the acceptance of form is closely related to human nature. This paper presents a network of state mixing solutions to address the significant differences between form recognition topics. The model aims to learn the functions of identity and the processes related to expression through two branches to include identical images. The bilinear model has been introduced to combine the two aspects and understand the relationship between them. Experimental results show that personality-related processes can improve the effectiveness of the recognition of the form factor. Our style is superior to most of the latest technologies. In the two popular facial expression data (CK + and Oulu-CASIA), our method reached 96.02% and 85.21%, respectively [13].

3 Facial Expression Database

For the attainment of facial expressions information sets in this paper, we use a Cohn-Kanade Dataset to obtain facial expressions. Then examine, remove the data in order, or lastly save data information. By capturing network image data, 35,887 facial appearance images were together, counting 28,709 images from the preparation set, 3589 images from the public test set, or 3589 images from the private test set. Each image consists of 400 * 500 grayscale images with fixed pixel sizes. There are seven expressions: anger, disgust, fear, contentment, sadness, surprise, or impartiality. We map them to digital marks 0–6, correspondingly. Figure 1 explains the types of dataset.

Therefore, the quality (such as completeness) (including facial expressions, facial details, changes in light) and accuracy (including aging status) of facial data set characteristics (different image file formats and colors/grayscale, face resolution, Limited/Unlimited Environment) are crucial to the AFR process. Also, several facial databases have been urbanized for public purposes and are in public obtainable [14].

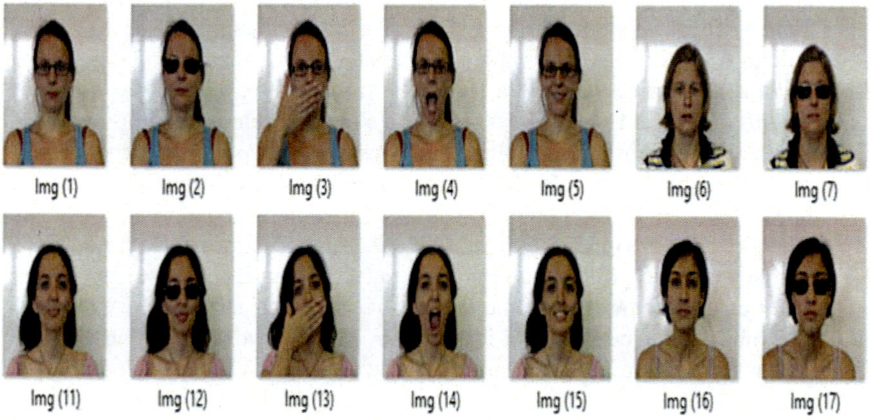

Fig. 1 Shows seven emoticons

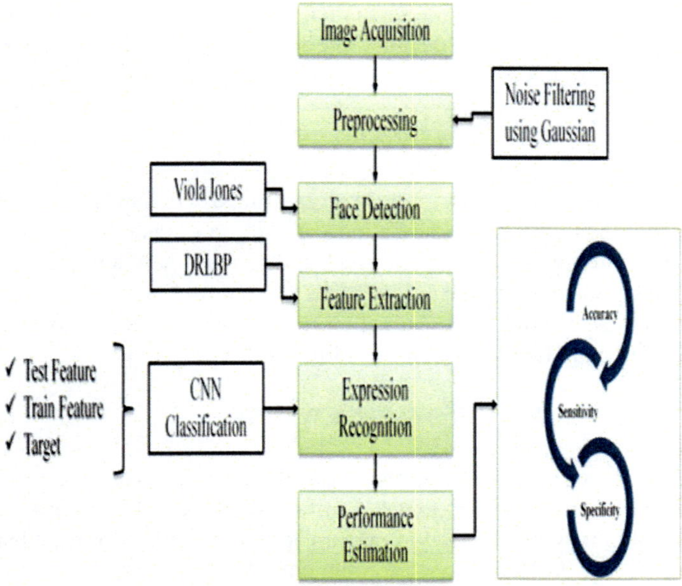

Fig. 2 Proposed flow chart

4 Proposed Work

FER technology requires a usable, reliable, and dependable face search database. Also, several facial databases have been residential for public purposes and are publicly obtainable. For the actual recognition step of the face recognition procedure, the organization is often used, as shown. This is a machine learning technology whose job is to learn first, and then use it to map a person's facial features on one of them. The predefined category marks, namely, category 1 (person's face) or category 2 (not person's face), the classifier can be applied to the whole set of unpacked facial skin texture or some detailed facial features, such as facial expressions. Gender, age, race, etc., lately, methods such as neural complex have been used as classifiers.

CNN-based expression classification algorithm—Due to the computer's excellent performance, deep learning has become widespread in many areas. A deep neural network based on CNN is used for the classification problem with expressions. Teachable Convolution cores involve high-level expression functions in each folding layer, high-dimensional expression function output to low convolution Floor. The calculation formula for the expression function card is as follows:

$$Gi = f(Gi + Wi + Bi) \tag{1}$$

In formula (1), Gi is the output of the i-th layer of neurons. Gi-1 is input of the i-1 layer of neurons, wireless is the weight vector of the one in the neuron's degradation nucleus, Bi the displacement vector, or f is foundation meaning. To some point, the pool layer can keep the function zoomed out or condense the function card's size. The combination formula is:

$$\mathrm{Gi} = \mathrm{Pooling}\ (\mathrm{Gi} - 1) \tag{2}$$

Among them, GI is a minor example layer or braid is a fusion function. A significant part of the neural system foundation can add non-linear factors to simulate the human body arrangement neural network and store or map items. The complete connecting layer attaches neurons in the upper layer or each neuron in this layer. To achieve the purpose of produce or extracting functions, it is also called a multilayer perception. The computation formula is as follows.

$$\mathrm{F(x)} \ominus \mathrm{f}\ (\mathrm{x} \ * \mathrm{W} \ominus \mathrm{B}) \tag{3}$$

F (x) is a fully associated layer, f is regularly called a creation purpose, W is typically called a weight vector, and B can be called a displacement vector. The SoftMax meaning is typically used to explain several organization troubles. In this article, the SoftMax function is used to map outputs from many neurons to values from 0 to 1.

System Implementation

- Image Acquisition
- Pre-processing
- Face Detection
- Feature Extraction
- Expression Recognition
- Performance Estimation.

Image Acquisition:
The read command is used to read the image into work area. This example reads one of the examples that comes with the toolbox, the image, and saves it in a matrix named I. Image read derives from the file that the graphic file format is "Tagged Image File Format (TIFF)" Im show function See the image in "Image Viewer" application.

Pre-processing:

- Noise reduction is the process of removing noise from the signal.
- All recording devices, including analog and digital recording devices, have characteristics that make them susceptible to noise. The noise can be random or white without coherence, or it can be coherent noise introduced by the device's processing mechanism or algorithm.

Face detection:
Face recognition is a computer technology used in applications to recognize faces in digital images. The Viola–Jones object detection framework is the first object

detection framework proposed by Paul Viola and Michael Jones in 2001, which aims to provide a competitive real-time object velocity. Although he may be trained in the discovery of different kinds of objects his primary motivation is facial recognition.

Feature extraction:
Pattern recognition is a branch of machine learning that focuses on pattern recognition and data regularity; although in some cases, it is considered almost synonymous with machine learning. Unlike LBP, it does not limit the pixel threshold to 0 and 1, but uses a threshold constant to limit the pixel threshold to three values. The result of using k as the threshold constant, c as the value of the central pixel, and using the adjacent pixel p as the threshold.

Convolutional Neural Network: Convolutional Neural Network (ConvNetal/CNN) is a deep learning algorithm that can absorb input images, assign meaning (learnable weights and deviations) to different aspects/objects in the image, and can distinguish them from each other. Compared to other classification algorithms, the pre-processing required in ConvNet is much lower. In the original method, filters are manually designed, and after adequate training, ConvNets can learn these filters/features. The architecture of ConvNet corresponds to the connection pattern of neurons in the human brain, which is inspired by the organization of the visual cortex. A single neuron responds only to stimuli in a limited field of vision called the receptive field. The collection of these fields overlaps to cover the entire visible area.

5 Experiments and Results

Acceptance of form can be divided into two stages: removal of structure and classification of form. This experiment obtained 150 imagery of each facial expression or used the expanded Cohn-Kanade data [3] as experimental data. We first divide this data into two sections, of which 50 imagery is used as a test case, or another part is used as a training series. On the other hand, as part of the face shape, we use the DRLBP mapping method to condense the dimensionality of work on the image or use neural networks to match the shape. On the other hand, we use two categories: CNN for organization and acceptance. In the experiment, the image must be placed in advance, and the Cohn-Kanade expression data contains a series of motion pictures from the beginning of the speech to the mature expression. For more accurate results, select the image for the face moving forward and not be blocked or shaken. This research introduces the facial expression recognition method, which can semi-automatically extract and classify facial expressions from an image dataset. In this study, the data set is first used, and the quality of the extracted expressions is evaluated. Then the digital representation of the extracted facial expressions is presented to the machine learning algorithm, and the ability to automatically predict is evaluated [14].

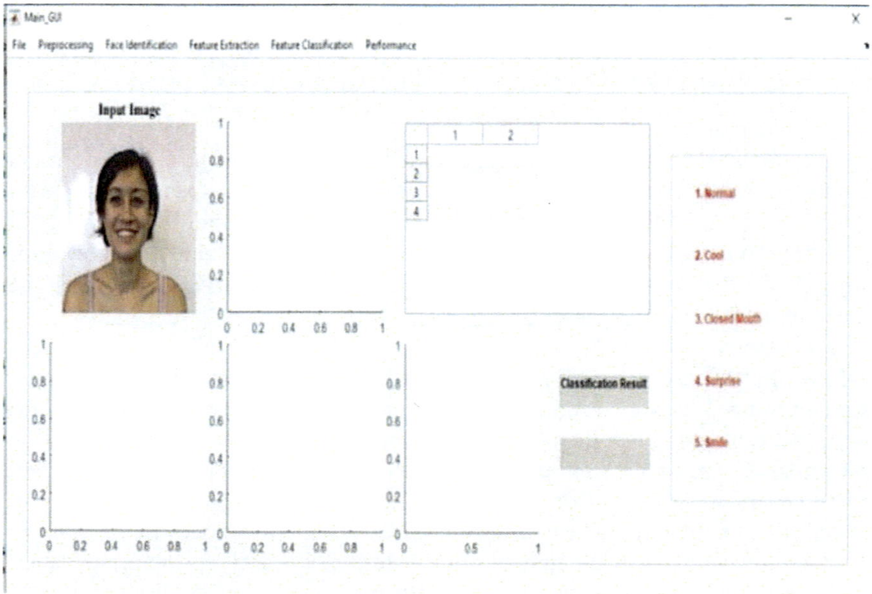

Fig. 3 GUI window input image

In the experiment, the image should be preprocessed initially, or the Cohn-Kanade phrase folder is poised of a series of exploit images from the opening of appearance to the phrase of mature as shown in Fig. 3.

Noise diminution is the procedure of confiscating noise from a signal. All recording equipment, counting analog, or digital cassette, has features that make them vulnerable to noise. The noise can be haphazard or white noise without consistency or rational noise established by the tool mechanism or indulgence algorithm shown in Fig. 4.

A lot of data is used to represent images, so image analysis needs a lot of memory and therefore requires more time. To reduce the amount of data, a series of functions are used to display the image. The exploitation of features is a kind of original pattern acceptance, which is very important in pattern recognition. This process brings out some features, such as color, shape, and structure. The feature contains the information related to the image and will be used in the image shown in Fig. 5.

This quality describes the structure of the image. This component also detects the object's face in the identify face county or changes or resizes the face. As shown in Fig. 6.

The feature extraction component is used to create face illustration, here, DRLBP is used to feature extract as shown in Fig. 7.

This means we need to reduce the size of the DRLBP code. In the high-dimensional space, we can use some mathematical transformation methods to convert the original high-dimensional space into a low-dimensional subspace. As this sample density

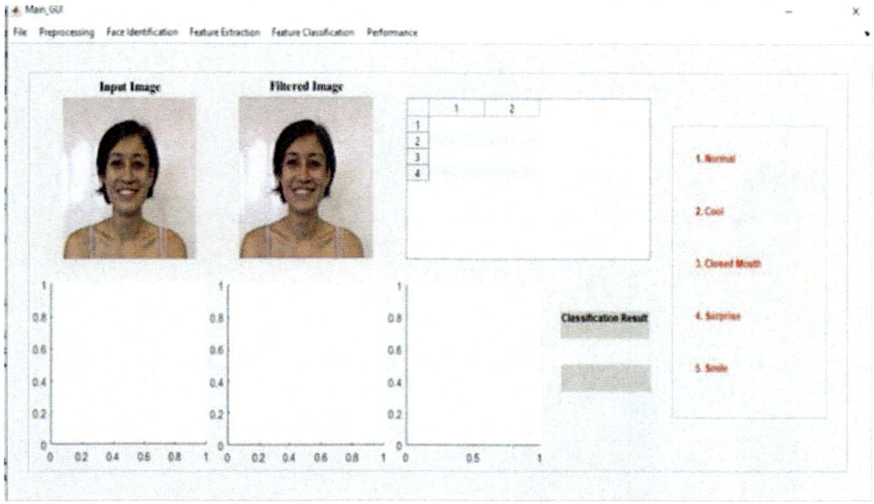

Fig. 4 GUI window input image

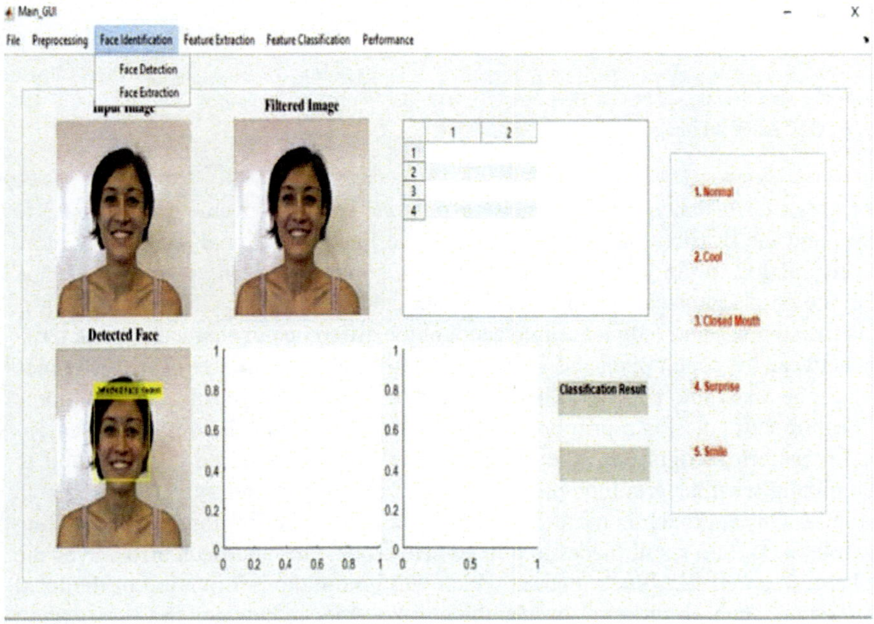

Fig. 5 GUI Window Filter Image

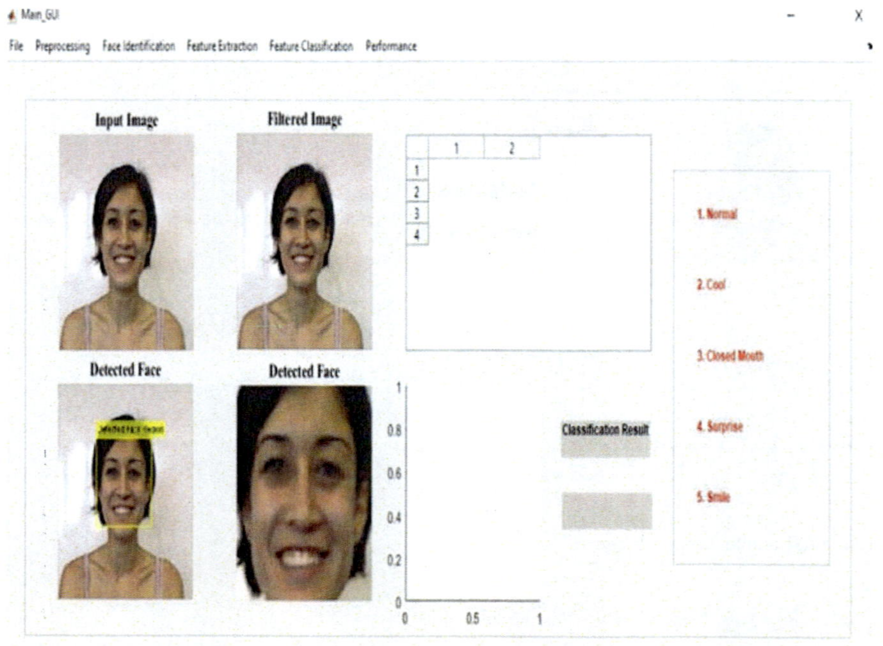

Fig. 6 GUI detected image

increases, the distance calculation is in this way more convenient and easier to learn, and the distribution or embedding of the low-dimensional sub-area is more closely related to the learning task. There are some limitations to the robustness of monotonous grayscale changes and computational efficiency. For example, it is sensitive to noise and only takes into account the difference in character characteristics between the center pixel and the neighboring pixel. And the size of the difference will not be taken into account. Therefore, it will inevitably lose some information.

Feature extraction is a primordial nature of pattern appreciation, or it is essential for pattern recognition. The last assurance compute module is committed to recognizing whether gratitude product is exact or not pursued by classifier shown in Fig. 8. Our approach is more reasonable and avoids the irrationality of dividing a whole organ into small pieces in a fixed-size block. Our approach also solves the problem of small size block method which will ignore the information of the local expression. Then, in order to discuss the correspondence between the local region and the facial features, we select different expression areas for the experiment. Under the condition of only retaining eyes and mouth regions and covering other regions, the specific situations of identification.

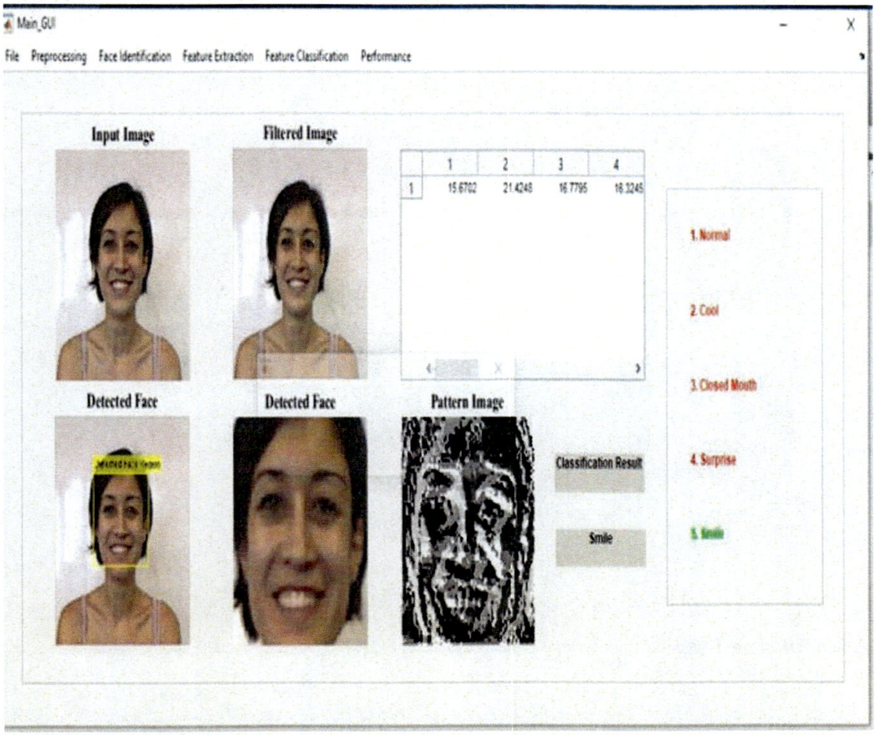

Fig. 7 GUI feature extraction

6 Performance Analysis

Process performance is measured by performance indicators such as accuracy, sensitivity, specificity, or time expenditure.

TP—is a total number of properly confidential prospects (true positive).

TN—is a total figure of pre-classified prospects (true negative numbers).

UN is a total number of false negatives and accounts for the wrong number of forefront pixels confidential as background (false negatives).

FP—is the entire number of false positives, which means that the pixel is imperfectly confidential as foreground (false positive). The presentation value is considered for each image of the input video based on the above indicators.

Accuracy: Accuracy in arrangement troubles is the number of accurate predictions made by the model over all kinds of calculations made.

$$\text{Accuracy} = (\text{TP} + \text{TN}) / (\text{TN} + \text{TP} + \text{FN} + \text{FP})$$

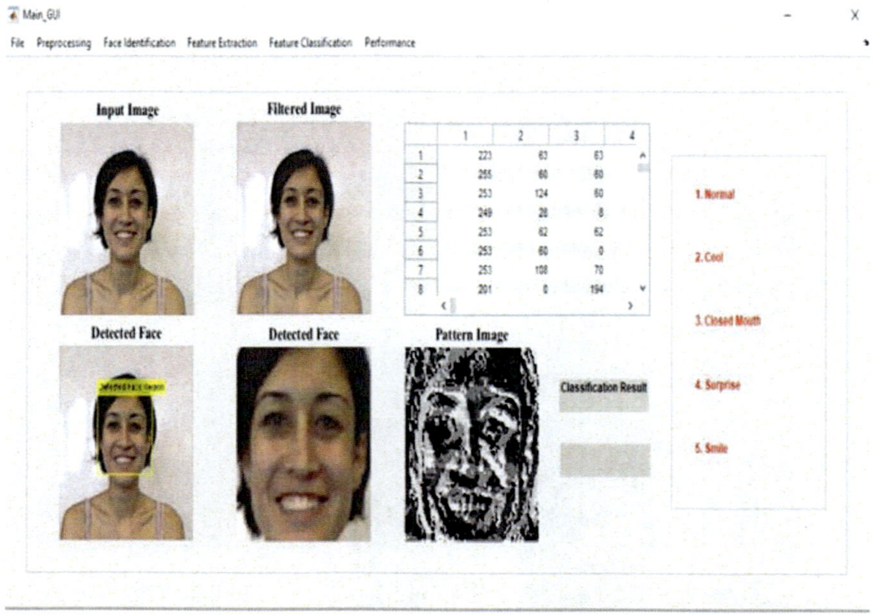

Fig. 8 GUI smile image

Sensitivity: A test's ability to correctly identify those with the disease (true positive rate). Measures the proportion of actual positives that are correctly identified.

$$\text{Sensitivity} = \text{TP}/(\text{TP} + \text{FN})$$

Specificity: The ability of a test to acceptably recognize those without infection (true negative rate). Measures percentage of actual negatives that are correctly identified.

$$\text{Specificity} = \text{TN}/(\text{TN} + \text{FP})$$

F-measure: F metric (F1 number or F-number) is measured by the test's validity and is defined as the mean of the similarity and memory of the test.

$$\text{f\_measure} = 2 * ((\text{precision} * \text{recall})/(\text{precision} + \text{recall}));$$

The results show that eyes and mouths can distinguish expressions better than other areas, especially when they are angry and disgusted. As shown in Table 1, the classification number of each facial expression after adding the area to the left cheek is described in detail. Similarly, Table 2 shows that when the forehead is expressed, the forehead area is different from the cheek area. This result also supports the

Table 1 Confusion matrix and automatic facial expression recognition performance

	Classification	Accuracy	Specificity	Sensitivity
Proposed work	SVM	85	99.83	99.63
Previous work	CNN	94	100	90.32

Table 2 Comparison of the results obtained with existing solutions

	Normal	Cool	Surprise	Closed Mouth	Smile
Normal	20	05	0	0	0
Cool	2	25	0	1	0
Surprise	0	1	25		
Closed Mouth	1	0	0	24	4
Smile	0	0	2	1	25

conclusion that people in cognitive neurology mainly recognize expression through their eyes and mouth.

This part of the expression is not among the seven discrete expressions. Volunteers say this is a complex outcome composed of different emotions. Table 2 shows that the forehead area is different from the cheek area in terms of expression of emotions. This result also supports the conclusion that people in cognitive neurology mainly recognize facial expressions through their eyes and mouth.

7 Conclusion

This paper mostly studies the technique of recognizing facial expressions based on difficult neural networks. By building a difficult neural network model, facial expressions can be standard or confidential. According to unique expression sets, face recognition applications can call singular model files for emotion recognition, providing theoretical and practical references to face recognition research. Experiments show that the satisfaction rate in this face shape is higher than 70% when used in training and higher than 80% when it is possible to apply a test with good results. The main emphasis has been given to face recognition problems via clarification variance in images, expression variance, and moderate pose variations in the present research work. As multiple images of a person may not be available for training in most practical scenarios, attempts have been made to design algorithms in such a way that a minimal number of images of a person is required for face recognition.

References

1. Lilly WS, Kamalaharidharini JT (2017) PNN-SIFT: an enhanced face recognition and classification system in image processing. In: 2017 4th international conference on electronics and communication systems (ICECS)
2. Lilly WS, Kamalaharidharini JT (2017) Robust face recognition and classification system based on SIFT and DCP techniques in image processing. In: International conference on intelligent computing and control (I2C2)
3. Satare RN, Khot SR (2018) Image matching with SIFT feature. In: 2nd international conference on inventive systems and control (ICISC)
4. Moghaddam B, Nastar C, Pentland A (2001) A bayesian similarity measure for deformable image matching. Image Vis Comput 19(5):235–244
5. Shan B (2010) A novel image correlation matching approach. JMM 5(3)
6. Lowe DG (2004) Distinctive image features from scale-invariant keypoints. Int J Comput Vis 50(2):91–110
7. Qi C, Li M, Wang Q, Zhang H, Xing J, Gao Z, Member, IEEE, Zhang H (2018) Facial expressions recognition based on cognition and mapped binary patterns. IEEE Access 2169–3536 (c) 2018 IEEE. https://doi.org/10.1109/ACCESS.2018.2816044
8. Jun H, Jian-feng C, Ling-zhi F, Zhong-wen H (2015) A method of facial expression recognition based on LBP fusion of key expressions areas. In: The 27th Chinese control and decision conference (2015 CCDC). Qingdao, China. ISBN: 978-1-4799-7017-9. https://doi.org/10.1109/IEEE
9. Agarwal S, Mukherjee DP (2019) Synthesis of realistic facial expressions using expression map. IEEE Trans Multimed. https://doi.org/10.1109/IEEE
10. Wang Y, Yu H, Dong J, Jian M, Liu H (2017) Cascade support vector regression-based facial expression-aware face frontalization. In: 2017 IEEE international conference on image processing (ICIP). IEEE Beijing, China. ISBN: 978-1-5090-2175-8. 10.1109/
11. Mal HP, Swarnalatha P (2017) Facial expression detection using facial expression model. In: 2017 international conference on energy, communication, data analytics and soft computing (ICECDS). IEEE Chennai, India. ISBN: 978-1-5386-1887-5. 10.1109/
12. Zhang H, Su W, Wang Z (2019) Expression-identity fusion network for facial expression recognition ICASSP 2019. In: 2019 IEEE international conference on acoustics, speech and signal processing (ICASSP). Brighton, United Kingdom, United Kingdom. ISBN: 978-1-4799-8131-1. https://doi.org/10.1109/IEEE
13. Sushama M, Rajinikanth E (2018) Face recognition using DRLBP and SIFT feature extraction. In: 2018 international conference on communication and signal processing (ICCSP)
14. Mokoena DN, Nair K (2018) Representation of pose invariant face images using SIFT. In: International conference on advances in big data, computing and data communication systems (icABCD)
15. Jebarani WSL, Kamalaharidharini T (2017) PNN-SIFT: an enhanced face recognition and classification system in image processing. In: 2017 4th international conference on electronics and communication systems (ICECS)
16. Karami E, Shehata M, Smith A (2015) Image identification using SIFT algorithm: performance analysis against different image deformations. In: Proceedings of the 2015 newfoundland electrical and computer engineering conference St. john's, Canada

Chapter 4
Post-Stroke Readmission Prediction Model Using Machine Learning Algorithms

Lavanya Kommina, Prasannavenkatesan Theerthagiri, Yogeshwari Payyavula, Prudvi Sai Vemula, and Gurram Deekshith Reddy

Abstract Heart Stroke or Ischemic Stroke occurs whenever the blood flow in the brain is interrupted. Stroke is one of the major causes of disability and even mortality in India. The patients who are already affected with the stroke have the chances of readmission. Readmissions after stroke leads to several problems to the patients and to the hospitals. The problems for patients can be anything like medical costs, health risks, any disabilities, or mortality to the patients. The main problem for the hospitals is availability of resources to treat patients. We were considering various risk factors to predict the chances of post-stroke readmission for the patients. The factors include age, gender, body mass index, hypertension, average glucose level, resident type. The total of 40,000 individuals was considered in the analysis. The main objective of this is to analyze the stroke patient's details, based on that we will predict which algorithm is giving the best result. The algorithms used for predicting this model are Naive Bayes classifier (NBC), K-Nearest Neighbor algorithm (KNN), Support Vector Machine algorithm (SVM), Logistic Regression algorithm (LR), Random Forest algorithm (RF). The algorithm which is giving highest accuracy will be considered as the best suitable algorithm for predicting this model. When compared to all the algorithms, SVM algorithm has given the highest accuracy which is 98.10%. And, the algorithm which is giving the least accuracy is Naive Bayes which is 93.90%. And, the scores of SVM for precision, recall, F1 score, and AUC score are 0.45, 0.5, 0.49, and 0.5. This analysis helps to reduce the mortality rate.

Keywords Stroke · Machine learning · Mortality · KNN · RF · SVM · LR · NBC

L. Kommina · P. Theerthagiri (✉) · Y. Payyavula · P. S. Vemula · G. D. Reddy
Department of Computer Science and Engineering, GITAM School of Technology, GITAM University, Bengaluru, India
e-mail: vprasann@gitam.edu

R. Mathur et al. (eds.), *Emerging Trends in Data Driven Computing and Communications*, Studies in Autonomic, Data-driven and Industrial Computing,
https://doi.org/10.1007/978-981-16-3915-9_4

1 Introduction

Stroke is one of the leading diseases which causes disability and even mortality to many of the people worldwide. In many of the stroke survivors, it leads to recurrence of stroke. This increases the burden on affected patients as well as on the health care system [1].

Stroke is similar to heart attack, but when the blood supply in the brain is interrupted, then it occurs. According to WHO, 15 million people are suffering from the stroke worldwide. And, 5 million people die every one year. So, after-stroke readmission not only leads to disability of the patients and mortality but also leads to an increase in their overall medical costs. So, predicting readmission after stroke is very important [2].

Also, for hospital performance, readmission and mortality are the important indicators. Stroke will last only for a few minutes. But several symptoms last only for one day. These include the paralysis in faces, legs, etc. Readmissions to different hospitals admission within 30 days or 90 days and also it can occur either at the same hospital or different hospital. So, the models should be implemented by using hospital-specific methods with hospital-specific information [7, 10].

For determining the efficiency and quality of hospital healthcare system, readmitting to the hospital after discharging is one of the important factor. Readmission rates are more in stroke patients worldwide. So, we have to concentrate on this [3]. So, reducing readmission has become one of the active areas in medical literature. And also, health care administrators have been developing several health care policies for proper decision making in the hospitals [4].

And, there are many machine learning algorithms which have been used to improve patients' health by early detection of the patient's condition and giving treatment to them [5].

And, there can be a number of variables which can be incorporated because all the calculations in machine learning can be performed by computer. So, these all advantages made machine learning most suitable for the medical field. This project makes use of machine learning algorithms in order to accomplish the accurate predictions [6]. Machine learning algorithms give accurate results.

Several risk factors are responsible for the heart stroke. Some of the risk factors can be controlled and some of the risk factors cannot be controlled. The risk factors which cannot be controlled are age, heredity, gender, etc. The factors which can be controlled are lifestyle choices and health conditions like high blood pressure, excess weight, smoking, alcohol consumption, etc.

And, also there are several risk factors that were identified from several years; they are socio economic factors, alcohol consumption, drug usage, etc. And recognizing all these factors can be used to decrease the stroke [7]. The main contribution in this paper is to develop a best suitable method for stroke prediction of patients.

2 Related Works

In recent years, there were different works published based on Machine Learning algorithms for stroke prediction. Ling-Chien Hung et al. [8] in the paper A Machine Learning Approach to Predict Readmission for mortality in patients hospitalized for stroke or ischemic attack stated that Now-a-days heart stroke is the major cause of death and it leads to disabilities for many people. So, they proposed readmission after stroke leads to several problems to the patients and to the hospitals. It increases the mortality and disability rate and also medical costs for patients. Predicting the risk for readmission and also the factors which are responsible for readmission are a must needed for quality treatment and resources required for that treatment. Several factors included in this research paper are age, visits, BMI, severity, etc. The model is predicted using several algorithms like CART, KNN, LR, MLP, NB, RF, and SVM. In many of the health-related data sets, the class is imbalanced, so many cost-sensitive learning and data resampling methods are used to balance. In this study, they compared several prediction models using various machine learning techniques. And, among these, data resampling methods under sampling method has given better performance for prediction of ML techniques. Smote has given lesser performance. While compared to other models, "Naive Bayes and logistic regression" algorithms have given more accuracy [8].

Yu-Ching Chen et al. proposed a model named Machine learning algorithms to predict 30-day readmission in patients with stroke. The main purpose of this research is to find the accuracy of the Artificial Neural Networks (ANN), K-nearest Neighbor (KNN), Naive Bayes Classifier (NBC), Support vector machine (SVM). Among all these algorithms, ANN has given the highest accuracy. And also, it suggested that the patients before discharging from the hospital should be known about their recovery factor and possible outcomes. This analysis helps to reduce the mortality rate as well as disability rate. The main idea is to find the risk factor of the person who is already affected with the stroke and also finding the accuracy of the algorithm. Previous analyses were mainly performed using a single medical center data set, but this study used data from several other medical institutions so that it provided a much more accurate result. Using the KNN model, we can get accurate results from noisy data sets as well. If we use several other predictive factors or attributes with the ANN model, we will get much more accurate models [4].

Joon Nyung et al. proposed a model named Machine learning-based model for prediction of outcomes in acute stroke. The main purpose of this research, which is to predict the outcomes of long-term Ischemic Stroke patients, helps us to make several decisions, so the machine learning techniques are adapted because they give high accuracy results. Favorable outcomes were mainly defined by ranking scale score. They developed three models by using machine learning techniques named Deep Neural networks, Random Forest, Logistic Regression. And, they should compare the predictable outputs of those and should find which has the highest accuracy. By this, it is proven that unpredictable solutions can also be predicted by using Machine Learning algorithms. Models are trained with all variables to classify patients to get favorable outcomes. Among all the three techniques, Deep Neural networks gave better performance when compared to other models. So, the main purpose of this study is to predict the outcome of Ischemic stroke patients by applying several Machine Learning techniques [6].

Pattanapong et al. proposed a model named Long Short-Term Memory Recurrent Neural Network for Stroke Prediction. EHR (Electronic Health Record) gives more details about patient's physical, mental, lab results, etc. And also, there is a code (ICO-10) for each and every patient. Several data mining techniques are used for analyzing patient's records. RNN (Recurrent Neural networks) is used in machine learning bioinformatics. Firstly, they included the (ICO-10) code and other important risk factors into the model for prediction. In this EHR, they consider the factors such as gender, DOB, clinical operation, operation date, diagnosis, code, diagnosis type. Here mainly, they predicted using EHR and Deep learning techniques. Accuracy rate mainly depends on the size of sample. The result is more reliable if there are large data sets. So, the LSTM algorithm is more suitable for predictive analysis of stroke [9].

Pattanapong Chantamit et al. proposed a model named A Case based reasoning framework for prediction of stroke. They stated Now-a-days stroke is the 3rd common disease of death all around the world. And patients who are surviving also face so many problems. There are several health conditions and lifestyle factors that have been identified as the risk factors. They can be changed, less well documented, and some are unchanged like age, heredity, race, gender, prior stroke etc. Recognizing these types of factors, like diabetes, smoking, alcohol consumption, location, either rural or urban, can reduce the occurrence of stroke. CBR (Case-Based Reasoning) Algorithm is used for solving this problem. It has four steps, retrieve, reuse, revise, retain. From this CBR, it will determine the cases that develop the stroke so that the burden can be reduced on patients. And also, it compares the stroke patient's risk factors like BP, smoking, etc. And also, it gives suggestions, warnings to patients, because strokes do not have warning signs. For complex data, CBR is applied by using Machine learning and data mining techniques. And, it uses KNN algorithm with weighted-feature technique to compare and retrieve previous cases and new cases. It has six main framework processes, clustering, retrieval, reusing, prediction, retain, revise [7].

Ashan Rao et al. have proposed a paper named Systematic Review of hospital Readmissions in Stroke Patients. This paper stated that the main aim of this study is identifying the common causes and patterns which are responsible for stroke patient's readmission by using hospital patients' stroke patient's data. They collected 2,126,627 individuals' data. They found that some of the common and important risks are age, infection, cardiac conditions, and history of heart disease, diabetes, and high blood pressure. This study investigated causes and their sequence of readmission among strokes. Kidney and urinary tract infections and other gastrointestinal discards are more common in stroke patients [1, 20].

Domenico et al. have proposed a paper named Machine Learning to predict mortality after rehabilitation among patients with severe stroke [21]. They stated that this stroke is the leading disease which causes death worldwide. The main aim of this model is to identify the performance of machine learning tree-based algorithms and by using a logistic regression algorithm, Random Forest. The model which has been developed by using LR has achieved 0.745 of AUC and total accuracy achieved was 75.7%. RF has achieved 0.928 of AUC and total accuracy of 86.3%. So, among all the three algorithms, RF algorithm has achieved the higher AUC and highest F measure. They stated that approximately (20–25)% of stroke-attacked patients have severe disabilities. And also, ML algorithms may also be used for screening, diagnostics, prognosis, purpose [3].

Damien Echevin et al. have proposed a model named Hospital Readmission is highly predictable from DL. They stated that readmission is very costly and existing models which have been developed are moderate in predicting readmission of stroke patients. They thought of developing a model which has a better chance for readmissions. They have taken 1,633,099 individual data between the year 1995 and 2012. They measured the performance by using AUC. They have developed the model using five algorithms, namely, Simple Logistic Regression, Decision Trees, Naive Bayes, Random Forest, Deep Learning. Among all the above algorithms, Deep learning has given the highest accuracy. Random Forest also almost reached the accuracy of Deep learning. The AUC reached 78% of hospital admission for these two algorithms [7].

3 Introduction

3.1 Data Sources

We have collected the stroke dataset from Kaggle repository [18]. Our dataset contains a total 43,400 individual's patient information of which 17,724 are male and 25,665 are female, and others are 11. The dataset contains a total of seven features. The list of attributes of our stroke dataset is given in Table 1. The attributes in our dataset are age, resident type, BMI, hypertension, glucose level, gender, and stroke. It is very difficult to find a symmetric dataset.

Table 1 List of attributes of the Stroke dataset

Attributes	Description
Age	Age of the patient
Hypertension	Blood pressure of the patient
BMI	Body mass index is a measure of body fat based on height and weight of the patient
Gender	Sex of the patient (Female or male)
Average glucose level	Glucose levels of the patient
Residence type	Where the patient lives (urban or rural)
Stroke	The patient is affected with stroke or not

3.2 Classifiers

In this proposed work, we used five algorithms to prepare a model and to predict whether the person is affected with stroke or not. The algorithms which we have used are KNN, LR, SVM, NB, RF.

KNN Algorithm (K-Nearest Neighbor) is a ML algorithm which is related to supervise learning. It is used for both classification, regression, and also for missing imputation [13]. It is the simplest one when compared to all other ML algorithms [5]. It is a simple, valid, and non-parametric algorithm [12]. It is also known as "Lazy learner algorithm." It stores the trained data, whenever a new instance is given to test based on the similarity, it will put the new instance into the most suitable categories. So, at last prediction can be done by using searching the entire training data [16]. There are several distance metrics that can be used for KNN. In our model, we used Euclidean distance. In our model, KNN algorithm is used to classify whether a person or patient is affected with stroke or not i.e., (yes/no). It takes the distance in test data and calculates its distance with trained instances by using the Euclidean distance formula.

Naive Bayes Classifier is a machine learning algorithm which is related to supervise learning. It is mainly used for solving the classifier problems. For making quick and fast predictions, this is the simplest and effective algorithm [11]. Based on the probability, it predicts the output. It is mainly related to Bayes theorem. The main advantage of using NB is it can handle the dataset which contains many attributes or columns [14]. In our model, it is used because it gives high scalability. And, Bayes theorem is mainly used to find out the conditional probability of occurring event to the probability of an event occurred [5]. Why do we choose NB to be our model? Because the data set is large and we want to get more effective output. In our model, we will give the new instance or record of the patient. Then, it will calculate the probability of each and every attribute in the record. Then it decides the record of the patient belongs to which class, i.e., (yes/or).

Logistic Regression is a ML algorithm which is related to supervise learning technique. The outcomes of LR are discrete or categorical values [5]. LR is mainly used for solving classification problems. There is a logistic function which is used in LR to predict which has the maximum value (either 0 or 1, yes/no, etc.). The curve is used to predict those outputs. In our model, the logistic function is used to predict whether the person is affected with stroke or not. Here, we use the concept of threshold value, the new instance, or record of what we have given will appear in the form of a curve; if the curve falls above, then it is yes, if it falls below then it is no, with respect to stroke [16].

Random Forest is a ML algorithm belonging to supervised learning. It is used for both classification and regression. It is an ensemble learning method [5]. It has many numbers of decision trees, and it takes an average of all to predict the accuracy of the dataset. RF tables a prediction from each tree and considers the votes of all. And which has more votes, then it will predict according to that. If there are many numbers of trees, then it gives the highest accuracy [16]. In our model, RF is used because it takes less time for training. RF has succeeded in many real-time applications [15]. As our dataset is very large, this algorithm predicts output with the highest accuracy. As our dataset is having many missing values, it works well for that. Firstly, we will give new instances, and then it will make the prediction for each decision tree, then it assigns the data points to the new category, then based on the votes of all decision trees, it predicts the output [16].

Support Vector Machine (SVM) is a supervised ML algorithm. It is used for both classification and regression problems. Here, hyperplanes are used. Hyperplanes are nothing but these are the decision boundaries which helps us to classify new instances [17, 19]. So, it finds a hyperplane in a N-dimensional space that classifies our data points. In our model, we will give the new instance for testing, then it outputs the hyperplane. This line is a decision boundary, it will classify whether the person is affected with stroke or not. If the new instance falls under one side, then it is classified as a person affected with stroke. If it falls on another side, then it is considered as a person not affected with stroke.

3.3 Data Preprocessing

Our dataset contains string values and integer values. For processing the data, the values should be only integers. We will convert all the string values to integer values, like in gender, we will replace "male" with "1" and "female" with "0," and in residence type, "rural" as "1" and "urban" as "0." We have also replaced null values with mean values. After preprocessing, we will use the preprocessed dataset to find the accuracy of machine learning algorithms. The given Fig. 1 represents the overall flow of the proposed model.

- Firstly, we have collected the dataset of patients who are already affected with stroke.
- And, we have done data preprocessing for the dataset what we have collected, i.e., cleaning and preparing the dataset for model prediction.
- We have trained five different prediction models using five different algorithms (KNN, SVM, LR, NB, RF).
- And, then we tested the training model.

And, we have calculated the accuracy of models by using different performance metrics to validate the models.

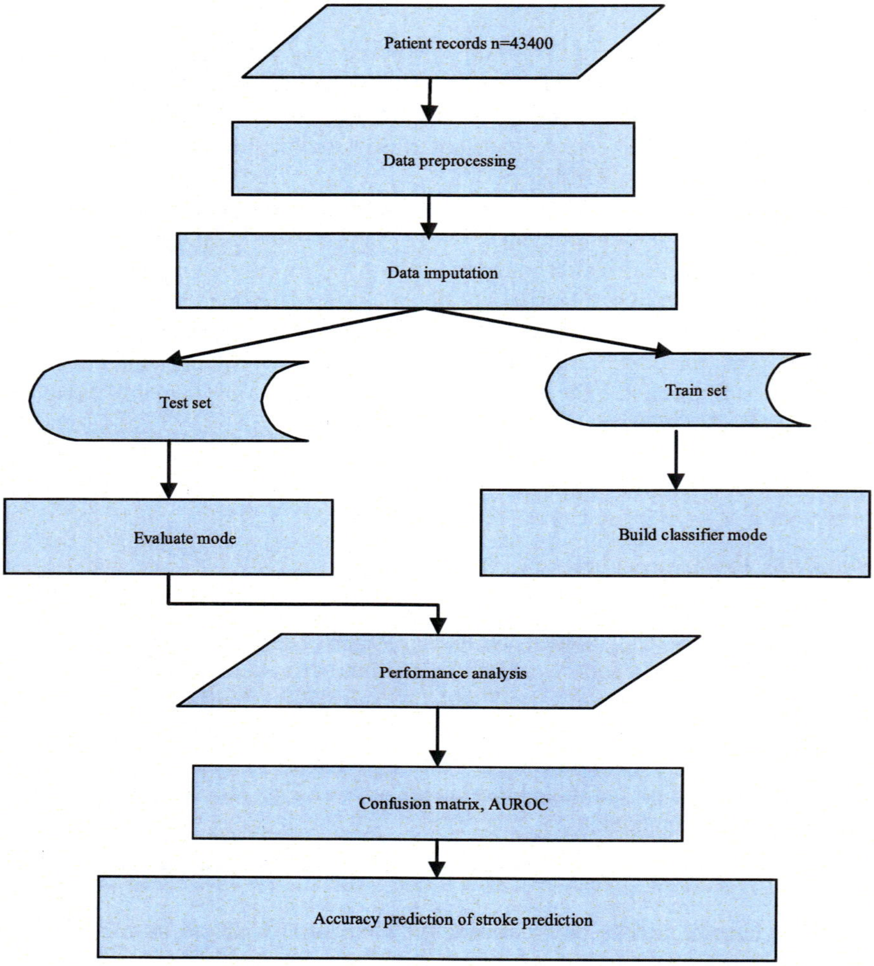

Fig. 1 Flow of proposed model

4 Experimental Results

To evaluate the performance of our model, we have used several performance metrics such as Accuracy, Precision, Recall, Confusion matrix, F1 score, and AUROC score. Classification accuracy is nothing but the number of correct predictions made as a ratio of all predictions made [17].

$$\text{Accuracy} = \text{TP} + \text{TNTP} + \text{FP} + \text{FN}$$

F1 score will give us the harmonic mean of precision and recall. Mathematically, F1 score is the weighted average of the precision and recall. The best value of F1 would be 1 and worst would be 0. We can calculate F1 score with the help of the following formula:

$$\text{F1 Score} = 2 * (\text{precision} * \text{recall})/(\text{precision} + \text{recall})$$

Precision, used in document retrievals, may be defined as the number of correct documents returned by our ML model.

$$\text{Precision} = \text{TP}/\ (\text{TP} + \text{FP})$$

Recall may be defined as the number of positives returned by our ML model.

$$\text{Recall} = \text{TP}/\ (\text{TP} + \text{FN})$$

A confusion matrix is nothing but a table with two dimensions viz. "Actual" and "Predicted," and furthermore, both the dimensions have "True Positives (TP)," "True Negatives (TN)," "False Positives (FP)," and "False Negatives (FN)" as shown below Actual Predicted. It is the easiest way to measure the performance of a classification problem where the output can be of two or more types of classes. Figure 2 illustrates the confusion matrix.

Performance of all the five algorithms is shown in Table 2. From the above table, we see that accuracy of KNN algorithm is 97.9%, random forest algorithm is 98.08%, NB algorithm is 93.90%, logistic regression and SVM have 98.10% (Fig. 3). Figure 3 illustrates AUROC score for the algorithms.

Confusion matrix of all algorithms:

Figures 4, 5, 6, 7 represent the confusion matrix of the algorithms SVM, LR, RF, KNN, and NB, respectively.

TP: Predicted result is positive and it is true. The proposed model predicted that the patient is affected with stroke, and actually, he is affected with stroke.

Fig. 2 Confusion matrix

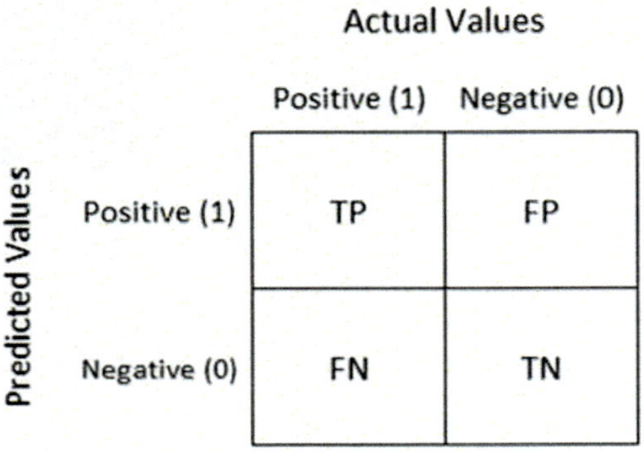

Table 2 Detailed performance of all five algorithms

Algorithm	Accuracy	Precision	Recall	F1 score	AUC score
SVM	98.10	0.532	0.5	0.49	0.54
KNN	97.99	0.536	0.502	0.501	0.502
RF	98.08	0.657	0.506	0.507	0.506
NB	93.99	0.539	0.604	0.552	0.604
LR	98.10	0.53	0.5	0.495	0.5

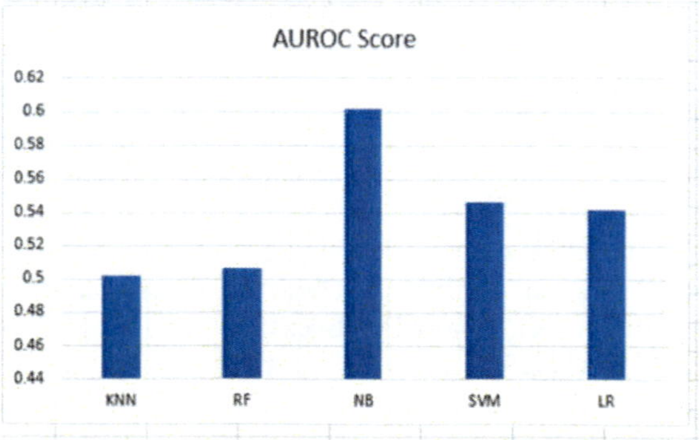

Fig. 3 AUROC score for all the algorithms

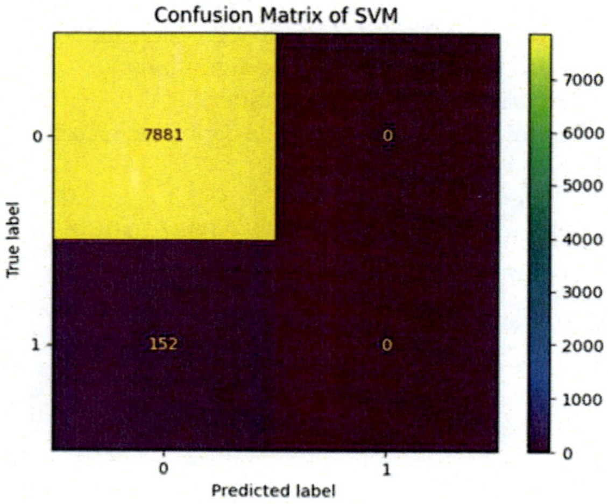

Fig. 4 Confusion matrix for support vector machine

TN: Predicted result is negative but actually it is true. The proposed model predicted that the patient is not affected with stroke but actually he is affected with stroke.

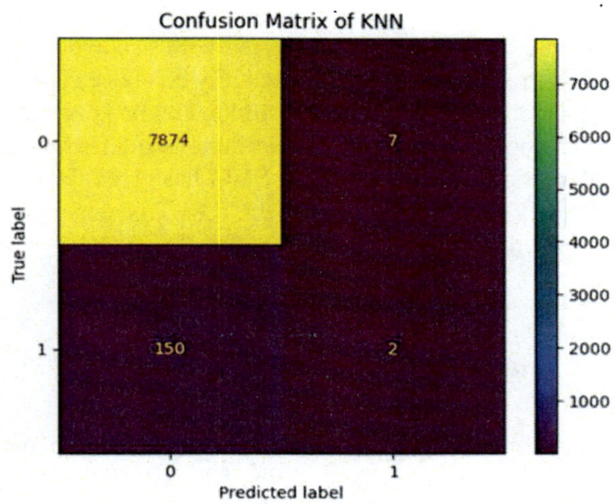

Fig. 5 Confusion matrix for K-nearest neighbor

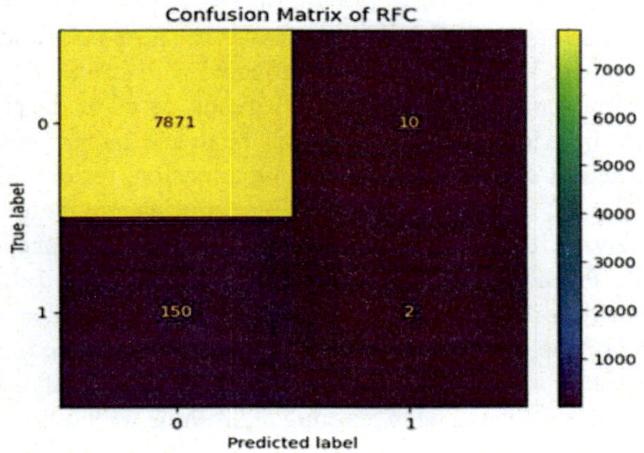

Fig. 6 Confusion matrix for random forest classifier

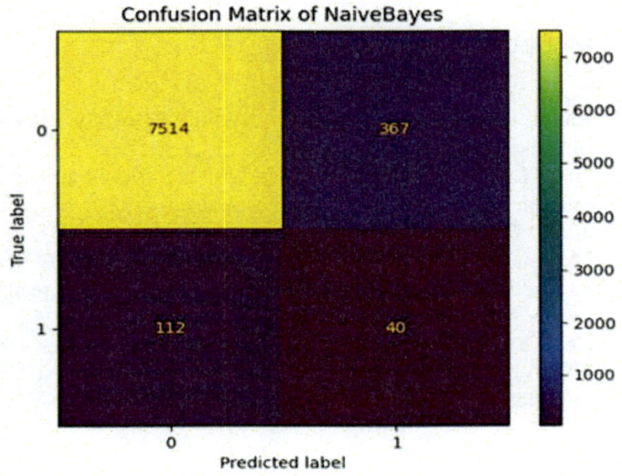

Fig. 7 Confusion matrix for Naive Bayes algorithm

FP: Predicted result is positive but actually it is false. The proposed model predicted that the patient is affected with stroke but actually he is not.

FN: Predicted negative and actually it is false. The proposed model predicted that the patient is not affected with stroke, and he is actually not affected with stroke.

True positive for all the algorithms is high when compared to TF, FP, and FN. And, among all the algorithms, SVM has given the highest result for True Positive.

The SVM algorithm predicts the 7881 True positives, and it is high when compared to all the other algorithms.

5 Conclusion

In this paper, a large dataset of stroke-attacked patients has been analyzed. Five different classifiers such as KNN, RF, NB, SVM, and LR were used for detection of the stroke disease. Based on the performance analysis of all the algorithms, SVM performed better than all the other models. And Naive Bayes has given the least accuracy among all the other models. The proposed model suggests that it might help people to know prior to being affected with stroke. In this paper, we have analyzed several parameters of the stroke patients, based on the patient details and past disease history. Moreover, this work has analyzed larger patient records with the patient attributes such as age, gender, hypertension, resident type, average glucose levels, and BMI. Even though a large number of patient records and several attributes were analyzed, the proposed model produces 98.10% of accuracy and higher true positive rate. Further, the result shows that among all analyzed algorithms, the SVM algorithm for stroke readmission gives accurate results. And, also this model helped the patients to find the risk factors which are mainly responsible for re-occurrence of the stroke. In future work, this model will be extended to analyze several other parameters of the patients using deep learning algorithms with optimization techniques.

References

1. Rao A, Barrow E, Vuik S, Darzi A, Aylin P (2016) Systematic review of hospital readmissions in stroke patients. Stroke Res Treat
2. Hung L-C, Sung S-F, Ya-Han Hu (2020) A Machine learning approach to predicting readmission or mortality in patients hospitalized for Stroke or transient ischemic attack. Appl Sci 10(18):6337
3. Lichtman JH, Leifheit-Limson EC, Jones SB, Watanabe E, Bernheim SM, Phipps MS, Bhat KR, Savage SV, Goldstein LB (2010) Predictors of hospital readmission after stroke: a systematic review. Stroke 41(11):2525–2533
4. Chen YC, Chung JH, Yeh YJ, Lin HF, Lin CH, Hsien HH, Hung KW, Yeh SC, SHI HY (2020) Machine learning algorithms to predict 30-day readmission in patients with stroke: a prospective cohort study

5. Shoily TI, Islam T, Jannat S, Tanna SA, Alif TM, Ema RR (2019) Detection of stroke disease using machine learning algorithms. In: 2019 10th international conference on computing, communication and networking technologies (ICCCNT). IEEE. pp 1–6
6. Heo J, Yoon JG, Park H, Kim YD, Nam HS, Heo JH (2019) Machine learning–based model for prediction of outcomes in acute stroke. Stroke 50(5):1263–1265
7. Échevin D, Li Q, Morin MA (2017) Hospital readmission is highly predictable from deep learning. Chaire de recherche Industrielle Alliance sur les enjeux économiques des changements démographiques
8. Chuang K-Y, Shwu-Chong Wu, Ma A-H, Chen Y-H, Chen-Long Wu (2005) Identifying factors associated with hospital readmissions among stroke patients in Taipei. J Nurs Res: JNR 13(2):117–128
9. Madhu G (2018) Long short-term memory recurrent neural network for stroke prediction. In: International conference on machine learning and data mining in pattern recognition. Springer, Cham, pp 312–323
10. Scrutinio D, Ricciardi C, Donisi L, Losavio E, Battista P, Guida P, Cesarelli M, Pagano G, D'Addio G (2020) Machine learning to predict mortality after rehabilitation among patients with severe stroke. Sci Rep 10(1):1–10
11. Singh MS, Choudhary P (2017) Stroke prediction using artificial intelligence. In: 2017 8th annual industrial automation and electromechanical engineering conference (IEMECON). IEEE, pp 158–161
12. Wang Y, Wang ZO (2007) A fast KNN algorithm for text categorization. In: 2007 international conference on machine learning and cybernetics, vol 6. IEEE, pp. 3436–3441
13. Zhang S, Cheng D, Deng Z, Zong M, Deng X (2018) A novel KNN algorithm with data-driven k parameter computation. Pattern Recogn Lett 109:44–54
14. Huang Y, L Lei (2011) Naive Bayes classification algorithm based on small sample set. In: 2011 IEEE international conference on cloud computing and intelligence systems. IEEE, pp 34–39
15. Prasannavenkatesan T, Jacob J, Usha Ruby U (2020) Prediction of COVID-19 possibilities using KNN classification algorithm. 03 Nov 2020, PREPRINT (Version 2) Research Square
16. Wu Q, Ye Y, Zhang H, Ng MK, Ho SS (2014) ForesTexter: an efficient random forest algorithm for imbalanced text categorization. Knowl-Based Syst 67, 105–116
17. Prasannavenkatesan T (2021) Probable forecasting of epidemic COVID-19 in using COCUDE model. EAI endorsed transactions on pervasive health and technology, Online First
18. Kaggle Webpage, https://www.kaggle.com/lirilkumaramal/heart-stroke, last accessed 2021/01/21.
19. Theerthagiri P (2021) Forecasting hyponatremia in hospitalized patients using multilayer perceptron and multivariate linear regression techniques. Concurr Comput Pract Exper e6248
20. Prasannavenkatesan T (2020) FUCEM: futuristic cooperation evaluation model using markov process for evaluating node reliability and link stability in mobile Ad hoc network. Wirel Netw 26(6):4173–4188
21. Theerthagiri P (2019) COFEE: context-aware futuristic energy estimation model for sensor nodes using Markov model and auto-regression. Int J Commun Syst e4248

Chapter 5
Handwriting Recognition Using Deep Learning

Shubh, Muskan Agarwal, Fardin Hassan, Garima Pandey, and Soumalya Ghosh

Abstract Handwritten figure recognition, whether it has characters or digits, has been a problem for a very long time in the department of design/pattern recognition and their classification. There are traditional techniques that have created Optical Character Recognition (OCR) systems in the 1950s, which were more costly, platform-dependent, and heavy in size. With the active contribution, it becomes possible to develop more efficient techniques. Now, it is also possible to use and contribute at the same time. One of the most prominent modern techniques is focused on segmenting individual handwritten text, i.e., Convolutional Neural Network (CNN), a technique of deep learning. In this experimental analysis, our motive is to spot the impression of the deep learning technique, and predict written digits and characters from 0 to 9 and a/A to z/Z when a set of lakhs of samples of EMNIST are given as an input. The major goal of this paper is to contribute to this topic, providing efficient and steady procedures for the identification of handwritten alphanumerals. As we all know that everyone has a different form of writing digits and characters, humans will acknowledge simply except for the computer; it's relatively a troublesome task. Therefore, the neural network-based approach comes into the picture, which is a technique based on how humans recognize any imagery. The machine can recognize too and become much more accurate from its experience. The model is trained by using convolutional neural networks. The general system accuracy obtained was 99.40% (digits) and 94.45% (characters).

Keywords Optical Character Recognition (OCR) · Handwriting recognition · Convolutional Neural Network (CNN) · Deep learning

---

Shubh · M. Agarwal · F. Hassan · G. Pandey (✉) · S. Ghosh
Galgotias University, Uttar Pradesh, Greater Noida 226001, India
e-mail: Garima.pandey@galgotiasuniversity.edu.in

S. Ghosh
e-mail: soumalya.ghosh@galgotiasuniversity.edu.in

R. Mathur et al. (eds.), *Emerging Trends in Data Driven Computing and Communications*,
Studies in Autonomic, Data-driven and Industrial Computing,
https://doi.org/10.1007/978-981-16-3915-9_5

1 Introduction

Handwritten characters or digit recognition can be used within the fields of medicine, schools or colleges, banking, and in other different fields in the form of the ability to store, read, write material, access, etc., and to ease up many processes. Earlier machine learning techniques such as Neural Network, Support Vector Machine (SVM), Random Forest, etc., are used as classification techniques. As the technology is advancing day-by-day, new advancements within the field of component and economic analysis in the computer science field also came to the limelight, where the booming field of data science gave emergence to Deep Learning. As per the latest published articles, the use of deep learning techniques for character/digit or pattern identification is becoming much more popular. The reason behind its popularity is the improvements in its performance within the field of pattern recognition.

There has been lots of research on the recognition of handwritten characters and digits, but still, this problem has persisted due to a variety of handwritings or data emerging day by day, which intern affects the effective accuracy. During this experimental analysis, the written character recognition is going to be done by employing a convolutional neural network and TensorFlow. Analysis within the handwriting recognition field is targeted with deep learning techniques as weapons helped many to achieve breakthrough performances, within the past few years. Still, the growth in the number of written facts and therefore the requirement or availability of better processing model power remains in demand, though for more improvement in recognition accuracy, it deserves further investigation. Convolutional neural network (CNN) is a special deep learning technique by which one can detect the design of the written symbol such as digits and characters. in ways that facilitate the automation in the field of extraction, giving distinct choices and producing the foremost appropriate solution for finding handwriting recognition problems. The goal is to test the CNN model on larger datasets to lookout different patterns and work on its accuracy conjointly.

2 Dataset

The MNIST or Modified National Institute of Standards and Technology [1] is the collection of handwritten numbers with many different handwritings in the form of structured data. Here, we have used EMNIST or Extended-MNIST, which is the bigger picture of the collection, having handwritten numbers as well as handwritten alphabets. These structured collections are derived from the NIST Special Database-19. In these collections, the image of a handwritten symbol is transformed into a size of 28×28 (length $\times$ breadth) layout, and a dataset shape that immediately suits the dataset is used. The set of structured data used in training a model has 697,932 images and to test the model it has 116,323 uppercase and lowercase alphabets and numeral images from 0 to 0039 that are labeled. The testing set and the training set

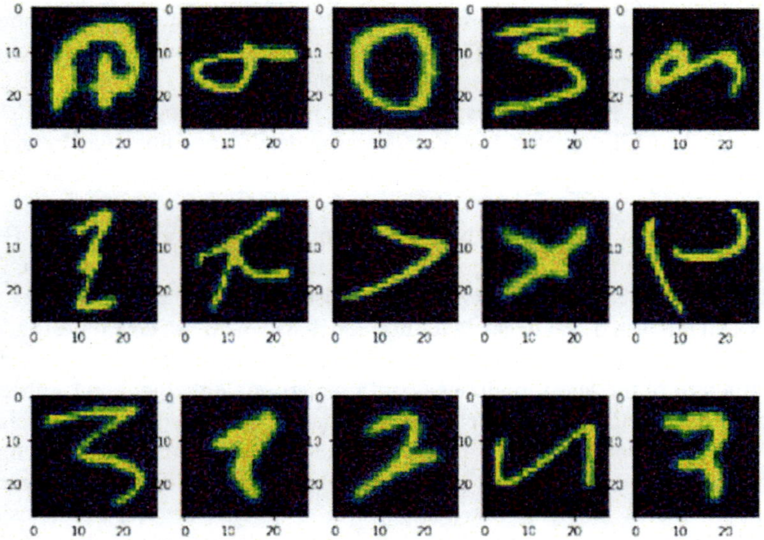

Fig. 1 Types of images used in this research

are to be had withinside the shape of the listing inside a listing. Each object of the outer listing depicts an image and the internal listing shows the depth value of 784 pixels (due to the fact that the scale of the image is 28 × 28 pixels) starting from 0 to 255. The testing images and training images are generally used as grayscale images. Both the testing images in addition to training images are turned horizontally by rotating at 90 degrees clockwise. Y train and Y test are formed arrays that comprise numbers ranging from 0 to 61 depicting a particular symbol. Here, 0–61 numbers represent 10 numbers from 0 to 9, plus 26 alphabets both in uppercase and lowercase (Fig. 1).

3 Methodology

3.1 Convolutional Neural Network (CNN)

CNN is a special category of deep learning that has attracted a great deal of interest in current studies. It can routinely extract high-degree features from the raw input, which is a great deal with greater effect. Hence, it further added bigger upgrades in some parts of its process—like image partitioning [2] and identification [3]. In the prediction part, CNN has already been hired in lots of working segments, and its performance has exceeded it in large part already.

Let's begin from the start, from where CNN came into the picture, a comparable framework which consists of learning from their labels (or teacher) commonly referred to as supervised, followed by the machine learning (ML) models. CNN was born from this learning pattern technique. Illustrative models like DNCON2 [4], DeepContact [5], RaptorX [6], etc. These models' capabilities can be divided dimensionally as the features that got extracted from both horizontal and vertical planes.

The former traits were related with the rest of the series and encompass HMM profile produced from Hhblits, expected another layout, and worthy readiness. Final capabilities have speculated effects of unsupervised learning models, together with PSICOV and mfDCA. Later to get a tridimensional matrix 'm' for convoluting action, the unidimensional attributes of remains, 'a' and 'b' withinside the series are merged to the structure of the latest unidimensional function tuple that is then localized in 'm(a, b)'. Through this, the 1-D features of remainder enclosed by the association of the remainder a and b are taken into consideration through the convolution operation while predicting whether a and b are in touch. Later, the bidimensional features are merged with 'm' alongside the passage size to produce the last tridimensional matrix. The deep learning architectures hired withinside the prediction models are specifically residual networks (ResNets) [7] that are broadly utilized in the disciplines like image analysis and NLP. Because of the top-notch capabilities of CNN in attribute detection and pattern identification, the overall improvement of these models is remarkable (Fig. 2).

3.2 Callback APIs

A callback [8] is an objective that may carry out different movements at diverse degrees of training (e.g., on the beginning or cease of an epoch, and earlier than or after a single batch).

You can use callbacks to

- Write TensorBoard logs after each batch of education to reveal your metrics;
- Periodically keep your version to disk;
- Do early stopping to get a view on inner states and data of a version at some stage in training...and more.

Here, we've got used to ModelCheckpoint, EarlyStopping, and ReduceLROnPlateau.

ModelCheckpoint [9]. It is a callback API, used to preserve the model that shows the best performance among the rest. Generally, the word ModelCheckpoint comprises two words model and checkpoint. In combination, we can say that it marks the checkpoints on the basis of the performance of the model. The checkpoint represents the best performance of the model recorded across multiple epochs. To maintain and to declare whether the model is good in comparison to others, this API

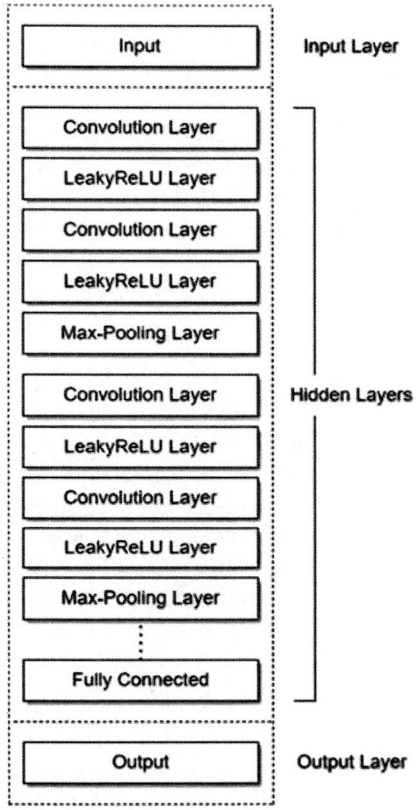

Fig. 2 CNN architecture used in this research

maintains a file to cross-check the best one out. A few alternatives this callback gives include

- Either it simply preserves the model that has shown great accuracy so far, or may not keep the model on the cease of each repetition of the learning cycle no matter what is efficient.
- The frequency has to store. Currently, the callback helps to save, on the cease of each repetition of the learning cycle, or after a set range of training blocks.
- Either best weights are preserved, or the entire version is preserved.

EarlyStopping [10]. It is a callback API used to stop the training of the model when it stops to improve in its judging terms such as validation accuracy in our case. Not only does it help in getting the output on time but also helps the model to not overfit the data, as the aim is to decrease the loss also. For instance, the metric to be monitored is 'loss', which should be minimum. When we train the model, it will make count of loss on each epoch of the phase. If it seems on further epochs the value is not changing or minimum value is already being evaluated, then it will terminate the training.

ReduceLROnPlateau [11]. It is a callback API used to lower the rate of learning while the model's performance seems to be constant or no change after a certain level of patience mentioned by the user. The word ReduceLROnPlateau consists of four words, namely reduce, learning (L), rate (R), on (On), and plateau, which in combination means that reduce the learning rate after reaching a certain condition. Generally, models get benefitted from lowering the learning rate, as it allows the training in a better manner. Here, we're going to monitor validation accuracy.

3.3 Confusion Matrix

A Confusion matrix [12] is a square matrix of an N x N dimensions, used for comparing the overall performance of a class model, wherein 'N' is the range of goal classes. The matrix compares the real target values with the ones anticipated via the way of means of the machine learning model. The rows constitute the anticipated values of the goal variable. Confusion matrices are usually used to verify and visualize important predictions to analyze recall, specificity, accuracy, and precision. Confusion matrices are beneficial as they supply direct comparisons of values between True-Positives (a), False-Positives (b), True-Negatives (c), and False-Negatives (d).

$$Accuracy = (a + c)/(a + b + c + d) \quad (1)$$

$$Precision = a/(a + b) \quad (2)$$

$$Recall = a/(a + b) \quad (3)$$

When an element in a matrix is not a part of its diagonal elements, then it leads to an error.

$$Error\,Percentage = (error\text{x}100)/(a + b + c + d) \quad (4)$$

As per the learnings, we have calculated accuracy and error percentage on the basis of the confusion matrix so that we can have a much clearer picture of our model. For the first one, we got on the basis of our original model and the second one is depicting its improvement. From Figs. 3 and 4, we can see that it has exceeded the accuracy from 99.0476 to 99.4047%, by addition of approximately 0.4%, and reduced the number of errors from the validation set.

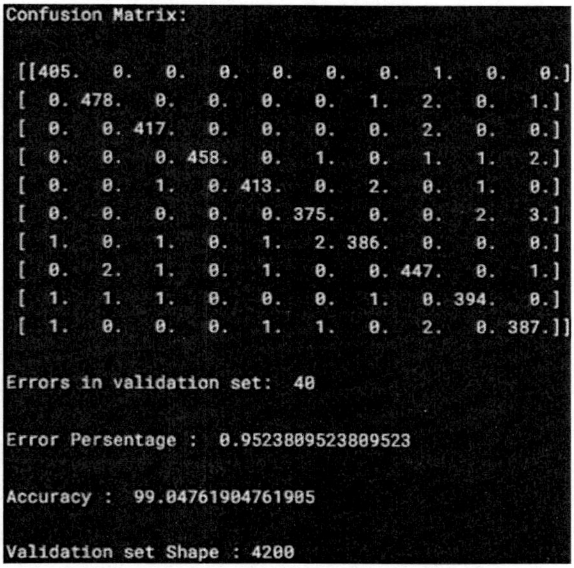

```
Confusion Matrix:

 [[405.   0.   0.   0.   0.   0.   0.   1.   0.   0.]
 [  0. 478.   0.   0.   0.   0.   1.   2.   0.   1.]
 [  0.   0. 417.   0.   0.   0.   0.   2.   0.   0.]
 [  0.   0.   0. 458.   0.   1.   0.   1.   1.   2.]
 [  0.   0.   1.   0. 413.   0.   2.   0.   1.   0.]
 [  0.   0.   0.   0.   0. 375.   0.   0.   2.   3.]
 [  1.   0.   1.   0.   1.   2. 386.   0.   0.   0.]
 [  0.   2.   1.   0.   1.   0.   0. 447.   0.   1.]
 [  1.   1.   1.   0.   0.   0.   1.   0. 394.   0.]
 [  1.   0.   0.   0.   1.   1.   0.   2.   0. 387.]]

Errors in validation set:  40

Error Persentage :  0.9523809523809523

Accuracy :  99.04761904761905

Validation set Shape : 4200
```

Fig. 3 Confusion matrix before image augmentation

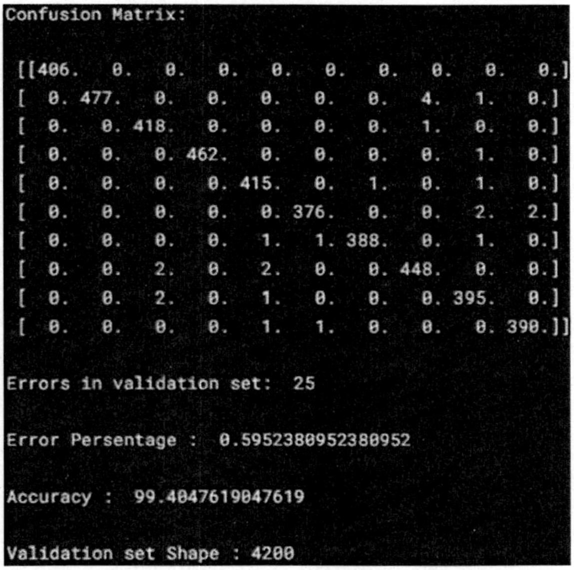

```
Confusion Matrix:

 [[406.   0.   0.   0.   0.   0.   0.   0.   0.   0.]
 [  0. 477.   0.   0.   0.   0.   0.   4.   1.   0.]
 [  0.   0. 418.   0.   0.   0.   0.   1.   0.   0.]
 [  0.   0.   0. 462.   0.   0.   0.   0.   1.   0.]
 [  0.   0.   0.   0. 415.   0.   1.   0.   1.   0.]
 [  0.   0.   0.   0.   0. 376.   0.   0.   2.   2.]
 [  0.   0.   0.   0.   1.   1. 388.   0.   1.   0.]
 [  0.   0.   2.   0.   2.   0.   0. 448.   0.   0.]
 [  0.   0.   2.   0.   1.   0.   0.   0. 395.   0.]
 [  0.   0.   0.   0.   1.   1.   0.   0.   0. 390.]]

Errors in validation set:  25

Error Persentage :  0.5952380952380952

Accuracy :  99.4047619047619

Validation set Shape : 4200
```

Fig. 4 Confusion matrix before image augmentation

3.4 RMSprop and Adam

As our purpose is to decrease the cost function with the aid of locating the optimized cost for weights, we additionally wanted to make sure that the set of rules generalizes properly. This will assist in making a higher prediction for the statistics that changed into now no longer visible before.

Optimizers replace the load parameters to decrease the loss function. Loss function acts as a route to the destination (value), tells the optimizer if it is shifting too far from the original value, to attain its minimum value. Here, we've used RMSprop and Adam optimizers.

RMSprop [13]. The core of this optimizer:

- Sustaining the active state of the mean of (gradients)^2.
- Dividing the gradient by the mean's root.

This application of this optimizer is to make use of undeniable momentum, now no longer Nesterov momentum. The targeted model moreover keeps a shifting common of the gradients and makes use of that common to estimate the variance [14].

$$e_t = \rho e_t - 1 + (1 - \rho) * st^2 \tag{5}$$

$$\Delta w_t = -(lr/(e_t + \varepsilon)1/2) * s_t \tag{6}$$

$$w_{t+1} = w_t + \Delta w_t \tag{7}$$

lr: learning rate

e_t: Exponential mean of quad of the slope

s_t: slope at time t along parameter w^j.

In Eq. (5), we calculated an exponential mean of the quad of the slope. Seeing that we perform it one after the other for every criterion, slope s_t right here correlates with the estimate or factor of the slope alongside the course acted for via the means of the criterion we're upgrading. To do that, we increase the radial growth mean calculated until the closing replaces from a hyper-parameter, depicted via the symbol 'z'. We then produce the quad of the latest slope with (1–z). After that, we upload them collectively to obtain the exponential mean until the modern time-step. The reason why we are using exponential mean is due to the fact as we are seeing, momentum example, it facilitates us to contemplate the extra latest slope updates other than the much least latest ones. The additives of the slope alongside weight-1 are much larger than those alongside weight-2. Since we make their quad and include them, they do not cancel out, and the exponential mean is massive for weight-2.

In Eq. (6), we determined our step length. We flow withinside the course of the gradient, however, our step length is laid low with the exponential mean. We selected a preliminary learning rate lr, after which we divide it with their mean. In our case, for the reason that the common of weight-1 is much larger than weight-2, the learning step for weight-1 is much lesser than that of weight-2. Hence, this could assist us to keep away from bouncing among the ridges and flow closer to the minima.

Equation (7) is simply the replace step. The hyper-parameter ρ is normally selected to be 0.9, however, one would possibly need to track it. The ε in Eq. (6) is to make sure that we no longer turn out to be dividing via way of means of zero and is usually selected to be 0.0000000001. It's additionally to be stated that the optimizer absolutely plays replicated tempering. Suppose if we are heading towards the minima, and we want to slow down so as to not to overshoot the minima. This optimizer robotically will lower the inclination of the slope closer to minima while the steps are too big.

Adam [15]. It is an optimizer, known for a hypothetical slope-drop approach that is primarily rooted in the supple assessment of first-order and second-order instant. The approach is computationally efficient, has little reminiscence requirement, is invariant to the crosswise resizing of the slope, and properly appropriate for troubles that can be big in terms of statistics/parameters.

$$e_t = b1 * e_{t-1} - (1 - b1) * s_t \tag{8}$$

$$c_t = b_2 * c_{t-1} - (1 - b_2) * s_t^2 \tag{9}$$

$$\Delta w_t = -(lr * e_t/(c_t + \varepsilon)^{1/2}) * s_t \tag{10}$$

$$w_{t+1} = w_t + \Delta w_t \tag{11}$$

lr: Initial Learning Rate
s_t: Slope at time t along parameter w^j
e_t: Exponential mean of slope along parameter w^j
c_t: Exponential mean of quad of slope along parameter w^j
b_1, b_2: Hyper-parameters.

Here, we have calculated the exponential mean of the slope, plus, the quad of the slope for every parameter (from Eqs (8) and (9)). To determine our getting to know the step, we multiply our learning charge mean of the slope and divide it by quad of the exponential mean of a quad of the slope [16].

4 Workflow

In this analysis, we have used methods that are already stated above and now let's understand the workflow of it. Figure 5 given below is illustrating about the flow of the work, beginning from data gathering to obtaining the results.

4.1 Image Acquisition

For this research, we had acquired the images from the NIST database and Kaggle website [17].

Fig. 5 Workflow

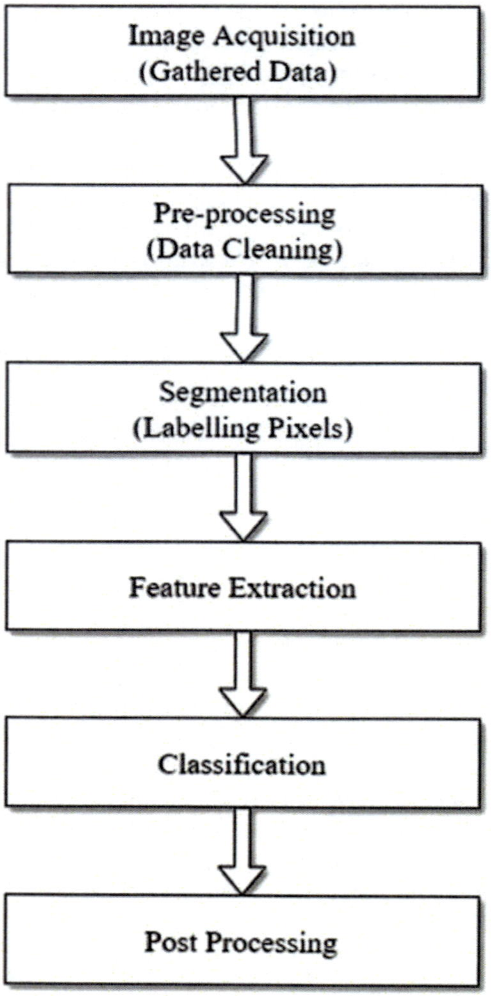

4.2 Pre-processing

In the lemman language, pre-processing [18] is a method of cleaning the data. Here, cleaning refers to the extraction of the required field out of the data gathered. Cleaning depends on the requirements of the person. For instance, in this analysis we wanted a 28×28 shape of images, but we have the bigger one, so in the stage of pre-processing we would reshape the image, in order to fulfill the requirement. Similarly, many other tasks can also be performed such as normalization and binary conversion as per the requirement.

4.3 Segmentation

This is a process in which the partition or segmentation [19] of images takes place. As a CNN model cannot process the whole image at once, it takes several repetitions for scanning an image, which at the end results in some features. There are some chances of skipping certain features which may lead to better results. So, to cope with this behavior, one can feed segments as an input so that there may be no lag in the result. In this process, various layers were created in order to filter out the prominent features of the image, in order to learn about it.

4.4 Feature Extraction

It is a process by which prominent features of an image are extracted out. To discover the prominent feature out of the image, one must understand the numeric nature of an image. Images are taken as an input in the form of an array, having number or pixel values depending upon the intensity of the pixel. A small filter in the form square matrix traverses over the array of an image and pools out of it the maximum number or intense pixel. Here, we used pooling layers and activation functions to get a clearer picture by extracting out the important features [20].

4.5 Classification

This is the process by which it determines whether the image is what, as per the training on the data. It is also called the decision-making stage. Here, CNN itself is a class of deep learning which works as a classifier [21] to make predictions.

4.6 Post-processing

This is the process of gathering results for a better lookout. Generally, in this final stage, the results are being finalized [22].

5 Result

The experimental outcomes on the EMNIST handwritten digit and character dataset and the usage of distinct parameters of CNN four-layer structure are recorded and analyzed in Table 1. An EMNIST dataset having sample images is made into the form

Table 1 Configuration details and accuracy achieved

Model	Layers	Kernel size	Stride	Dilation	Padding	Input size	Output size	Recognition accuracy of digits (%)	Recognition accuracy of characters (%)
Case-1	Layer-1	3	2	1	1	28	14	98.5	90.5
	Layer-2	3	2	1	1	14	7		
	Layer-3	3	2	1	1	7	4		
	Layer-4	3	2	1	1	4	2		
Case-2	Layer-1	3	2	2	2	28	14	97.3	89.1
	Layer-2	3	2	2	2	14	7		
	Layer-3	3	2	1	1	7	4		
	Layer-4	3	2	1	1	4	2		
Case-3	Layer-1	5	2	2	2	28	14	99.1	92.4
	Layer-2	3	2	2	2	14	7		
	Layer-3	3	2	1	1	7	4		
	Layer-4	3	2	1	1	4	2		
Case-4	Layer-1	5	1	2	2	28	17	98.7	91.4
	Layer-2	3	2	2	2	17	7		
	Layer-3	3	2	1	1	7	4		
	Layer-4	3	2	1	1	4	2		
Case-5	Layer-1	5	1	2	2	28	28	99.4	94.45
	Layer-2	5	2	1	2	28	14		
	Layer-3	3	2	1	1	14	7		
	Layer-4	3	2	1	1	7	4		

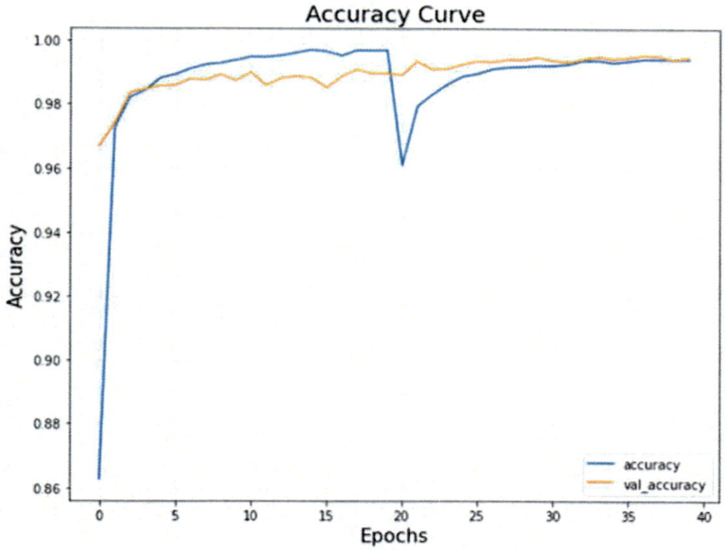

Fig. 6 Accuracy curve

of a 1-D array of 784 floating values representing 28 X 28 size of an image, ranging values from 0 to 1, where 0 stands for black and 1 stands for white. The calculation of the receptive area is completely rooted in the fields, input, output, and kernel length of the function map, stride, dilation, padding, and the accuracy to recognize, which is displayed in Table 1, using CNN structure with 4 layers. Learning rate of 0.01 and epoch counts of 20 are being used. The maximum recognition accuracy completed in case 5 is 99.4% for digits and 94.45% for characters.

In Fig. 6, Accuracy Curve is having a plot of learning accuracy (in blue), and validation accuracy (in yellow). In Fig. 7, Loss Curve is having a plot of learning loss (in blue), and validation loss (in yellow). After 30 epochs, it is visible that both the accuracies would be the same, which tends to be a good fit.

6 Conclusion

In this research, we practiced different techniques using deep learning and trained different models in order to get better recognition of handwritten text and better accuracy in predicting these alphanumeral symbols. Handwritten character recognition is a more complex problem than recognizing handwritten digits because of a variety of characters in different styles as different people have different handwriting styles. The complex architecture of characters is another major reason that makes the handwritten character recognition task tough. Research in this direction focuses on segmentation procedures, feature extraction procedure, and classification

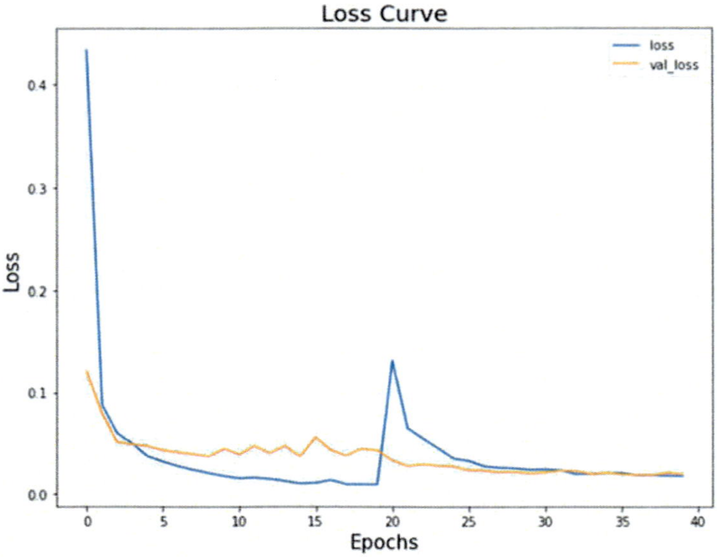

Fig. 7 Loss curve

algorithms. With the goal of bringing improvement in the discipline of recognition methodologies, we had carried out multiple evaluations on different variants of CNN models and as a result to avoid complex methods of a traditional recognition system. Further, we have verified this accuracy by extensively evaluating through changing different layers of the CNN architecture, and finally obtained 99.40% of accuracy for digits and 94.45% of accuracy for characters.

References

1. https://www.nist.gov/itl/products-and-services/emnist-dataset
2. https://en.wikipedia.org/wiki/Convolutional_neural_network
3. https://www.ijcaonline.org/archives/volume69/number19/12076-7656
4. Hanson J, Paliwal K, Litfin T, Yang Y, Zhou Y (2018) Accurate prediction of protein contact maps by coupling residual two-dimensional bidirectional long short-term memory with convolutional neural networks Bioinforma (Oxford, England). https://doi.org/10.1093/bioinformatics/bty481
5. Arora S (2008) Combining multiple feature extraction techniques for hand-written devnagari character recognition. In: IEEE region 10 colloquium and the third ICIIS. Kharagpur, India
6. http://raptorx.uchicago.edu/ContactMap/
7. Hinton GE, Salakhutdinov RR (2006) Reducing the dimensionality of data with neural networks. Science
8. https://keras.io/api/callbacks/
9. https://keras.io/api/callbacks/model_checkpoint/
10. https://keras.io/api/callbacks/early_stopping/

11. https://keras.io/api/callbacks/reduce_lr_on_plateau/
12. https://towardsdatascience.com/understanding-confusion-matrix-a9ad42dcfd62
13. https://keras.io/api/optimizers/rmsprop/
14. https://blog.paperspace.com/intro-to-optimization-momentum-rmsprop-adam/
15. https://keras.io/api/optimizers/adam/
16. https://ruder.io/optimizing-gradient-descent/
17. https://www.kaggle.com/crawford/emnist
18. https://en.wikipedia.org/wiki/Data_pre-processing
19. https://en.wikipedia.org/wiki/Image_segmentation
20. https://en.wikipedia.org/wiki/Feature_extraction
21. https://www.analytixlabs.co.in/blog/classification-in-machine-learning/
22. https://www.kdd.org/exploration_files/KDD2000PostWkshp.pdf

Chapter 6
A Theoretical Approach to Reinforcement Learning Algorithm and Nash Equilibrium to Maximize the Score in Imperfect Information Game

Ruchi Vyas and Alok Singh Gahlot

Abstract Computers have just overshadowed the degree of human play in imperfect information games like Scrabble, yet there remains an opportunity to get better. Specifically, there is a lot to be acquired about the rival's tiles and moves. Reinforcement learning is one of the parts of Machine Learning. It is tied in with making an appropriate move to increase reward in a specific circumstance. The Nash Equilibrium is a hypothesis in game theory that communicates that a player can achieve the ideal outcome by not wandering from their initial strategy. Reinforcement learning algorithm and game theory would be best technique for training the agent. It will learn from as much as it plays, such that our agent can play and win the game with different persons with their different playing strategy they use for Scrabble game. To do this, our agents need to find out about other players' strategies and win every time they play against the human.

Keywords Reinforcement learning · Nash equilibrium · Decision-making · Machine learning

1 Introduction

Scrabble is an imperfect information game, that is, the current player is unknown of the rack of other player, making it elusive out of the rival's best course of action until the finish of the game. Additionally, there is innate arbitrariness present in Scrabble as arbitrary letters are being chosen from the pack to the current player's rack at each round [1]. Our main motive is to provide training to the agent every time it plays against the human and gets itself skilled and ready for the next game.

In supervised learning, an agent is trained to play by gaining knowledge from the data sets provided by the trainer but in reinforcement learning an agent gets rewards whenever it wins so the main objective of the agent is to maximize its rewards and according to our implementation, the agent will be learning how to win the game

R. Vyas (✉) · A. S. Gahlot
M.B.M. Engineering College, Jodhpur, India

R. Mathur et al. (eds.), *Emerging Trends in Data Driven Computing and Communications*, Studies in Autonomic, Data-driven and Industrial Computing,
https://doi.org/10.1007/978-981-16-3915-9_6

irrespective of the letters of the opponent's rack; our agent will be concerned about the words being placed at the board and which combination will give the maximum score, and whenever it plays it will learn new words from its opponent so that it gets trained and becomes more powerful with its dictionary words scoring more rewards [2].

2 Scrabble Overview

A standard Scrabble comprises cells that are situated in a huge square framework which is a 15*15 matrix. There is a bag of 100 tiles that are being used in the game where 98 of them contain letters and points, 2 blank tiles can be used as wild tiles to be used in place of any required letter [3]. At the point when a blank is played, it will stay in the game as the letter it fills in for various letters in the game will have different points, and this will rely upon how uncommon the letter is and how troublesome it could be to lay that letter. Blank tiles will have no points. The essential test of Scrabble is it is a round of defective data as each rack of tiles is stowed away from the rival. Because of this missing data, it is difficult to foresee the specific progressive move of the adversary as its state is obscure to the player; what's more, the ideal move isn't characterized [4]. Furthermore, the machine ought to rapidly create promising up-and-comer moves for a given rack and current board state (Fig. 1).

2.1 Basic Strategy

We have a bag of 100 letters with 98 letters with point values and two blank tiles which could be made any letter as per the requirement of the player; there are n racks

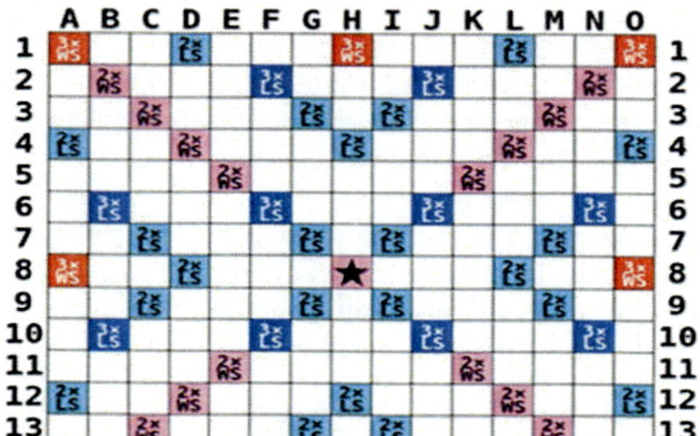

Fig. 1 A Scrabble board game. The red-colored tiles specify triple word score, dark blue specifies triple letter score, turquoise specifies double letter score, and pink specifies a double word [14]

for n players where seven letters could be placed, and rearranging of those letters could be done for making a meaningful word as per the Scrabble dictionary. There are total 3 modes for playing the game which are easy, medium, and hard; when a human wants to play the game, he/she is asked for the age and on the basis of age out of 3 one mode gets activated; after entering the game, the player is asked whether to play against a single player or multiple players on the basis of selection [11]. Nash equilibrium has to be used or not will be decided as if human selects multiplayer (multi agents), the equilibrium stage would be required to take a common decision so that every time it would be required so that agents can achieve desired outcome without changing the strategy. We will be using Reinforcement learning to identify the strategy of our agent to give the best possible move every time it plays with humans. Since we are using Reinforcement learning, our agent has no idea of what to play so when our agent is having his first turn to play the Scrabble game, he makes any word from the combination of letters on the rack and letters present on the board, while checking that word in Scrabble dictionary whether it is a valid word or not.

When our agent plays several games, he will be learning through the experience which word has to be played against which word to maximize the score; after gaining experience while playing several games, the agent will now be able to differentiate between all the three modes that he will be placing the word with a minimum score if he is in the easy mode; similarly, if he is in the medium mode, he will be playing the word with median score word, while if he is in hard mode, he will be placing the word with a maximum score or trying to make a word from all the letters in the rack for getting bonus points. It will also compare maximum score word and bonus point word as agent will keep word with maximum score.

One more thing our agent will be learning through experience is the word which the human makes; our agent will be placing all the words in its dictionary so that it can use those words next time; as initially our agent's dictionary is having no words, he just makes random words from the letters of rack and while checking them against Scrabble dictionary [10]. Through experience, our agent also gathers information as to which place will give more score as Scrabble has a double letter score, double word score, triple letter score, and triple word score. So, this implementation requires a model-free Q learning which trains the agent through a network which is built by playing as many times as it can to gain better knowledge on what word to place and where to place to maximize the score.

2.2 Reinforcement Learning

Reinforcement learning (RL) is a zone of Artificial Intelligence which is concerned about how smart agents should make moves in an environment to amplify the total reward. Reinforcement learning is one of the three Machine Learning techniques; two other techniques are supervised learning and unsupervised learning [5] (Fig. 2).

Some of the terms used in reinforcement learning are as follows.

Fig. 2 Describes reinforcement learning mechanism [14]

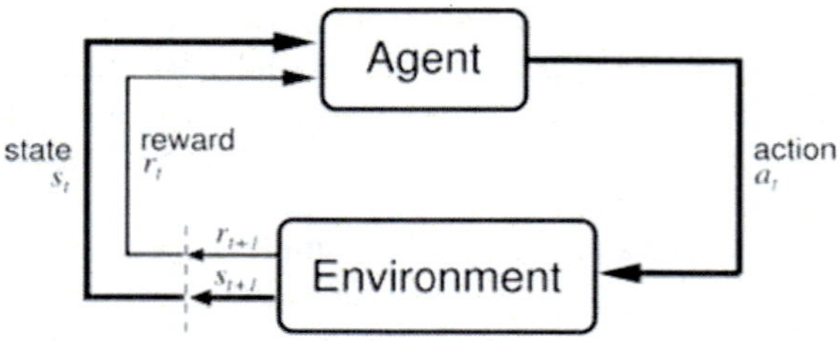

Agent takes decisions so it has to be trained for taking decisions which has to be trained for taking decisions. Environment refers to the place where the agent gets trained and takes decisions according to the strategy. Action is the step agent takes to fulfill the task. State of the agent is the position of the agent in a particular environment. Reward for each action performed by the agent in the current situation provides a reward [6]. It is a scalar value usually. Policy is the decision-making function of the agent, which maps situations to actions. Value function maps from states to numbers, where the value of a state is the prolonged reward achieved starting from the initial state, and by using a specific policy. Function approximator refers to the issue of a function value from training examples [12].

3 Literature Review

There are several research being done on imperfect information games like Scrabble to date, and some of them have already been recognized and are being used by gaming platforms.

While Computers are undeniably the most capable Scrabble players, it isn't by and large realized which Scrabble-playing program is awesome. The Brian Sheppard's Maven program unequivocally vanquished human World Champion Ben Logan in a 1998 display coordinate. Since that time, the National Scrabble Association has utilized Maven to clarify title games. Expert's design is laid out in (Sheppard 2002). The program partitions the game into three stages: the endgame, the pre-endgame, and the midgame. The endgame begins when the last tile is drawn [4]. Expert uses B*-search (Berliner 1979) to handle this stage and is probably almost ideal. Little data is accessible about the strategies utilized in the pre-endgame stage; however, the objective of that module is to accomplish a good endgame circumstance.

Most of the game is played under the direction of the midgame module. In its turn, Maven creates all conceivable lawful moves and positions them as per their prompt esteem (focuses scored on this turn) and on the capability of the leave. The qualities used to rank the leaves are processed disconnected through broad re-enactment [11]. For instance, the estimation of the space QU is controlled by estimating the distinction in future scoring between a player with that space and his rival; what's more, averaging that esteem more than a huge number of games in which it is experienced.

When all lawful moves have been created and positioned by the static assessment work, Maven utilizes reproductions to assess the value of those moves concerning the current board arrangement and the excess inconspicuous tiles. Since it isn't exceptional to have a few hundred lawful plays to browse on each turn, a profound inquiry isn't manageable. Sheppard recommends that a profound hunt may not be essential for superb play. Since master players utilize a normal of 3–4 tiles in each turn, the total turnover of a rack can be considered typical for each two to four turns [6]. Reproductions past that level are of problematic worth, particularly if the pack actually contains numerous letters. Experts by and large use a two-to four-employ look in their recreations. After the distribution of (Sheppard 2002), rights to Maven were bought by Hasbro, and it is presently appropriated with that organization's Scrabble programming item. Since its commercialization, extra insights regarding its procedures and calculations have not been freely accessible.

Jim Homan's CrossWise is another business programming bundle that can be arranged to play Scrabble. In 1990 and 1991, CrossWise won the PC Scrabble rivalry at the Computer Olympiad [9]. (In ensuing Olympiad rivalries, Scrabble has not been challenged.) The algorithmic subtleties of CrossWise are not promptly available. Shockingly, Expert and Crosswise have not been set in opposition to one another in an authority rivalry, so it isn't realized which program is unrivaled. In light of freely accessible data, Maven would likely have the edge. Homan claims that CrossWise created more than US $3 million in deals, which shows that there is an extraordinary interest in amazing Scrabble PC programs [13] (Table 1).

The second recognized work done was on five unique AIs having diverse move age and reproduction strategies. Expert AI with Q-Sticking lethargic endgame methodology was required to outscore the Quackle AI, and it ended up being the best PC player AI in our undertaking. It can impede the problem areas just as procuring extra focuses by catching an adversary utilizing the lethargic endgame procedure. Expert Q- Sticking AI additionally had an important move age heuristic and outscored the Quackle adversary. In any case, the heuristics were not as ideal as Maven Q-Sticking sluggish endgame AI.

Expert sluggish endgame AI didn't perform well for endgames, and the AI lost to Quackle commonly. Quackle AI could just outscore Maven moderate endgame AI [7]. At last, No Q-Sticking AI was discovered to be a promising AI during midgames. Be that as it may, this AI didn't perform well during endgame situations. The significant downside of our task was the execution time taken by the three-handle look-ahead re- enactment. For instance, our venture burns through 100–255 s to complete

Table 1 Comparative study of Quackle with inference [14]

	With inferences	Quackle
Wins	324	306
Mean Score	427	422
Biggest Win	279	262

Table 2 Average mean score [14]

Player	Mean score
Maven Q-Sticking slow-endgame	479.9
Without Q-Sricking	415.1

100 reproductions of Quackle while initially the Quackle title player just requires around 19–50 s for a move age. If there should be an occurrence of Maven AI, the execution season of a solitary move age was around 0–2000 ms; in any case, initially Maven AI devoured just 0–30 ms. The time utilization of unique Maven AI was processed utilizing a stopwatch. As a future upgrade, we will decrease this execution season of the recreation motor and will actualize an intelligent graphical UI for the task. Besides, we understood that Quackle isn't only a Scrabble play motor; it is additionally an investigation apparatus that can recreate up to multiple times and gives an alternative to the player to admire 6 utilizes profound which could be tedious. So, we will attempt to consolidate these highlights into our AI [8] (Table 2).

4 Conclusion and Future Scope

The empirical implementation suggested above will improve the intelligence of computers playing imperfect information games like Scrabble and will definitely increase the number of times our agent wins against humans as when we are applying reinforcement learning, our agent will become more proficient as much as it plays against humans. As we have seen many of the previous researches on Scrabble board game are working on predicting opponent rack letters while we are focusing on learning experience and making our agent more skillful and winner as many times it plays against human or against agents.

Since our agent is rewarded whenever it wins, the main goal of the agent is to maximize its rewards and according to our implementation, the agent will be learning how to win the game irrespective of the letters of the opponent's rack and our agent will be more focused about the words being placed on the board and which combination will give the maximum score, and whenever it plays it will learn new words from its opponent so that it gets trained and becomes more powerful with its dictionary words scoring more rewards.

References

1. Artificially intelligent game bots pass the turing test on turing's centenary (2015). http://news.utexas.edu/2012/09/26/artificially-intelligent-game-botspass-the-turing-test-on-turings-centenary
2. Barriga NA, Stanescu M, Buro M (2014) Parallel UCT search on GPUs. In: IEEE conference on computatonal intelligence and games, CIG

3. Arneson B, Hayward RB, Henderson P (2010) Monte Carlo tree search in hex. IEEE Trans Comp Intell AI Games 2(4):251–258
4. Benbassat A, Sipper M (2013) Evo MCTS: enhancing MCTS-based players through genetic programming. In: 2013 IEEE conference on computational intelligence in games. Niagara Falls, Ontario, Canada, pp 57–64
5. Shapiro SC, Smith HR (1988) A scrabble crossword game-playing program computer games I, pp 403–419
6. Aiolli F, Palazzi CE (2019) Enhancing artificial intelligence in games by learning the opponent's playing style. In: IFIP international federation for information processing, [online] pp 1–10
7. Cheng C, Zhu Z, Xin B, Chen C (2017) A multi-agent reinforcement learning algorithm based on Stackelberg game. In: Conference: 2017 6th data driven control and learning systems (DDCLS), p 727.https://doi.org/10.1109/DDCLS.2017.8068163
8. Vrabie D, Lewis FL (2011) Integral reinforcement learning for finding online the feedback nash equilibrium of nonzero-sum differential games. Adv Reinf Learn. https://doi.org/10.5772/13804
9. Whitehouse D, Powley EJ, Cowling PI (2011) Determinization and informationset Monte Carlo tree search for the card game Dou Di Zhu. In: Proceedings IEEE conference computer intelligence games. Seoul, South Korea, pp 87–94
10. Whitehouse D, Cowling PI, Powley EJ, Rollason J (2013) Integrating MonteCarlo tree search with knowledge-based methods to create engaging play in a commercial mobile game. In: Proceedings artificial intelligence and interactive digital entertainment conference. Boston, Massachusetts, pp 100–106
11. Peng J, Williams RJ (1994) Incremental multi-step Q-learning. In: Machine learning, pp 226–232. Morgan Kaufmann
12. Dahl FA (2001) A reinforcement learning algorithm applied to simplified two- playertexas hold'em poker. In: Proceedings of the 12th European conference on machine learning. Springer
13. Miikkulainen R, Bryant B, Cornelius R, Karpov I, Stanley K, Yong CH (2006) Computational intelligence in games. IEEE Computational Intelligence Society, Piscataway, NJ
14. He K, Zhang X, Ren S, Sun J (2016) Deep residual learning for image recognition. In: Proceedings of the IEEE conference on computer vision and pattern recognition, pp 770–778, 64–67

Chapter 7
A Review Paper on Sign Language Recognition Using Machine Learning Techniques

Ritesh Kumar Jain and S. K. Rathi

Abstract Correspondence is a significant piece of our lives. Hearing disabled individuals who can't talk experience various issues while bantering with regular people. These individuals locally can't speak with others without any problem. There are numerous manners by which individuals with incapacities attempt to convey. Sign language is one of the most natural and sensible ways for the disabled. Since deaf people cannot speak like normal people, they often have to rely on some form of visual communication. It has been shown that they sometimes find it difficult to communicate with ordinary people with their hands, as very few of them are recognized by most people. Communication through signing is a significant method for correspondence with the hearing impaired community. Gestures come in numerous structures, for example, hand gestures, two hands, and face movements. Gesture-based communication can be partitioned into two classifications: static and dynamic. Static signs are a hand setup and a particular shape, addressed by a single image. The dynamic sign is a moving action, represented by a sequence of images. Accordingly, the requirement for an intelligent computer-based program is in desperate need for deaf and dumb people that will empower them to convey all the more successfully with any remaining individuals utilizing their normal hands. This paper presents a top-to-bottom audit of Sign Language recognition techniques, a report related to those methods, and identifies challenges. The various methods and algorithms that have been used in sign language recognition projects have been refined and compared in terms of their advantages and disadvantages.

Keywords Sign language recognition · Convoluted neural networks · SVM · Hand gestures · Hand detection · Classification

R. K. Jain (✉) · S. K. Rathi
Sobhasaria Engineering College, Sikar, Rajasthan, India

R. Mathur et al. (eds.), *Emerging Trends in Data Driven Computing and Communications*, Studies in Autonomic, Data-driven and Industrial Computing,
https://doi.org/10.1007/978-981-16-3915-9_7

1 Introduction

The gesture is an activity that can be characterized as movement, hand or face frequency that communicates thoughts, sentiments, or feelings, for example, raised eyebrows; shoulder movements are a portion of the activities we use in our day-to-day lives.

Gesture-based communication is a more formal and expressive type of correspondence where all words or letters are given a particular activity.

Gesture-based communication is an all-around coded demonstration of code; each activity has an appointed significance. Communication through signing is the solitary method for correspondence for the hard of hearing. There are different classifications of communications via gestures: accessible, Indian Sign Language (ISL), British Sign Language (BSL), American Sign Language (ASL), and so on. In American Sign Language (ASL), every English letter in order, A–Z, is given an alternate action word. Correspondence through signs is most commonly used by deaf people.

Two techniques are utilized for gesture-based communication acknowledgment; one depends on vision and the other depends on the sensor. The advantage of systems that support the idea more than their counterpart is that users do not have to use sophisticated equipment but in the pre-processing phase, it requires sufficient calculation instead of cameras used in vision-based systems, sensory systems use sensor-enabled gloves tool.

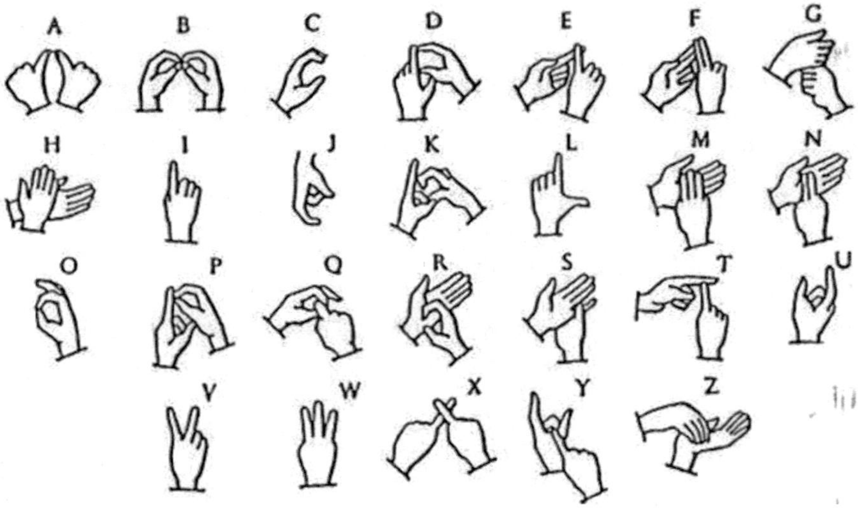

Fig. 1 Indian sign language

Nonetheless, there are a large number of hard of hearing and deaf individuals around the planet, and verbal correspondence is incomprehensible for these individuals. Likewise, speaking with others is a significant test in their everyday lives because a great many people don't comprehend gesture-based communication.

It is our social obligation to make this general public more autonomous in life with the goal that it also can be a piece of this becoming a technological world. With the progression of science and innovation, numerous strategies have been created not exclusively to reduce the issue of hard of hearing yet additionally to apply it to different organizations.

Not at all like speaking with a discourse that utilizations sound to communicate one's considerations; communication through signing utilizes outward appearances and lips developments, motions and non-verbal communication, and signals. Like communicated in language, communication via gestures differs all around like Indian Sign Language (ISL), British Sign Language (BSL), and American Sign Language (ASL).

2 Literature Review

A great deal of work has been done on creating frameworks for various communications via sign languages. In [1], the author used Convolution Neural Network to recognize American Sign Language. They worked on 26 English Alphabets and 10 Numerals. The dataset consists of 2524 Images. The author achieved 96% recognition rate.

In [2], the creator utilized Hidden Markov Model to perceive communication via gestures utilizing various sensors. The creator utilized a multi-sensor system for Sign Language acknowledgment utilizing the joint alignment of two profundity sensors, in particular, Leap movement and Kinect. The dataset comprises 25 powerful hand motions. This model doesn't chip away at different hand information. The creator accomplished 90.80% recognition rate.

In [3], the creator utilized versatile skin shading demonstrating and a help vector machine to perceive American Sign Language (ASL) and Chinese Sign Language (CSL). The creators purposed a vision-based gesture recognition system which can be utilized in conditions with complex background. The dataset comprises 39,000 pictures of 26 letter sets. The creator accomplished a 100% acknowledgment rate for CSL and 94% acknowledgment rate for ASL.

In [4], the author used Neural Networks and KNN Classifier to recognize Indian Sign Language. The authors purposed a gesture recognition system that is capable of only numerical Indian Sign Language (ISL) static signs. The dataset consists of 5000 images of numerical signs (0–10). The authors used Direct Pixel value and Hierarchical Centroid Feature extraction methods to extract the features. The author achieved 78.63% recognition rate using KNN and 97.10% using Neural Network.

In [5], the creator utilized Artificial Neural Network to perceive Indian Sign Language. The creator proposed a vision-based methodology for naturally perceiving static and dynamic hand signals utilizing an Artificial Neural Network. They worked on 10 Static and 10 Dynamic hand gestures. The author achieved 99.98% recognition rate for static hand gestures and 99.51% for dynamic hand gestures.

In [6], the creator utilized the Support Vector Machine to perceive Vietnamese Sign Language. The creators purposed Vietnamese Sign Language Recognition System (VSLRS) can perceive completely Vietnamese letters in order communication through signing including static and dynamic signals. They worked on 23 dynamic hand gestures and 6 dynamic gestures. The creator accomplished 90% recognition rate.

In [7], the creator utilized Artificial Neural Networks and Support Vector Machine Indian Sign Language. The creators built up an acknowledgment framework for Indian Sign Language (ISL) numerals (0–9). The creator accomplished 93% acknowledgment rate utilizing SVM and 99% recognition rate using ANN.

In [8], the creator utilized KNN and Back Propagation Technique to perceive Indian Sign Language. The creators purposed a gesture recognition system for single- and double-handed ISL to perceive English letters and numbers. They worked on 26 letters (A–Z). The creator accomplished a 100% recognition rate utilizing KNN and 94–96% recognition rate utilizing the Back Propagation Technique.

In [9], the creator utilized Principal Component Analysis (PCA) and Support Vector Machine (SVM) to perceive Sign Language. The creators purposed a proficient arrangement of sign acknowledgment of 15 hand motions and the dataset comprises 200 pictures. The creator accomplished 94.5% acknowledgment rate.

In [10], the creator utilized the Hierarchical Clustering Classification Method to perceive ongoing Sign Language. The creators purposed a vision-based motion acknowledgment framework which can be utilized in conditions with a complex foundation. The authors plan a system to adaptively revive the skin concealing model for different customers and distinctive lighting conditions. The dataset comprises 7800 pictures of 26 letter sets and 5200 pictures of 26 letter sets of ASL. The creator accomplished a 99.8% acknowledgment rate for CSL and a 94% acknowledgment rate for ASL.

3 Steps of Sign Language Recognition

The flowchart in Fig. 2 is clarifying the total framework in which all the important advances are thought of. The flowchart has the accompanying advances: image acquiring and pre-preparing, skin color detection, background removal, and afterward edge detection and separating an important motion and in the wake of perceiving and removing the specific motion, it changed over into an indication of that specific signal has been effectively perceived.

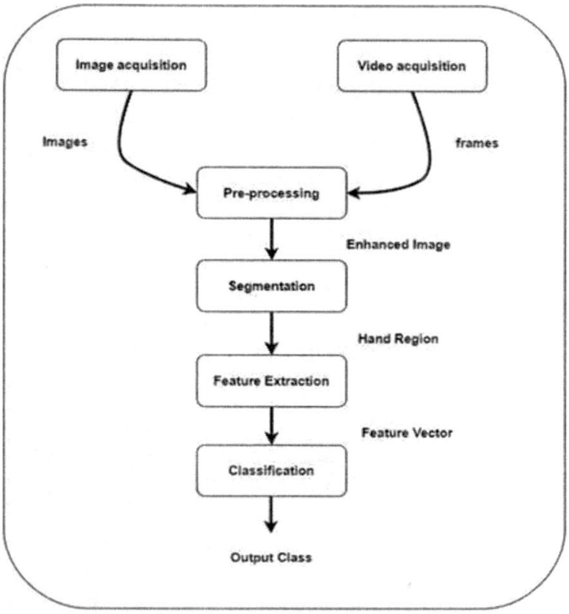

Fig. 2 Steps of Sign language recognition

3.1 Data Acquisition

In a vision-based touch see, the data obtained is an image layout. The commitment of such a program is assembled using pictures that get contraptions like a standard camcorder, a webcam, a sound framework camera, a hot camera, or more powerful techniques like Kinect and LMC. Sound system, Kinect, and LMC cameras with 3D cameras can accumulate all-around nuances.

3.2 Pre-processing

This stage is done to change the image or video commitment to improving the overall presentation of the system. As the data is acquired and made under different lighting conditions, pictures have assorted comparative characteristics.

Consequently, there is a need to change the pixel thickness esteems before continuing. The histogram of the picture with the low difference is little and has bars confronting the focal point of the grayscale. Dispersing the histogram to a wide scope of Median channel quality and Gaussian channel is probably the most broadly utilized techniques to diminish the commotion in pictures or video found.

3.3 Segmentation

Segmentation is the path toward isolating pictures into different parts. It is a stage where the Region of Interest (ROI) is detached from the rest of the image. The procedure for a course of action can be content or non-content.

Relevant partition considers spatial connections between factors, for example, edge-discovering methodologies. While non-all-out developments don't consider nearby connections yet bunch pixels depend on worldwide images.

3.4 Feature Extraction

The image classification gives us a binary image that contains a hand shape addressing a specific image. In the partition of this isolated activity, there is a need to avoid certain parts of the picture that might be needed to distinguish the image. A functioning shape can be an indispensable piece of the picture, as make-up is the fundamental resource of any article. There are just two kinds of status descriptions: region-based descriptions and string-based depictions.

In the proposed study, Hu-fixed occasions were utilized to include extraction. It is a well-known kind of string-based depiction that was first acquainted by Hu in 1962 to the pattern recognition community. A bunch of transitory symmetrical irregular installers (for example, Revolution) were acquired from the impacts of mathematical aggressors' vision. This is determined by making ordinary orders up to three orders. This can be utilized to recognize the scale, position, and revolution design.

- Translation consistency is gotten at typical PC times about the territory of gravity with the goal that the focal point of dissemination size is of the source (transitional occasions).
- Minimum size times are gotten from arithmetical data sources, however, these can be demonstrated to be the consequence of typical size.
- From the second- and third-request orders for the standard middle-of-the-road times, a bunch of seven variable occasions can be determined freely thus.

3.5 Classification

Classification can be isolated into supervised and unsupervised ML methods. Supervised machine learning technique is a procedure that shows a framework to see a specific information design, which is used to envision future data. It takes a lot of striking getting ready data and is used to incorporate work from data-checked planning. Untreated machine perusing is utilized to draw non-existent from informational indexes with non-marked info information. Since no name reaction is remembered for the separator, there is no prize or punishment weight where the information ought to have a place with classes.

4 Conclusion and Future Scope

Hand acknowledgment has been the ceaseless investigation of its extraordinary force of utilizations, for example, communication through signing recognition, controller robots, and human interaction with the PC. Notwithstanding, the obstructions to acquiring an exact and hearty framework remain, in particular hand conclusion, the presence of relative factors, dataset debasement, unmistakable foundation lighting, and high computational expenses.

There are a developing number of arising advancements, for example, EMG, LMC, and Kinect, that catch activity subtleties without any problem. The standard preparing technique utilized is the Median and Gaussian channel and picture decreases before the following stages.

Skin shading partition is perhaps the most broadly utilized division technique. The strongest shading space in the lighting climate is CIE Lab, YCbCr, and HSV. Ongoing exploration utilizes a mix of explicit spatial highlights and displaying strategies to improve part usefulness.

Normal component-based insight strategies fuse SIFT, SURF, PCA, LDA, and DWT. Model-based approaches fuse both imaging and solid showing systems and convexity failure modes. The Hybrid method for the extraction technique has been comprehensively used to give a solid affirmation feature.

From past works, HMMs have been distinguished as promising techniques in agreement hand signals as they have been utilized effectively in numerous examinations. From a static point of view, SVM is the most well-known strategy as it has appeared to work best in a few examinations. There are a few varieties proposed about the current strategy, and financing costs are generally utilized as they can beat the deficiencies of a solitary technique. There are significant holes that should be filled for contact recognition to be successful. Future tasks utilizing benchmark data are encouraged to permit direct correlations between the calculations utilized.

References

1. Masood S, Thuwal H, Srivastava A (2017) American sign language character recognition using convolution neural network. Smart Comput Inform, 403–412. https://doi.org/10.1007/978-981-10-5547-8_42
2. Kumar P, Gauba H, Roy P, Dogra D (2017) Coupled HMM-based multi-sensor data fusion for sign language recognition. Pattern Recogn Lett 86:1–8. https://doi.org/10.1016/j.patrec.2016.12

3. Pan T, Lo L, Yeh C, Li J, Liu H, Hu M (2018) Sign language recognition in complex background scene based on adaptive skin color modeling and support vector machine. Int J Big Data Intell 5(1/2):21–30
4. Sharma M, Pal R, Sahoo AK (2014) Indian sign language recognition using neural networks and KNN classifier. ARPN J Eng Appl Sci 9(8):1255–1259
5. Kaur C, Gill N (2015) An automated system for Indian sign language recognition 5(5):1037–1043
6. Hai PT, Thinh HC, Phuc BV, Kha HH (2018) Automatic feature extraction for vietnamese sign language recognition using support vector machine. In: 2018 IEEE 2nd international conference on recent advances in signal processing, telecommunication & computing (SigTelCom), pp 146–151
7. Ekbote J, Joshi M (2017) Indian sign language recognition using ANN and SVM classifiers. In: 2017 IEEE international conference on innovations in information embedded and communication systems (ICIIECS)
8. Dutta KK, Bellary SAS (2017) 2017 IEEE international conference on current trends in computer, electrical, electronics and communication (ICCTCEEC-2017), pp 333–336
9. Kumar S, Yadav G, Singh HP, Gupta A (2018) SSVM classifier and hand gesture based sign language recognition. In: 2018 IEEE international conference on intelligent circuits and systems, pp 456–460
10. Pan T, Lo L, Yeh C, Li J, Liu H, Hu M (2016) Real-time sign language recognition in complex background scene based on a hierarchical clustering and support vector machine. In: 2016 IEEE second conference on multimedia big data, pp 64–67

Chapter 8
Deep Learning Methods for the Prediction of Chronic Diseases: A Systematic Review

Gunjan Sahni and Soniya Lalwani

Abstract Deep learning (DL) is a machine learning optimization technique that has represented amazing performance in identifying obscure structures in high-dimensional data to find the most optimal settings. Thus, DL has been effectively applied in many diverse fields in image and speech recognition, visual art, natural language processing, and bioinformatics. Other than this, lots more are still needed to be investigated. This paper systematically reviews publications used in deep learning methods for the prediction of chronic diseases more accurately. In medical science, it is always challenging to analyze chronic diseases before the major damages. Some of the chronic diseases cannot be recognized in primary diagnosis until they put a drastic impact on health, as some of them have no treatment. Hence, to avoid such an awful condition, there is a sturdy requirement of some models that can predict disease more accurately in an early stage. Different models have been designed using deep learning's multilayer approach and provide better result in the prediction of some chronic diseases that comprises Coronary Heart disease, Alzheimer disease, labeling of multiple chronic disease, Diabetic Retinopathy, Breast cancer, Autoimmune disease, and skin diseases. This paper summarizes all of these DL models for predicting the mentioned diseases.

Keywords Deep learning methods · Chronic diseases · Autoencoders

1 Introduction

Deep learning (DL) provides efficient methods to process raw or unstructured data using neural networks. Therefore, DL is an eminent trend in the present research period. Technical growth is futile until it works for concern of humanities. In this

G. Sahni (✉)
Career Point University, Kota, India

S. Lalwani
Bal Krisha Institute of Technology, Kota, India
e-mail: slalwani.pdf@rtu.ac.in

R. Mathur et al. (eds.), *Emerging Trends in Data Driven Computing and Communications*,
Studies in Autonomic, Data-driven and Industrial Computing,
https://doi.org/10.1007/978-981-16-3915-9_8

contaminated and overexerted lifestyle, people are more prone to suffer from chronic disease, some diseases do not have any effective treatment yet. Early diagnosis can reduce the impact of the disease. Artificial intelligence (AI) and machine learning (ML) brought incredible performance in predicting chronic diseases, simultaneously improving related algorithms and processes for more accurate results. So there is a requirement for more accuracy in the prediction process due to the high-dimensional data of medical images. Since different techniques have evolved through DL, researches provide a great impact on signal processing, information processing, speech recognition and audio recognition processes, language modeling, and natural language processing.

This paper addresses different researches that proposed the models using DL network, supervised or unsupervised algorithms to process unlabeled or raw pathological data with the purpose of predicting chronic disease. Performance evaluation of the proposed method is also discussed.

The classification of the manuscript is as follows: Sect. 2 presents the details of DL and its comparison with ML. In this subsection, different algorithms and functions of DL are discussed. Section 3 is based on defining the summary of studies performed on different research models, along with presenting the different techniques of DL, used for analyzing data and providing results as possibilities of diseases. Lastly, comparative analysis represents the performance of the proposed models in Sect. 4 followed by the conclusion of the presented work in Sect. 5.

2 Deep Learning

As the name states, DL is itself learning of system in bottomless. Discussion about automation is never complete without Artificial Intelligence (AI), and neural network (NN) is the step towards achieving AI. A neural network works similarly to a human brain, as it has series of different algorithms to find out relationships in a set of data. NN can adjust for varying input values. In this way, it can generate the best possible results without the need for any output criteria. Whereas ML can construct machines that can process data and learn itself with or without supervision. So the main features of DL are:

1. DL depends on the layers of artificial neural network (ANN).
2. DL networks do not require human intrusion because of the nested layers present in the neural networks, which set data through hierarchies of concepts, which have self learning properties as they learn through their own fault.
3. In DL network, the quality of input data finally decides the quality of the outcome.
4. DL networks are best suited for complex data and complex calculations with a huge amount of data. DL networks can recognize edges or concepts, and also the differences within layers of neural networks when exposed to over a million data points (Fig. 1).

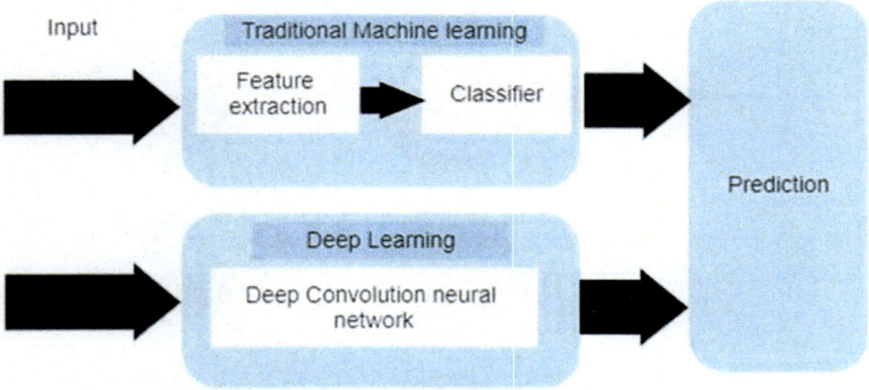

Fig. 1 Traditional machine learning versus deep learning [11]

Some of the important features related to DL are illustrated below.

2.1 Sparse Autoencoders

Autoencoders (AE) are techniques of unsupervised learning. They influence neural networks to achieve representation learning. As there is a need to design a neural network architecture that can enforce a bottleneck in the network, basically work to force a compressed knowledge image of the original input. This compression and reconstruction would be a tedious job if input features were independent. But if any correlation exists between features, then it can be learned and accordingly controlled at the time of forcing input through the bottleneck of the network as in Fig. 2. The sparse auto encoder provides alternative methods for introducing information bottleneck without reducing counts of a node in the hidden layer as it is a type of competitive learning. It is a neural network method whose target output is input. This encoding technique provides linear and nonlinear transformation.

Sparse AE are types of encoders that contain a greater number of hidden units than input units. But the type of input decides which hidden units will work and which will remain dumb [12]. By using this sparsity, the constraint model is forced to react to the exclusive arithmetical properties of the input data, which is applied for training.

So, for sparse AE, training criteria contain a sparsity penalty $\Omega(g)$ on layer g.

$$\mathscr{L}\left(y, y'\right) + \Omega(g),$$

Here, $g = f(\mathrm{C}y + \mathrm{d})$, where C represents the weight matrix and d represents the bias vector.

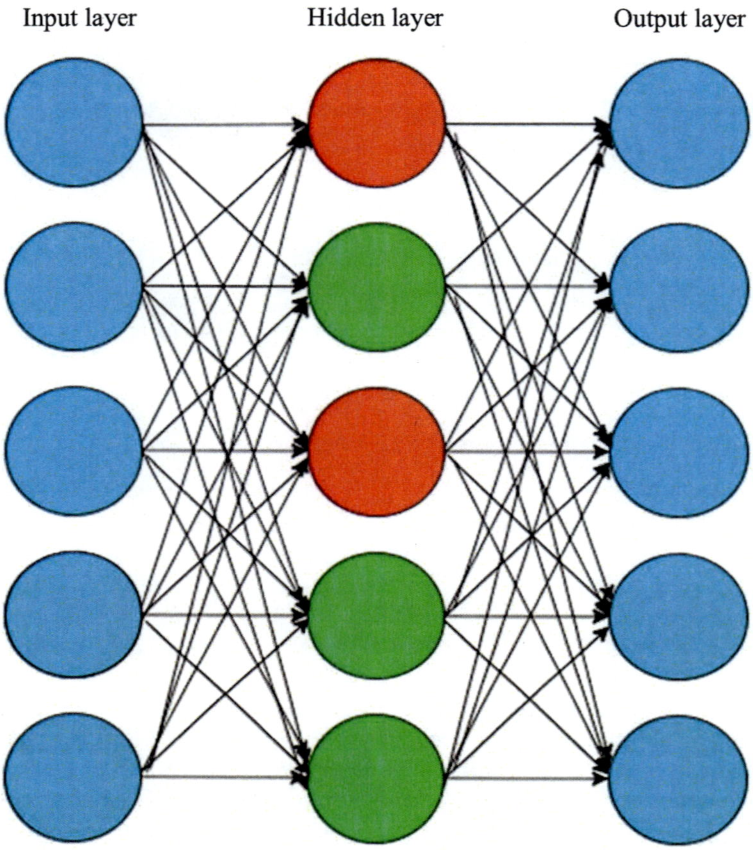

Fig. 2 Basic diagram of a single layered sparse AE (Nodes in bright red represents activated nodes, and nodes in the green are dumb) [8]

2.2 *Softmax Regression*

Softmax regression (or multinomial logistic regression) is a simplification of logistic regression in situations for which there is a need to process multiple classes. In the case of general logistic regression, the labels were binary k(i) ϵ {0,1}. Softmax regression has an extraordinary property as has a redundant set of parameters. K(i) ϵ {1.....M}, where M represents a number of classes.

2.3 Convolution Auto-encoders

Convolution auto encoder (CAE) is an AE neural network. It applies convolution layers and pooling layers to take out the hidden patterns of input features in encoding and to reconstruct the hidden pattern's features applies deconvolution layers and unpooling layers. Convolution of the continuous case classifies as the integral operation applied on the product of two functions or signals after one of them is reversed and shifted. It is a commutative operation, therefore, used in encoding and decoding.

2.4 Sparse Filtering (SF)

It is an efficient algorithm having one hyperparameter that has basically multiple features to learn. SF optimizes a simple cost function. SF uses scaling to handle high-dimensional inputs, and the greedy layer-wise stacking technique, used for learning major features [13].

2.5 Stochastic Gradient Descent (SGD)

SGD is basically used for optimizing the target function. It provides apposite smoothness properties as differentiable or sub differentiable. It is an iterative process. It can be assumed as an arbitrary estimation of gradient descent optimization, which is calculated from the entire data set, and here SGD is calculated from a randomly selected subset of the data. This approach results in the reduction of computational complexity and provides faster iteration in big data applications [2].

2.6 Gradient Boosting (GB)

GB is a machine learning technique. It can be applied for regression and classification problems [10]. It provides prediction model as a group of weak prediction models, normally decision trees. Here, the models can be built in a stage-wise manner. At the time of generalization, it allows optimization of a random differentiable loss function.

2.7 Decision Trees Classifier

These are a type of supervised machine learning. The main process of this is to divide data continuously based on some parameters. It contains basically two entities: decision nodes and leaves. Here, leaves represent final result, whereas nodes are the points where the data is divided [16].

2.8 Spatial Pooling

Spatial pooling is the process when local features are grouped mutually from spatially adjacent pixels. This is usually done to recover robustness to slight deformations of objects, and is now one of the standard "tricks" of computer vision. Pooling is one of the most important features of convolution neural network, split the input map into a set of rectangles, and outputs the maximum for nonlinear down-sampling. The pooling layer decreases the resolution of the feature map. It saves map's features that are required for cataloging through translational and rotational invariants [18].

3 Deep Learning for Prediction of Diseases

3.1 Alzheimer Disease (AD)

Alzheimer disease's accurate diagnosis in the early stage can be very beneficial in effective treatment. Siqi Liu1 et al. [9] worked on an early diagnosis model for AD, which was based on the deep learning model. This research method having stacked sparse autoencoders and softmax regression in the network, can reduce the dependence on the previous idea about the data. By using a semi-supervised approach, unlabeled training samples can also be used. The AE achieve deep demonstration of the actual input, whereas the softmax regression layer categorized the instance by choosing the maximum probability of the labels that are predicted. Training of the hidden layers of sparse AE was performed once and then piling of layers was performed to figure out an absolute neural network, by eliminating the provisional output layer.

Ehsan Hosseini-Asl et al. [7] applied a deep 3D convolution neural network (3DCNN) in the prediction of AD. The proposed method was for unsupervised generic and transferable feature extraction based on 3DCAE. The proposed network combines a pertained 3DCAE in the source domain, such adaptation of prelearned generic features allows for calling the proposed classifier a 3DACNN.

FirouzehRazavi et al. [15] proposed a two-stage method to provide solutions for the high dimension by processing the limited number of samples in the investigation of brain images. At an early stage of sparse filtering learning, a two-layer network like system was used. It learns the features that represent AD. This algorithm minimizes the impact of features that consist of low or no information at all. In the first step, the sparse filtering was skilled to obtain the weight matrix. In the second step, sparse filtering was processed to collect the local features from each of the samples. At last, averaging of the features was performed. Now, after the features collection from each sample, the next step of learning was SoftMax regression method, performed so that it can classify healthy and unhealthy conditions automatically.

3.2 *Coronary Heart Disease (CHD)*

To predict coronary heart disease, different researches have been taken place. Juan Wang, Huanjun Ding et al. [17] proposed the method of analyzing mammograms by DL, to predict CHD in women. Here, breast arterial classification (BACs) was used as a risk marker. They formulated the predicament as a pixel-wise, patch-based two-class classification problem. An image patch around the pixel was used to analyze the pixel under concern. These image patches are input to a deep CNN. This deep CNN had the capability to dig out features automatically. These were already trained to categorize the central pixel, whether it belongs to the BAC class or not.

Ryan Poplin1, Avinash V. Varadarajan et al. [14] proposed a novel approach of predicting CHD from retinal fundus photographs by applying DL. They used Inception-v3 neural-network architecture37 to calculate the labels of classes.

3.3 *Diabetic Retinopathy (DR)*

DR is also an impact of chronic disease that needs accurate and timely diagnosis. After a study of different models for predicting DR using the multilayered structure of the DL network, one of the findings is that most of them are considering retinal fundus photographs as raw data.

Varun Gulshan et al. [5] proposed a system to automate the DR screening process that can be very useful in diabetic retinopathy. The proposed algorithm does not explicitly detect graze, but it learns about it by learning from local features as DL concepts. This system used a convolution neural network, it utilized a function that first grouped nearby pixels into local features, and then combined those into global features. Here, stochastic gradient descent implementation is used to train the network weights. The training process can be speeded up by applying batch normalization and pre initialization. Through this training process, a single network was ready to perform multiple binary predictions.

Gargeyaand Leng [4] proposed a method of automated prediction of DR by using the theory of deep residual learning to build up a convolution network (CCN), which involve learning discriminative features for DR prediction. In the proposed model, a convolution visualization layer is also implemented at the end of the network to check whether the network learning takes place as per process. The average pooling layer and a traditional softmax layer was introduced after the visualization layer. In the process of generating a final diagnosis, a second-level gradient boosting classifier was trained on a feature vector of 1027 values.

3.4 Breast Cancer (BC)

Akselrod-Ballin, A., et al. [1] proposed a novel method to predict BC by analyzing mammography (MG). MG analysis is always exigent because of not only the faint close-grained visual grouping, as well as large inconsistency in the appearance of anomalies, making them complex to detect and classify. The tenderness of the breast and unstable viewing setting result in considerable intra-expert and inter-expert inconsistency. They used here deep fully convolution network to provide a solution. The proposed system was composed of two main components: A region proposal network (RPN) and fast region convolutional neural network (R-CNN) detection network. RPN was trained to detect windows on the input image and capture objects of interest, at the same time, it works to predict object bounds and abjectness scores. Fast R-CNN detection network was also skilled to categorize candidate object of windows, output of RPN. These windows were categorized either as a class of interest or refused class.

3.5 Autoimmune Disease

Indirect immunofluorescence (IIF) is a common tactic to find out the occurrence of any type of autoimmune disease by discovering antibodies in the patient serum. Han et al. [6] proposed a DL model supposed to automate prediction and analyze IIF. In the proposed method, the convolution layers in the deep CNNs was used. This method opted for the use of a sliding window that essentially convoluting their input with a much smaller size kernel when compared to actual input, so for these layers, fixed size of input was avoided. K-support pooling was used in CNN. It was performed by combining the maximum activated magnitude and magnitude of response for the comparatively stimulated patterns of a neuron within a spatial region together.

3.6 Skin Disease

Deep neural network‘s application in the categorization of skin lesions has considerable performance, as the programmed categorization of skin lesions using images is a challenging task due to the fine-grained inconsistency in the different types of skin lesions. Esteva et al. [3] proposed a novel technique for the categorization of skin cancer using deep learning. A recursive algorithm was proposed to control the taxonomy that provides training classes for diseases that are difficult to identify or differentiate. At the time of implication, the deep CNN gave outputs as a probability distribution over the different classes of defined categories. The probability of inference node is the summation of offsprings training classes' probabilities.

4 Result and Discussion

About 25 research papers related to different models of DL in disease prediction have been studied. Most of the models are taking pathological images as raw data that have multiple features to extract, but complex algorithms of computer vision's provide efficient performance in converting visual images into raw data. All of the proposed models are dealing with features extraction and labeling of features. AD prediction by using the DL approach [9] applied a semi-supervised approach and created an average accuracy of 87.76%, which is better than SVM. AD prediction based on unsupervised learning as 3DCNN, achieved performance greater than 90% for AD and other brain diseases. In the proposed method [15], DL-based AD prediction performed better than the % value of 3.4, 4.8, and 3.9. For the proposed research [17], TP rate = 57.43% and FPs = 0.4198 cm2, which is quite better results achieved through DL. Performance sensitivity of 97.5% to 96.1% was achieved in the method [5] of predicting diabetic retinopathy using DL. The model of research [4] achieved a sensitivity of 90% with a specificity of 97% using external data. In the breast tumor detection work research [1] using the DL approach, the performance is about 77–78%. DL-based work for the prediction of autoimmune disease [6] attains 99–96% performance. Prediction of different types of skin cancer using DL related proposed research [3] accomplished performance about 72.1% in case of three way accuracy and 55.4% in case of nine way accuracy (Table 1).

5 Conclusion and Future Scope

The exotic lifestyle increases the chances of chronic disease and they need to be treated straight away. Most of the pathological images cannot directly be analyzed by a machine because of the lack of absolute structure of data. A single image data contains multiple features that need to be extracted and categorized. DL multilayer structure provides a magnificent model that can analyze multidimensional data of enormous size without requiring any specific format of input data. This has encouraged researchers to apply DL models in the prediction of diseases and the results of these researches astonished the medico technical system.

In the concurrence of the predefined results, this paper involves the work that has been done in the prediction of most common diseases or sub diseases due to that chronic diseases. Then, all the related algorithms or mathematical approaches used

Table 1 Analysis of publications on deep learning in prediction of diseases

S. no	Year	Author	Algorithm	Objective	Performance
1	2014	Liu, S., Liu, S., Cai, W., Pujol, S., Kikinis, R., & Feng, D	Sparse auto encoders, Softmax regression	Diagnosis t AD using DL	87.76
2	2016	Hosseini-Asl, E., Keynton, R., & El-Baz, A	3DCAE(3D Convolution autoencoder)	Diagnosis of AD with 3D-CNN	90%
3	2019	Razavi, Firouzeh, Mohammad JafarTarokh, and MahmoodAlborzi	Sparse auto encoders, Softmax regression	Diagnosis of AD using unsupervised feature learning	More by 3.4%, 4.8% and 3.9% than ML
4	2017	Wang, J., Ding, H., Bidgoli, F. A., Zhou, B., Iribarren, C., Molloi, S., & Baldi, P	Convolution layer followed batch normalization layer and a nonlinearity layer	Detecting Cardiovascular Disease from Mammograms	TP rate = 57.43% and FPs = 0.4198 cm^2
5	2018	Poplin, R., Varadarajan, A.V., Blumer, K., Liu, Y., McConnell, M.V., Corrado, G.S., Peng, L. and Webster, D.R	Stochastic gradient descent, Inception-v3 NNA	Prediction of cardiovascular risk factors from retinal fundus photographs	Sensitivity 97.5% with specificity 93.4%
6	2016	Gulshan, V., Peng, L., Coram, M., Stumpe, M. C., Wu, D., Narayanaswamy, A., … & Kim, R	Stochastic gradient descent	Detection of Diabetic Retinopathy	97.5% to 96.1%
7	2017	Gargeya, Rishab, and Theodore Leng	Gradient boosting, Spatial pooling,tree based classifier	Automated identification of diabetic retinopathy	Sensitivity of 90% with specificity of 97%
8	2016	Akselrod-Ballin, A., Karlinsky, L., Alpert, S., Hasoul, S., Ben-Ari, R., & Barkan, E	Faster-R-CNN	Breast tumor detection and classification using R-CNN	77–78%
9	2016	Han, Xian-Hua, Jianmei Lei, and Yen-Wei Chen	K-support spatial pooling, Slide window	Predict autoimmune disease	99–96%
10	2017	Esteva, A., Kuprel, B., Novoa, R. A., Ko, J., Swetter, S. M., Blau, H. M., & Thrun, S	Inference algorithm	Classification of skin cancer with deep neural networks	72.10%

in DL in the addressed work, are addressed with their specific role in the proposed model. This article provides an outlook to the researchers about the main functions used in DL related medical solutions.

In future, there is a need to look for more applications of DL in the medicinal field. Data processing time is needed to be reduced as the data mass incessantly increases. More researches are needed to be formulated, that can provide better performance in automation of prediction of disease, using multilayered structure of DL having complex function and ML algorithms. There may be the scope of using DL network in the cloud to work on vast data clasp by the cloud, due to the availability of pathological data on it. It is now the beginning of an exploration of DL network application and exercises it for civilization.

References

1. Akselrod-Ballin A, Karlinsky L, Alpert S, Hasoul S, Ben-Ari R, Barkan E (2016) A region based convolutional network for tumor detection and classification in breast mammography. In: Deep learning and data labeling for medical applications. Springer, Cham, pp 197–205
2. Bottou L (2012) Stochastic gradient descent tricks. In: Neural networks: Tricks of the trade. Springer, Berlin, Heidelberg, pp 421–436
3. Esteva A, Kuprel B, Novoa RA, Ko J, Swetter SM, Blau HM, Thrun S (2017) Dermatologist-level classification of skin cancer with deep neural networks. Nature 542(7639):115
4. Gargeya R, Leng T (2017) Automated identification of diabetic retinopathy using deep learning. Ophthalmology 124(7):962–969
5. Gulshan V, Peng L, Coram M, Stumpe MC, Wu D, Narayanaswamy A, Kim R (2016) Development and validation of a deep learning algorithm for detection of diabetic retinopathy in retinal fundus photographs. Jama 316(22):2402–2410
6. Han XH, Lei J, Chen YW (2016) HEp-2 cell classification using K-support spatial pooling in deep CNNs. In: Deep learning and data labeling for medical applications. Springer, Cham, pp 3–11
7. Hosseini-Asl E, Keynton R, El-Baz A (2016) Alzheimer's disease diagnostics by adaptation of 3D convolutional network. In: 2016 IEEE international conference on image processing (ICIP). IEEE, pp 126–130
8. https://www.wikiwand.com/en/Autoencoder
9. Liu S, Liu S, Cai W, Pujol S, Kikinis R, Feng D (2014) Early diagnosis of Alzheimer's disease with deep learning. In: 2014 IEEE 11th international symposium on biomedical imaging (ISBI). IEEE, pp 1015–1018
10. Mason L, Baxter J, Bartlett PL, Frean MR (2000) Boosting algorithms as gradient descent. In: Advances in neural information processing systems, pp 512–518
11. Mazurowski MA, Buda M, Saha A, Bashir MR (2018) Deep learning in radiology: an overview of the concepts and a survey of the state of the art. arXiv preprint arXiv:1802.08717
12. Ng A (2011) Sparse autoencoder. CS294A Lecture notes 72(2011):1–19
13. Ngiam J, Chen Z, Bhaskar SA, Koh PW, Ng AY (2011) Sparse filtering. In: Advances in neural information processing systems, pp 1125–1133
14. Poplin R, Varadarajan AV, Blumer K, Liu Y, McConnell MV, Corrado GS, Peng L, Webster DR (2018) Prediction of cardiovascular risk factors from retinal fundus photographs via deep learning. Nat Biomed Eng 2(3):158
15. Razavi F, Tarokh MJ, Alborzi M (2019) An intelligent Alzheimer's disease diagnosis method using unsupervised feature learning. J Big Data 6.1, 32

16. Swain PH, Hauska H (1977) The decision tree classifier: design and potential. IEEE Trans Geosci Electron 15(3):142–147
17. Wang J, Ding H, Bidgoli FA, Zhou B, Iribarren C, Molloi S, Baldi P (2017) Detecting cardiovascular disease from mammograms with deep learning. IEEE Trans Med Imaging 36(5):1172–1181
18. Wang Z, Shang X (2006) Spatial pooling strategies for perceptual image quality assessment. In: 2006 international conference on image processing. IEEE, pp 2945–2948

Chapter 9
Cyberinfrastructure for Advanced Research with High Performance Edge Computing

Dharm Singh Jat, Arpit Jain, and Kamwayamunzi Fortune Sepo

Abstract The technological revolution brings rapid change in scientific and computational approaches. A huge amount of data is generated due to concurrent request-response processes and computation on real-time data. And this data requires to store as historical data for future reference and data analysis. For the computational work with intelligent and dynamic algorithm processes in communication networks, the infrastructure requires many scientific and networking equipment such as High Performance Computing (HPC) cyberinfrastructure. Researchers working with the complex problem need a small, state-of-the-art HPC system for their research. Researchers also require HPC administration expertise and identify and install the required tools, system software. Most of the time, the researcher would install the required tools and software that will be expensive. Undoubtedly, there is a need for a fast and low-cost ready-to-use HPC system that can be straightway put to utilisation by researchers and users. This paper presents a comprehensive literature review about the high performance edge computing (HPEC) technologies of cyberinfrastructure and other existing related initiatives around the world. Further, the paper also presents a case study of an affordable supercomputing solution named PARAM !ARUB which offers ready-to-use supercomputing facility based on edge computing and AI technologies hardware resources for complex problems. This provides a support research tool for analysis, design and development.

Keywords HPC · Edge computing · High performance edge computing · HPEC

D. S. Jat · A. Jain (✉) · K. F. Sepo
Namibia University of Science and Technology, Windhoek, Namibia
e-mail: ajain@nust.na

D. S. Jat
e-mail: dsingh@nust.na

R. Mathur et al. (eds.), *Emerging Trends in Data Driven Computing and Communications*, Studies in Autonomic, Data-driven and Industrial Computing,
https://doi.org/10.1007/978-981-16-3915-9_9

1 Introduction

High Performance Computing (HPC) Systems are usually extensive systems that require big space and specialised infrastructure. Due to several reasons HPC systems are unaffordable and inaccessible for researchers when needed for short duration [1]. The HPC will strengthen the application development of packages in smart, secure environments such as Climatology, Bioinformatics, Agriculture, Preventive health care system, Astrophysics and Transportation systems required in smart cities. This will accelerate research and development (R&D) in the war against global diseases such as COVID-19 through faster simulations, medical imaging and forecasting.

High Performance Computing (HPC) can process data and perform complex calculations at extremely high speeds [2]. The HPC is to combine multiple processors to create a unified system that can process a tremendous amount of work within a short period. To cope up with the fourth industrial revolution, nations must build infrastructures that will handle the big data associated with the resulting technologies. However, for most developing countries, the HPC facilities that support artificial intelligence and big data analytics are limited or non-existent. The lack of HPC facilities poses a severe problem in the research fraternity and the entire community. As we approach the era of the Internet of Things (IoT), more and more data will be generated daily. Figure 1 shows a prediction by the international telecommunication union (2015) for the expected data traffic in exabytes per month from 2020 to 2030 [3].

The figure shows the traffic will be growing at an annual rate of around 55% in 2020–2030. The global traffic per month is estimated to reach 607 exabytes (EB) in 2025 and 5016 EB in 2030. Thus, the major challenges are the facilities that will store and process such a large traffic volume. It is required to build widespread available, affordable and accessible communication networks with enhanced hardware HPC

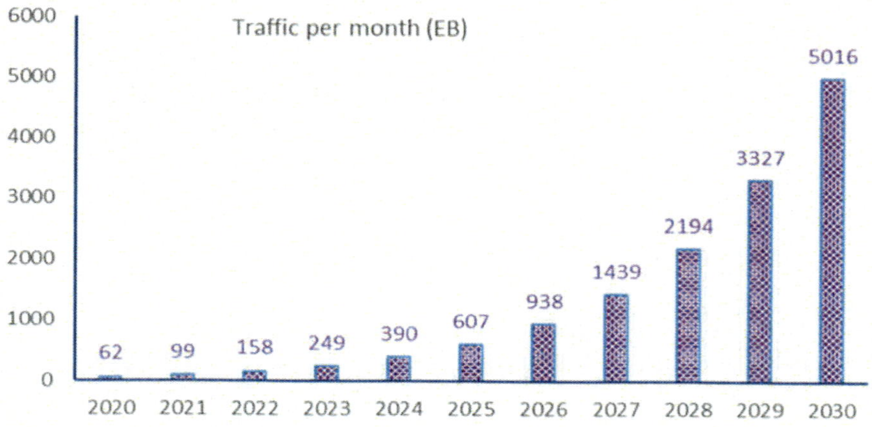

Fig. 1 Estimates of global traffic from the year 2020–2030 [3]

infrastructure to meet edge's current demand for intelligent services. High Performance computing (HPC) takes on challenges that will not fit in a single desktop/server configuration or be solved by a simple system in an acceptable time frame. In the HPC setup, multiple computers known as cores are clustered together to function in unison as a single high processing machine. This setup is highly needed for a researcher who collects and processes big data that requires large storage and rapid processing power.

There is multiple software that makes an HPC run effectively, efficiently and securely. HPC software's can be broken into system software and application software. The system software is used to build and manages the HPC system, while the application software supports the various application running on the HPC system. Mostly HPC infrastructure are developed to serve multiple and concurrent users in a cloud platform, thus there are unique softwares required for each set of users.

The general profile of HPC applications is constituted by an extensive collection of compute-intensive tasks that need to be processed in a short period of time. Historically, HPC was only available and used by key institutions such as scientific laboratories, university research and the military. Today, HPC is used to solve a range of problems, including applications in Bioinformatics, Climatology, Astrophysics, Computational Fluid dynamics, Computational Chemistry, Molecular Dynamics, Finite Element Analysis, Agriculture, Big data analytics, Artificial Intelligence, Internet of Things, etc. HPC applications are always provisioned using cloud computing.

2 Related Works

In this study, autotuning tools are emphasised to facilitate High performance computing, which helps the programmer write a set of codes efficiently without putting many efforts. It eliminates the manual efforts in the implementation of low-level architecture, which results in programmer productivity. The study analyses the powerful features of autotuning such as forecasting, predictability and integrational aspect into the applications running in HPC. The autotuning tools are much suitable for heavy computational aspects like deep learning, data exploration, data visualisation and data analytics for a huge number of datasets [3].

As the concurrent processes and computational overhead are executing in High performance computing components like clusters, base nodes, master nodes and clouds, where complex problem-solving and decision-making algorithms consumes much energy in continuous or longer execution. To optimise the energy consumption, the study proposed a novel energy scheduling algorithm based on the concept of maximum execution time with respect to minimum energy consumption. In this study, the optimised scheduling algorithm is tested with statistical analysis of variance followed by post hoc analysis to verify the energy efficiency and effectiveness [4].

The study analyses the challenges and techniques to overcome these issues in communication. It also introduces the efficient technique in edge AI communication system with an intelligent algorithm and inferences computational task at edge network. Further, the study presents an efficient solution in edge AI applications to train the AI models at the edge layer with several order optimisation algorithms. It also summarises the types of edge-based AI model architecture includes data partition and model division training approach. The study concludes with an analysis of computational latency, offloading and inference at edge node with an AI edge framework [5]. In this study, the intelligent system, coupled with edge computing, is analysed in the cyber human evolution and concludes with the Edge AI environment's challenges. The study analyses Edge Intelligence's fabric with critical components like sensing substrate, edge network with AI functionalities complimented by HPC. This study is also focussed on the investigation gaps in edge intelligence infrastructure of smart-city scale. Also shown are the negative aspects due to complexity in AI applications and machine learning processes for which new efficient techniques are required [6].

In this study, the singularity container-based technology is investigated to analyse the CPU's performance, memory and bandwidth of the network. Through this container-based technology, user and HPC data centres have the flexibility to utilise and distribute software environment. According to the given analysis, the study shows compatibility, mobility and security in the computing environment [7].

The study shows the vulnerability and security challenges in High Performance Computing systems, which have increased due to advanced integration and computing techniques. Further, the study employs three-dimensional integrated circuits with case studies to analyse the latest security threats. The study also highlighted the security measure to cover up the loopholes and provide approximations on computing. 3D integration plays a vital role in HPC with the added advantage of high device density, low power consumption and high bandwidth. The study shows the security threats on the hardware like 3D chips, specifically hardware trojans, with many other types of attacks [8].

The study proposed a framework named Edgent for Deep Neural Network (DNN) inferences with Edge Computing. To optimise the DNN inference latency, the study introduced two forms of design. One is the partition between device and edge computing. Another is the right-sizing design to describe the exit mechanism. The study designed collaboration aspects with different modes of networks. The fluctuation in bandwidth can be stable using a regression-based forecasting or prediction model known as a static network environment. The bandwidth changes rapidly, called dynamic. Further, the study implements to achieve the low latency edge intelligence with the help of a prototype developed in Raspberry pi [9].

The study shows the integration of blockchain technology with edge computing and implemented in blockchain-based architecture, which keeps track of the users accessing the data stored in distributed databases in the form of data analysis. This data analysis is concluded with edge artificial intelligence using the Ethereum blockchain technology. Further, the study emphasises the secure database and distributed trust with AI at the edge with the reduction in resource consumption [10].

The study analyses the limitations and advantages of working with edge architectures compared to cloud computing to execute AI algorithms or applications. With the help of a hardware edge accelerator, the edge-based AI workload is tested with benefit analysis of distributed or split processing, including model splitting and model compression. Split processing enables the deep learning process to be split across multiple nodes, whereas compression is the alternative method that provides the smaller compressed form of split processing with the lowest memory and least resources. Further, the study concludes with two analyses: firstly the edge accelerator can serve the concurrency of multitenant applications and secondly the disadvantage of isolation mechanism required for the edge computing environment [11].

3 HPC Edge and AI

The study shows the deployment of edge-based AI applications characterised by face recognition video analytics using machine learning and open-source infrastructure. The application is built using machine learning and artificial intelligence algorithm with many accelerators, which has limited software infrastructure. Further, the study shows the pre and post-processing code integration with multiple inferences stage. It reveals the system level implications on AI application which expose AI tax for overall computing for CPU performance. The study concludes with the analysis of upcoming challenges of accelerated AI [12].

Figure 2 shows the edge computing environment in which the High Performance Computing device interacts with edge devices and cloud computing. The HPC edge node is equipped with the data analysis machine learning algorithm, through which the real-time decision-making and resource management task can be achieved. The collaboration of edge computing and AI concludes the term edge AI, which enhances the system by learning the intelligence with data and programs. The various benefits of edge computing like optimised bandwidth and low latency make the added advantage to Edge nodes' designed algorithm. With the help of many datasets, AI can take out the inferences and make decisions according to the business requirements.

Due to the high demand for applications using IoT devices, the cloud extension in the form of edge computing can serve with AI components. The real-time data is used for predictive analysis from the sensors and IoT devices through this computing enhancement. To ensure the quality of Edge AI model inference results, it is required to evaluate the performance with such parameters as communication delay or latency, privacy and integrity of data, the accuracy of the transaction and communication overhead. The advancement in technology results in benefits and some challenges associated with the current system [13].

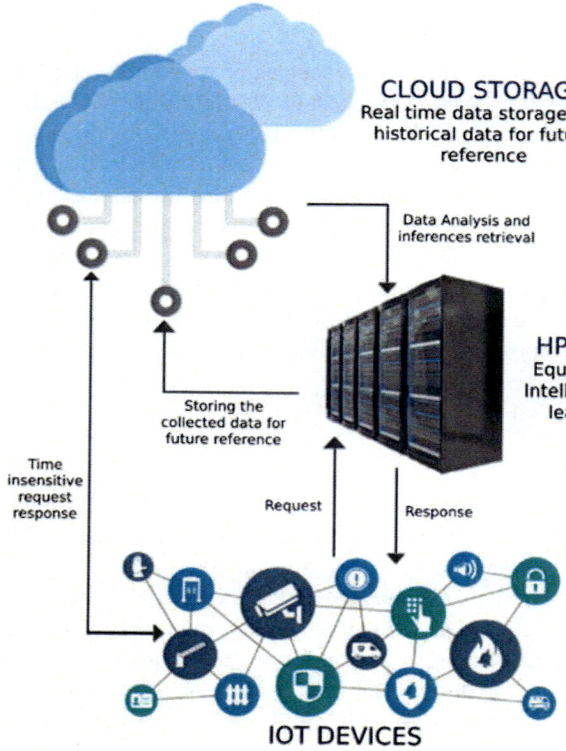

Fig. 2 HPC edge computing environment

4 PARAM !ARUB Architecture

An affordable supercomputing solution for a complex problem named PARAM !ARUB [14], which offers ready-to-use supercomputing facilities based on edge computing and AI technologies hardware resources is hosted in the Namibia University of Science and Technology (NUST). This provides a support research tool for analysis, design and development. PARAM !arub system is based on processor Intel Xeon Gold 6130. The cluster consists of compute nodes connected with INTEL OPA 100Gbps interconnect. The PARAM !ARUB Supercomputer aims to provide a computational resource in Namibia with edge artificial intelligence and other advanced technologies to perform complex tasks for industry, academic, scientific, technology, engineering programmes. PARAM !ARUB high performance edge computing (HPEC) is an affordable supercomputer facility with pre-installed required software, and the various applications from selected engineering and scientific domains are ready-to-use. Edge and artificial intelligence (AI) based PARAM !ARUB HPC system is designed at an affordable cost to perform complex and high-end computations for the scientific, technology, engineering and academic programmes to solve the complex problem by using modelling, simulation and data analysis.

Most researchers working with complex problem need a small state-of-the-art HPC system for their research. Researchers also require HPC administration expertise and identify and install the required tools, system software. Most of the time, the researcher would consume time to install the required tools and software and be expensive. Undoubtedly, there is a need for a fast and low-cost ready-to-use HPC system that can be straightway put to utilisation by researchers and users. The PARAM !ARUB supercomputing solution provides scalability and power efficiency. This paper presents a comprehensive literature review about the HPEC and other existing related initiatives around the world.

4.1 Machine Learning/Deep Learning Application Development

Most of the popular python based machine learning/deep learning libraries are installed on PARAM !arub system. While developing and testing their applications, users have the option to choose virtual environment-based python libraries or conda runtime based python libraries [14] (Fig. 3).

Fig. 3 PARAM !ARUB architecture at NUST, Namibia [15]

5 Conclusion

Today, a huge amount of data is generated from the Internet and other various sources. For the experimental work in communication networks research, the experimental setup requires many scientific and networking equipment such as high performance computing cyberinfrastructure. These equipments are normally geographically scattered with different capabilities around the nation, region and globe. For most of the researchers, this experimental setup is unavailable or hardly accessible. This paper explores how to enable wider accessibility of the HPC platform to potential researchers and learners and provide an affordable HPEC system named PARAM !ARUB.

Acknowledgements The authors would like to acknowledge the Centre for Development of Advanced Computing (C-DAC), Ministry of Electronics & Information Technology, Government of India, particularly the HPEC-Technologies Group, who have contributed to the development.

References

1. Agrawal S, Das S, Valmiki M, Wandhekar S, Moona R (2017) A case for PARAM shavak: ready-to-use and affordable supercomputing solution. In: 2017 international conference on high performance computing & simulation (HPCS), Genoa, 2017, pp 396–401. https://doi.org/10.1109/HPCS.2017.66. https://ieeexplore.ieee.org/stamp/stamp.jsp?tp=&arnumber=8035105
2. Cappello F, Gentzsch W, Valero M, Nygård M (2013) HPCS 2013 panel: the era of exascale sciences: challenges, needs and requirements. In: 2013 international conference on high performance computing & simulation (HPCS), Helsinki, Finland, 2013, pp 1–12. https://doi.org/10.1109/HPCSim.2013.6641380. https://ieeexplore.ieee.org/stamp/stamp.jsp?tp=&arnumber=6641380
3. ITU 2015.IMT traffic estimates for the years 2020–2030. https://www.itu.int/dms_pub/itu-r/opb/rep/R-REP-M.2370-2015-PDF-E.pdf
4. Orris BON, Uduc RIV, IEEE M (2018) Autotuning in high- performance computing applications 106(11)
5. Biswas T, Kuila P, Ray AK (2021) A novel energy efficient scheduling for high performance computing systems
6. Shi Y, Yang K, Member GS (2020) Communication-efficient edge AI: algorithms and systems 22(4):2167–2191
7. Rausch T (2019) Edge intelligence: the convergence of humans, things, and AI.https://doi.org/10.1109/IC2E.2019.00022
8. Wkh S, Ri S, Ljk IRU (2019) ([sorulqj wkh shuirupdqfh ri 6lqjxodulw\ iru +ljk 3huirupdqfh &rpsxwlqj 6fhqdulrv. https://doi.org/10.1109/HPCC/SmartCity/DSS.2019.00362
9. Yellu P, Zhang Z, Mezanur M, Monjur R, Abeysinghe R, Yu Q (2019) Emerging applications of 3D integration and approximate computing in high-performance computing systems: unique security vulnerabilities.
10. Li E, Zeng L, Zhou Z, Chen X (2020) Edge AI: on-demand accelerating deep neural network inference via edge computing 19(1):447–457
11. Nawaz A, Gia TN, Pe J (n.d.) Edge AI and blockchain for privacy-critical and data-sensitive applications, 2–3
12. Liang Q, Irwin D (2020) AI on the edge: characterizing AI-based IoT applications using specialized edge architectures, 145–156. https://doi.org/10.1109/IISWC50251.2020.00023

13. Richins D, Doshi D, Blackmore M, Nair AT, Pathapati N, Patel A (2020) Missing the forest for the trees: end-to-end AI application performance in edge data centers, 515–528. https://doi.org/10.1109/HPCA47549.2020.00049
14. Zhou Z, Chen X, Li E, Zeng L, Luo K, Zhang J (2019) Edge intelligence: paving the last mile of artificial intelligence with edge computing, 107(8)
15. Centre for Development of Advanced Computing (C-DAC) (2021) PARAM !Arub—user guide

Chapter 10
A Literature Review on Machine Learning Techniques and Strategies Applied to Stock Market Price Prediction

Pankaj Rambhau Patil, Deepa Parasar, and Shrikant Charhate

Abstract Forecast of financial exchange price and patterns is considered as a significant challenging task and is of extraordinary consideration as anticipating stock costs effectively may prompt alluring benefits by settling on appropriate choices. Because of non-stationary, blasting, and noisy information, stock market price prediction is one of the major challenges, and the expectation among financial specialists is therefore difficult to contribute to profit making. This work of literature review presents the survey of more than 100 research papers proposing the various philosophies of machine learning methods as well as SVM classifier, Random forest, Neural Network, Bayesian model, Fuzzy classifier, Artificial Neural Networks, etc., in view of stock market price prediction. Among the cutting-edge techniques, various machine learning approaches and models due to their ability to identify complex patterns in a number of applications are studied. The review works are investigated utilizing certain dataset parameters, forecast technique used, and results achieved by various procedures. It can be observed that the stock market forecast is a very complicated activity. Various parameters should be considered to predict the market more accurately and effectively. This review shows that in financial market forecasting, machine learning algorithms continue to outperform most conventional stochastic methods.

Keywords Machine learning · Stock market · Artificial neural networks · Support vector machine · Time series forecasting · Stock price

P. R. Patil (✉) · D. Parasar · S. Charhate
Amity University, Mumbai, India

D. Parasar
e-mail: dparasar@mum.amity.edu

S. Charhate
e-mail: scharhate@mum.amity.edu

R. Mathur et al. (eds.), *Emerging Trends in Data Driven Computing and Communications*,
Studies in Autonomic, Data-driven and Industrial Computing,
https://doi.org/10.1007/978-981-16-3915-9_10

1 Introduction

One of the most interesting inventions of our time is the finance markets. They had a massive effect on many sectors, such as production, education, housing, infrastructure, and hence the economy Hiransha [1]. Investors and analysts have focused on researching and developing models of stock price performance over the years Fama [2]. Investors have aimed to achieve a strategic gain over other market players since the early days of the financial markets, and a persistent area of concern for market participants is the ability to reliably predict time series [37].

The study of stock market movements and price behaviors is highly difficult due to the complex, non-linear, non-stationary, non-parametric, noisy, and disorderly existence of the markets [3]. As per Zhong and Enke [4], several highly interrelated factors influence financial markets. More essential than else is constant and consistent decision-making.

Although many strategic analysts and research efforts are made, no method has proved to correctly predict the movement of stock prices. Machine learning algorithms provide an extremely attractive choice for forecasting the financial time series [14, 44]. The difficulty of prediction lies in the complex and dynamic nature of market. They got some successes, with a lack of consistent methods. Two important investing methods, fundamental and technical methods, usually comprise of stock market research. In fundamental research, price fluctuations in the stock market are assumed to originate from the relative data of stability. As per Hu et al. [5], the three major elements of fundamental analysis are focused primarily on.

1. Macroeconomic data.
2. Analysis which estimates the company's value focused on status and prospects of the industry.
3. Study of a company's current activity and financial position to determine its intrinsic worth.

There are various approaches to valuation for fundamental analysis. In order to assess future predictions, fundamentalists use numeric details such as profits, balances, and organizational development. Market timing is crucial and an important element in technical analysis. Specialists use pattern charts and different analytical models to define stock price movements. Despite a growing number of papers being published in this research field, the literature is struggling to have a convincing overview of various algorithms and their findings. Our literature research therefore conducts a thorough, systematic analysis of current trade forecasting algorithms work to close the gap in research. The majority of this paper is arranged as follows. Section 2 identifies a need for stock market forecasting. Section 3 enlists different analysis methods. The literature review is mentioned in Sect. 4. Abstract analysis on analytical approaches is given in Sect. 5. Challenges and open problems are addressed in Sect. 6. The paper concludes with Sect. 7.

2 Need of Stock Market Prediction

Due to various advanced techniques, the stock market has gained investor and traders appeal, where prediction can contribute to a strong economic projection. The forecast of stock trends is entirely dependent on trading stock data. The methods of determining market will analyze, forecast, and monitor the market which can be used to make the appropriate decisions [6]. Sale transactions and delivery activities along with various stock information are to be adjusted by experts. An effective way of collecting data is to extract valuable information [7]. A predictive algorithm capable of steadily gaining profits above market indices over time would not only be significant proof against the EMH, but will also allow substantial financial operating profits to be made. This concept will help several investors and organizations invest their money in the right way as well as getting a comprehensive look at their assets and their potential investment scope.

3 Stock Market Analysis Approaches

Various theories are available for predicting the stock market prices. Specifically, there are two important theories of stock market prediction.

4 Efficient Market Hypothesis (EMH)

This hypothesis states that share prices reflect all information and consistent alpha generation is impossible. According to the EMH by Malkiel and Fama [38], stocks are still trading on exchanges at their market price, making it difficult for investors to buy undervalued stocks. Adherents of the Efficient Market Theory argue that investors will do well by investing in a low-cost, passive portfolio. Three forms of EMH:

1. Weak Form: all past information is considered.
2. Semi-Strong Form: all public information is used in the calculation of a stock's current price.
3. Strong Form: both public and private information is used in the calculation of a stock's current price.

4.1 Random Walk Theory

The random walk hypothesis argues that stocks take a random and unpredictable course. This makes it useless in the long term for all methods of forecasting stock prices. Statistics, pattern recognition, ML, and sentiment analysis are four types of stock analysis technologies. These categories include two conventional approaches.

- Technical Analysis
- Fundamental Analysis

Many machine learning approaches blend technical analysis with fundamental analysis.

4.2 *Review of Machine Learning Techniques*

ML is an aspect of artificial intelligence (AI) that allows systems the ability to automatically learn and enhance from experience without being programmed directly. Commonly, ML Techniques are divided into three categories [42].

Supervised Learning: Supervised learning is where you have input variables (x) and output variables (Y) and use an algorithm Y = f(x) to learn the mapping feature from input to output (X). Linear regression, logistic regression, and neural networks comprise the algorithms [43].

Unsupervised Learning: Unsupervised learning is where an algorithm is supplied with input, as a training set, without equivalent output values. There are no performance measurement values or trainers, as with supervised learning. Instead, algorithms are able to work and learn about the data and results independently.

Reinforced Learning: It is the training of ML models to make a sequence of decisions. The agent learns to achieve a goal in an uncertain, potentially complex environment.

5 Literature Survey

Supervised learning methods such as SVM, Decision Trees predict on the basis of historical prices and patterns and give effective historical price evaluation [45]. Echo State Networks (ESN), the subset of RNN was introduced by Bernal et al. [8, 84] to forecast S&P 500 stock prices using size, simple moving averages, and volume as characteristics [8]. The method surpasses with a marginal 0.0027 check error by Kalman Filter technique. Milosevic (2016) applied a set of manual features among financial ratios to understand how optimum features are found [9]. He conducted a manual collection of functions, chose 11 basic ratios, and applied ML algorithms to market prediction. It implies that the highest F-Score of 0.7511 against methods like SVM and Nave Bayes was achieved by Random Forest. Dey et al. [10] used eXtreme Gradient Boosting (XGBoost) to predict direction of stocks [10]. LSTM network has made other expectations for predicting the time series; Roondiwala et al. [11] have applied the LSTM network to forecast Nifty prices with attributes such as OHLC. Here, in view of frequent percentage changes, LSTM produces an RMSE of 0.00859 for the test results. Di Parsio and Honchar (2017) introduced three NN models, RNN, LSTM, and Gated Recurrent Unit (GRU) to Google stock prices. Here, LSTM surpassed other models on a five-day timeframe with 72% precision [12]. Ensemble of Chinese stock prediction feed-forward networks that used training algorithms such as back propagation and Adam was suggested by Yang et al. (2017). It achieves satisfactory precision, consistency, and recall [13]. Zhang et al. (2018) [14] suggested the movement of stock prices and their growth interval (or reduction) rate under predefined forecast periods using the stock market trend prediction method. Hybrid model based on deep learning was trained using S&P 500 time series consisting of LSTM and GRU architecture by Hossain et al. (2018) [15]. In expectation, the authors obtained 0.00098 MSE, outperforming past NN methods. Lv et al. (2019) studied separate ML algorithms and examined in depth the S&P 500 Index and CSI 300 Index Component Stocks (CSICS) stocks' daily trading outcomes between 2010 and 2017 [16].

Unsupervised learning aims to classify associations including financial markets in an unorganized dataset. Powell et al. in 2008 differentiate between the SVM and the K-means by the Principal Component Analysis for feature reduction, the data tested on S&P 500, and the analysis shows that both methods have close similarity, SVM reaches 89.1 percent and K-means reaches 85.6%, respectively [17]. In order to quantify the short-term impact after the release of financial results, Babu et al. (2012) introduce a clustering method called HAK through a mixture of Hierarchical Agglomerative Clustering and reverse K-means clustering [18]. The results indicate that the approach suggested exceeds SVM in terms of precision. Peachavanish [20] suggests a clustering approach to classify the category of stocks with the strongest patterns and traction at a given time [20].

Sentiments are a significant portion of capital markets and can induce instability in the short-term market. Schumaker and Chen (2009) assessed, rely on textual interpretations, the effect on stock markets of breaking news. To analyze stock dynamics, Bollen et al. [22] analyzed Twitter info. In order to explain the correlations and estimate DJIA closing prices, the Twitter data was analyzed using the Google Profile of Mood States (GPOMS) and Opinion Finder. The Emotion Finder and GPOMS time series are also cross-validated against typical occurrences such as presidential elections. A causative study of Granger was done. When calculating daily values [22], they achieved an accuracy of 87.6%. Based on the work of [22], Mittal and [23] implemented their process, but opted for a massively greater datasets. Their strategy obtained a precision of 75% [23].

Lee et al. in (2014) suggested a method to forecast when the stock price will increase, decline, or remain as it is by conducting 8-K data sentiment analysis of securities, which would boost the prediction accuracy by 10%, but would be sufficient for short-term predictions [24]. To extract sentiment from news stories and feed output into two ML algorithms, [25] suggested a model based on sentiment analysis [25]. Results have shown that with an accuracy of 60%, the ML model using gradient descent has predicted the feeling of a news article. The ML model, in reality, was 81.82% correct.

Cakra et al. (2015) predict the price, cost variance, and percentage of margin from Indonesian stocks [26], using a basic estimation investigation, combined with classification methods and a linear regression forecasting model. StockNet architecture for estimating market volatility from tweets and historical prices was introduced by Xu and Cohen (2018) [27]. Markowska-Kaczmar and Dziedzic (2008) developed a supervised NN feed-forward technique to classify patterns of stock data and minimize attributes using PIP [28]. A fascinating hybrid model was introduced by Tiwari et al. [29], combining the Hidden Markov Predictive Model with the 92.1% accuracy of the supervised learning approach [29]. An algorithm with mathematical and SVM hybrid methods was suggested by Shen et al. [30]. The findings of this analysis indicate a prediction performance of 77.6% on the DJIA and up to 85 percent longer-term [30]. The Hybrid Model (PHM) suggested by Wang et al. [31] involves the ESM, Backpropagation Neural Network, and ARIMA. This shows 70.16 percent directional precision [31]. Focused on Bollen's (2011) work, Mittal and [23] suggested their solution, but preferred a much bigger dataset. A precision of 75% was obtained by their methodology. Ding et al. (2015) on the S&P 500 index, introduced a technique that combines NN, sentiment analysis, and simulations that achieves 64.21 percent accuracy and 65.48% accuracy on the specific stock price forecast [33]. Rather et al. (2015) presented a hybrid model of ARIMA, ESN, and Recurrent NN models combining both linear and non-linear approaches [34]. The initial research of [31] was generalized by Creighton and Zulkernineine's use of the hybrid method for daily prediction [35] (2017).

Models such as the ARIMA and ESM, like the hybrid PHM ensemble, can be best predicted over longer periods of time, but, when applied to regular predictions, they rely on market volatility. Table 1 summarizes in depth all the alternative methods. Most of the literature articles apply neural network. How- ever, the SVM used by Kim [46] and Random Forest is another algorithm that are very current in prediction articles. Kumar and Thenmozhi [47] predict frequent returns by combining ARIMA, ANN, SVM, and RF.

Sentiment research based on Twitter data has recently been given a specific emphasis. Social network results can be unreliable and impossible to manage. The quarterly or annual reports submitted by the organizations would be an acceptable alternative to these or complementary tools. Even if it is suggested that preprocessing improves the predictive efficiency of neural networks in the collection of variables. Chen et al. [49] introduced the Fuzzy logic theory for time series in human linguistic terms [49]. From the overall literature, it can be noted that Neural Networks, RF/Decision Trees, SVM/SVR, Sentiment Analysis, kNN, ARIMA/GARCH, Fuzzy Logic, Fuzzy Logic, are the key and relevant forecasting strategies used by analysts and researchers as a priority and quality of performance.

Table 1 Literature review

Authors	Methods	Dataset	Result Accuracy (%)	Metrics	Time series
Kimoto et al. [74]	NN, GA	Japan	65–70	Direction	Monthly
Kamstra and Donaldson [71]	NN	Multiple	60–70	Volatility	Daily
Fernandez-Rodriguez et al. [66]	Neural Networks	Spain	70	Direction	Daily
Kim and Han [73]	Neural Networks	Korea	60–70	Direction	Weekly
Leung et al. [78]	NN, LDA	USA, UK, Japan	65–70	Return	Weekly, monthly
Leigh et al. [77]	NN, GA	USA	70–75	Prices	Daily
Chen et al. [60]	NN, GMM	Taiwan	80–90	Return	Weekly, monthly
Cao et al. [56, 57]	Neural networks, CAPM	China	65–75	Return	Weekly, monthly
Kyoung-jae [46], 89]	SVM	Koreacom-positestock priceindex (KOSPI)	52–54	Accuracy	Daily
Kyoung-jae Kim, Won BooLee [90]	GAand ANNs	Koreacom-positestock price index	60–64	hit ratio	Daily
Armano et al. [52]	NN, GA	USA & Italy	80	Sharperate, Prices	Daily
Enkeand Thawornwong [65]	Neural networks	USA	60–70	Direction	Weekly, monthly
Pai and Lin [82]	SVM	USA	85–90	Prices	Daily
Angand Quek [51]	NN, TA	Singapore	89.5	Return, prices	Weekly, monthly
Bo Qian [91]	ANN, KNN, DT	Dow Jones Industrial Average index	60–65	Accuracy	Daily

(continued)

Table 1 (continued)

Authors	Methods	Dataset	Result Accuracy (%)	Metrics	Time series
Hassan et al. [69]	Analysisof sentiment, Neural networks, GA	USA	80–85	Prices	Daily
Leigh et al. [85]	Template Matching	NYSE	3.14.59	Average profits	Daily
Changand Fan [58]	kNN, DWT, fuzzy logic	Taiwan	90	Prices	Weekly, Monthly
Li and Kuo [79]	regressions DWT, SOM	Taiwan	85–90	Prices	Daily
Chang et al. [59]	Neural networks, CBR	Taiwan	80–90	Direction	Weekly, Monthly
Huangand Tsai [70]	SVR	Taiwan	60–70	Prices	Daily
Schumaker and Chen [21]	Bag of words , noun phrases, and noun entities to SVM	News articles, S&P 500	2.57 (Noun phrases)	Returns, DA	Daily
Tiwari et al. [29]	HHMM + Decision Trees	Sensex + 3 stocks	92.1	Accuracy	Daily
Kara et al. [72]	Neural networks, SVM	Turkey	80–90	Direction	Weekly, Monthly
Bollen et al. [22]	Mood Indicator to SOFNN	DJIA, Twitter data	87.14	Accuracy	Daily
Bernal et al. [86]	ESN RNN	S&P 500	0.0027	Test error	Daily
Wang et al. [31]	ESM + BPNN + ARIMA	DJIA and SJI Index	70.16	Directional accuracy	Weekly
Shen et al. [30]	Auto,cross correlation + SVM	Indices, commodities	77.6	Accuracy	Daily, monthly

(continued)

Table 1 (continued)

Authors	Methods	Dataset	Result Accuracy (%)	Metrics	Time series
Hjek et al. (2013)[68]	Neural networks, SVR,	USA	80–90	Return	Weekly, Monthly
Lee et al. [24]	Ngramto Random Forest	8-K Reports, Stock prices, volatility	10 (Increase in accuracy)	Accuracy	Daily long term
Kalyanaraman et al. [25]	Dictionary Approach toLinear Regression	News articles (Bing API)	81.82	Accuracy	Daily
Rather et al. [34]	ARIMA + ESM + RNN + GA	NSE stocks	0.0009, 0.0127	Avg MSE, MAE	Daily
Ding et al [33]	NN(event embeddings) + CNN	S&P 500	64.21	Accuracy& MCC	Weekly, Monthly
AlNasseri [50]	Analysisof sentiment	USA	sentiment	Direction	Return
Ballings et al. [53]	NN,SVM, kNN, and RF	Europe	65–75	Accuracy, Direction	Weekly, Monthly
Laboissiere et al. [76]	decision treesNeural networks	Brazil	75–80	Maximums, minimums	Daily
Patel et al. [84]	NN,SVM, RF, NB	India	75	Direction	Weekly, Monthly
Milosevic [87]	RF, SVM, NB, Logistic Regression	1700 + individual stocks	0.751(Random Forest)	Precision, Recall, and F-score	Classification (good vs. bad)
Deyetal [10]	XGBoost vs SVM vs. ANN	Apple, Yahoo	80–85(XG-Boost)	Accuracy	Daily
Chen and Chen [62]	Pattern recognition	USA, Taiwan	APEranged between 1.5% and 9.6%	Return	Weekly, Monthly
Chiang et al. [64]	Neuralnetworks	Multiple	86	Direction	Weekly, Monthly
Gorenc Novak and Veluk [67]	SVM	USA	80–90	Direction	Weekly, Monthly

(continued)

Table 1 (continued)

Authors	Methods	Dataset	Result Accuracy (%)	Metrics	Time series
Pagolu et al. [88]	Ngram + wordvec ToRandom Forest	MSFT price, Twitter data	70.1	Accuracy	Daily
DiPersio and Honchar [12]	RNNvs LSTMvs. GRU	Google Stock	72, 5 day (LSTM)	Log loss, accuracy	Daily, Weekly
Yang [13]	DNN Ensemble	Shanghai composite index	71.34	Accuracy, relative error	Daily
Barak et al. [54]	NN, SVM, decision trees	Iran	65	Return and risk,Accuracy	Weekly, Monthly
Bezerra and Albuquerque [55]	decision trees,SVR, GARCH	Brazil, Japan	80–90	Volatility	Daily
Chen et al. [61]	SVM, kNN	China	90	Direction	Weekly, Monthly
Krauss et al. [75]	NN, RF	USA	80–90	Returns	Weekly, Monthly
Zhang [14]	Random Forest	Shenzhen GE Market	75.1	Returnper trade	Classification (up,down, flat)
Mo and Wang [80]	NN	China, USA	85–90	Correlation	Weekly, Monthly
Oliveira et al. [81]	NN, SVM, RF	Multiple	85–90	Return, volume, volatility	Daily, weekly
Panetal [83]	SVM	USA	85–90	Prices	Daily
Mojtaba Nabipour, Pooyan Nayyeri [92]	RNNand LSTM	Tehran stock exchange	80–88	Volatility	Daily
Sahil Bhatia, Srishti Kishore [93]	LSTM	GOOGL	50	epochs	Daily

6 Discussion

Economic benefits may not be seen as a metric of progress, as not all practice relies on implementing market plans. Some of the identified works suggest and check models for predicting factors in the financial market, regardless of whether those forecasts can be used for benefit. These models clearly turn into trading strategies. Direction prediction can definitely be viewed as a tactic for buying or selling signals. Models are designed, built, updated, configured, or hybridized. Creating a model for machine learning does not inherently require designing an entirely new algorithm. Customizing and refining possibly the best models will contribute to better results in forecasting.

7 Challenges and Open Problems

Share price analysis and prediction is still a significant and complicated task. When more details are released, in order to generate information and determine market values, we face new obstacles in the gathering and distribution of data. These challenges include problems relating to live monitoring, algorithmic trading, self-defeating, long-term estimates, and study of opinion for market filing. The analysis of these algorithms and efficiency poses yet another problem, with new algorithms targeting the markets on a daily basis. In the medium term, capital prices are usually more predictable. Several recent ANN methods, such as LSTM and RNN, SVM, RF, etc., are being experimented [36].

8 Conclusion

Stock markets have a strong exchange and investing environment in which any computer that can connect to the Internet can be exchanged. With many positives, such advances bring fresh opportunities and machine learning strategies, but they also bring a whole new multitude of problems. This article presents a categorization of computational methods of stock market research and evaluation, and also presents a comprehensive literature summary of the existing algorithms and techniques used to forecast stock markets. A fruitful direction for study is to investigate algorithms' effectiveness to estimate for the longer term. It is possible that hybrid techniques that incorporate statistical and machine learning methods would be more effective for market prediction.

References

1. Hiransha M et al (2018) NSE stock market prediction using deep-learning models. Procardia Comput Sci 132, 1351–1362
2. Fama EF (1995) Random walks in stock market prices. Financ Anal J 51:75–80
3. Abu-Mustafa YS (1996) Introduction to financial forecasting. Appl Intell 6, 205–213
4. Zhong X, Enke D (2017) Forecasting daily stock market return using dimensionality reduction. Expert Syst Appl 67:126–139

5. Hu Y, Liu K et al (2015) Application of evolutionary computation for rule discovery in stock algorithmic trading: a literature review. Appl Soft Comput 36, 534–51
6. Majumder M, Hussian MDA (2009) Forecasting of Indian stock market index using artificial neural network. National stock exchange of India Lim- ited
7. Kumar A Artificial neural networks for data mining, IASRI, Library Avenue, Pusa, New Delhi Google Scholar
8. Bernal et al (2012) Financial market time series prediction with recurrent neural networks. State College: Citeseer
9. Milosevic N (2016) Equity forecast: predicting long term stock price movement using machine learning
10. Dey S et al (2016) Forecasting to classification: predicting the direction of stock market price using xtreme gradient boosting
11. Roondiwala M et al (2017) Predicting stock prices using LSTM. Int J Sci Res (IJSR) 6, 1754–1756
12. Di Persio L, Honchar O (2017) Recurrent neural networks approach to the financial forecast of google assets. Int J Math Comput Simul 11, 7–13
13. Yang B et al (2017) Stock market index prediction using deep neural network ensemble. In: 36th Chinese control conference (CCC), Dalian, China. 26–28 July 2017
14. Zhang J, Cui S et al (2018) A novel data-driven stock price trend prediction system. Expert Syst Appl 97, 60–69
15. Hossain MA et al (2018) Hybrid deep learning model for stock price prediction. In: IEEE symposium series on computational intelligence (SSCI), Bangalore, India, and 18–21 Nov 2018
16. Lv D, Yuan S et al (2019) An empirical study of machine learning algorithms for stock daily trading strategy. Mathematical problems in engineering
17. Powell N, Foo SY (2008) Supervised and unsupervised methods for stock trend forecasting. In: 40th southeastern symposium on system theory (SSST 2008). New Orleans, LA, USA, 16–18 Mar 2008
18. Babu MS et al (2012) Clustering approach to stock market prediction. Int J Adv Netw Appl 3, 1281
19. Wu KP et al (2014) Stock trend prediction by using K-means and aprioriall algorithm for sequential chart pattern mining. J Inf Sci Eng 30, 669–86
20. Peachavanish R (2016) Stock selection and trading based on cluster analysis of trend and momentum indicators. In: International multi conference of engineers and computer scientists. Hong Kong, China, 16–18 Mar 2016
21. Schumaker RP, Chen H (2009) Textual analysis of stock market pre- diction using breaking financial news: the Azfin text system. ACM Trans Inf Syst (TOIS) 27, 12
22. Bollen J et al (2011) Twitter mood predicts the stock market. J Comput Sci 2, 1–8
23. Mittal A, Goel A (2012) Stock prediction using twitter sentiment analysis. Stanford University, CS229
24. Lee H et al (2014) On the importance of text analysis for stock price prediction. In: 9th international conference on language resources and evaluation, LREC 2014, Reykjavik, Iceland, 26–31 May 2014
25. Kalyanaraman V et al (2014) Sentiment analysis on news articles for stocks. In: 8th asia modelling symposium (AMS). Taipei, Taiwan, 23–25 Sept 2014
26. Cakra YE et al (2015) Stock price prediction using linear regression based on sentiment analysis. In: International conference on advanced computer science and information systems (ICACSIS), Depok, Indonesia, 10–11 Oct 2015
27. Xu Y, Cohen SB (2018) Stock movement prediction from tweets and historical prices. In: 56th annual meeting of the association for computational linguistics. Melbourne, Australia, 15–20 July 2018
28. Markowska-Kaczmar U et al (2008) Discovery of technical analysis patterns. In: International multi conference on computer science and information technology, 2008, IMCSIT 2008. Wisia, Poland, 20–22 Oct 2008

29. Tiwari S et al (2010) Predicting future trends in stock market by decision tree rough-set based hybrid system with HHMM. Int J Electron Comput Sci Eng 1, 1578–1587
30. Shen S et al (2012) Stock market forecasting using machine learning algorithms. Department of Electrical Engineering, Stanford University, Stanford, pp 1–5
31. Wang JJ et al (2012) Stock index forecasting based on a hybrid model. Omega 40, 758–66
32. Yoshihara A et al (2014) Predicting stock market trends by recurrent deep neural networks. In: Pacific RIM international conference on artificial intelligence, gold coast. Australia, 1–5 Dec 1–5
33. Ding X et al (2015) Deep learning for event-driven stock prediction. In: 24th international conference on artificial intelligence (IJCAI). Buenos Aires, Argentina, 25–31 July 2015
34. Rather AM et al (2015) Recurrent neural network and a hybrid model for prediction of stock returns. Expert Syst Appl 42, 3234–41
35. Creighton J et al (2017) Towards building a hybrid model for predicting stock indexes. IEEE international conference on big data (big data). Boston, MA, USA, 11–14 Dec 2017
36. Shah D, Isah H, Zulkernine F (2019) Stock market analysis: a review and taxonomy of prediction techniques. Int J Financ Stud 7, 26. 10.3390, 27 May 2019
37. Chen H, Xiao et al (2017) A double-layer neural network framework for high- frequency forecasting. ACM Trans Manag Inf Syst (TMIS) 7(4):11:2–11:17
38. Fama EF (1991) Efficient capital markets: II. J Financ 46(5):1575–1617
39. Weng B et al (2017) Stock market one-day ahead movement prediction using disparate data sources. Expert Syst Appl 79(1):153–163
40. Tay FE, Cao L (2001) Application of support vector machines in financial time series forecasting. Omega 29(4):309–317
41. Kumar D et al (2016) Proximal support vector machine based hybrid prediction models for trend forecasting in financial markets. J Comput Sci 17(1):1–13
42. Xiao Y, Xiao J et al (2013) Ensemble ANNs-PSO-GA approach for day-ahead stock e-exchange prices forecasting. Int J Comput Intell Syst 6(1):96–114
43. Tsaih R, Hsu Y, Lai CC (1998) Forecasting S&P 500 stock index futures with a hybrid AI system. Decis Support Syst 23(2):161–174
44. Kara Y et al (2011) Predicting direction of stock price index movement using artificial neural networks and support vector machines: the sample of the Istanbul stock exchange. Expert Syst Appl 38(5):5311–5319
45. Barak S et al (2017) Fusion of multiple diverse predictors in stock market. Inf Fusion 36(1):90–102
46. Kim K (2003) Financial time series forecasting using support vector machines. Neurocomputing 55(1–2):307–319
47. Kumar M, Thenmozhi M (2014) Forecasting stock index returns using ARIMA-SVM, ARIMA-ANN, and ARIMA-random forest hybrid models. Int J Bank Account Financ 5(3):284–308
48. Zhong X, Enke D (2017) Forecasting daily stock market return using dimensionality reduction. Expert Syst Appl 67(1):126–139
49. Chen YS et al (2014) Modeling fitting-function-based fuzzy time series patterns for evolving stock index forecasting. Appl Intell 41(2):327–347
50. Al Nasseri A et al (2015) Quantifying stock twits semantic terms' trading behavior in financial markets: an effective application of decision tree algorithms. Expert Syst Appl 42(23):9192–9210
51. Ang KK, Quek C (2006) Stock trading using RSPOP: a novel rough set-based neuro-fuzzy approach. IEEE Trans Neural Networks 17(5):1301–1315
52. Armano G et al (2005) A hybrid genetic-neural architecture for stock indexes fore-casting. Inf Sci 170(1):3–33
53. Ballings M et al (2015) Evaluating multiple classifiers for stock price direction pre- diction. Expert Syst Appl 42(20):7046–7056
54. Barak S, Arjmand A, Ortobelli S, Batagelj V (2017) Fusion of multiple diverse predictors in stock market. Inf Fusion 36(1):90–102

55. Bezerra PCS, Albuquerque PHM (2017) Volatility forecasting via SVR- GARCH with mixture of Gaussian kernels. CMS 14(2):179–196
56. Cao L (2003) Support vector machines experts for time series forecasting. Neuro Comput 51(1):321–339
57. Cao Q at al (2005) A comparison between Fama and French's model and artificial neural networks in predicting the Chinese stock market. Comput Oper Res 32(10):2499–2512
58. Chang P-C, Fan CY (2008) A hybrid system integrating a wavelet and TSK fuzzy rules for stock price forecasting. IEEE Trans Syst (Appl Rev) 38(6):802–815
59. Chang P-C et al (2009) A neural network with a case based dynamic window for stock trading prediction. Expert Syst Appl 36(3, Part 2):6 889–6 898
60. Chen A-S et al (2003) Application of neural networks to an emerging financial market: forecasting and trading the Taiwan stock index. Comput Oper Res 30(6):901–923
61. Chen H et al (2017) A double-layer neural network frame- work for high-frequency forecasting. ACM Trans Manag Inf Syst (TMIS) 7(4):11:2–11:17
62. Chen T-l, Chen F (2016) An intelligent pattern recognition model for sup- porting investment decisions in stock market. Inf Sci 346(1):261–274
63. Chiang WC et al (2016) An adaptive stock index trading decision support system. Expert Syst Appl 59(1):195–207
64. Enke D, Thawornwong S (2005) The use of data mining and neural networks for forecasting stock market returns. Expert Syst Appl 29(4):927–940
65. Fernandez-Rodriguez F et al (2000) On the profitability of technical trading rules based on artificial neural networks: evidence from the Madrid stock market. Econ Lett 69(1):89–94
66. Novak G et al (2016) Prediction of stock price movement based on daily high prices. Quant Financ 16(5):793–826
67. Hjek P, Olej V et al (2013) Forecasting stock prices using sentiment information in annual reports-A neural network and support vector regression approach. WSEAS Trans Bus Econ 10(4):293–305
68. Hassan MR et al (2007) A fusion model of HMM, ANN and GA for stock market forecasting. Expert Syst Appl 33(1):171–180
69. Huang CL, Tsai CY (2009) A hybrid SOFM-SVR with a filter-based feature selection for stock market forecasting. Expert Syst Appl 36(2):1529–1539
70. Kamstra M, Donaldson G (1996) Forecasting combined with neural networks. J Forecast 15(1):49–61
71. Kim KJ, Han I (2000) Genetic algorithms approach to feature discretization in artificial neural network for the prediction of stock price index. Expert Syst Appl 19(2):125–132
72. Kimoto T, Asakawa K et al (1990) Market prediction system with modular neural networks. In: IJCNN international joint conference on. IEEE, pp 1–6
73. Krauss C et al (2017) Deep neural networks, gradient-boosted trees, random forests: statistical arbitrage on the S&P 500. Eur J Oper Res 259(2):689–702
74. Laboissiere LA et al (2015) Maximum and minimum stock price forecasting of Brazilian power distribution companies based on artificial neural networks. Appl Soft Comput 35(1):66–74
75. Leigh W, Purvis R, Ragusa JM (2002) Forecasting the NYSE composite index with technical analysis, pattern recognizer, neural network, and genetic algorithm: a case study in romantic decision support. Decis Support Syst 32(4):361–377
76. Leung MT et al (2000) Forecasting stock indices: a comparison of classification and level estimation models. Int J Forecast 16(2):173–190
77. Li ST, Kuo SC (2008) Knowledge discovery in financial investment for forecasting and trading strategy through wavelet-based SOM networks. Expert Syst Appl 34(2):935–951
78. Mo H, Wang J (2017) Return scaling cross-correlation forecasting by stochastic time strength neural network in financial market dynamics. Soft Comput 1(1):1–13
79. Oliveira N et al (2017) The impact of microblogging data for stock market prediction: using twitter to predict returns, volatility, trading volume and survey sentiment indices. Expert Syst Appl 73(1):125–144

80. Pai PF, Lin CS (2005) A hybrid ARIMA and support vector machines model in stock price forecasting. Omega 33(6):497–505
81. Pan Y et al (2017) A multiple support vector machine approach to stock index fore- casting with mixed frequency sampling. Knowl-Based Syst 122(1):90–102
82. Patel J, Shah S et al (2015) Predicting stock and stock price index movement using trend deterministic data preparation and machine learning techniques. Expert Syst Appl 42(1):259–268
83. Leigh W, Frohlich CJ et al (2008) Trading with a stock chart heuristic. IEEE Trans Syst, Man, Cybern 38, 93–104
84. Bernal A et al (2012) Financial market time series prediction with recurrent neural networks. State College, Citeseerseer
85. Pagolu VS, Reddy KN et al (2016) Sentiment analysis of twitter data for predicting stock market movements. In: International conference on signal processing, communication, power and embedded system (SCOPES), Paralakhemundi, India, 3–5 Oct 2016
86. Kim K (2003) Financial time series forecasting using support vector machines", 0925–2312/03. Elsevier. https://doi.org/10.1016/S0925-2312(03)00372-2
87. Kim K, Lee WB (2004) Stock market prediction using artificial neural networks with optimal feature transformation. Springer
88. Qian B, Rasheed K (2006) Stock market prediction with multiple classifiers. Springer
89. Nabipour M, Nayyeri P (2020) Predicting stock market trends using machine learning and deep learning algorithms via continuous and binary data: a comparative analysis. IEEE Access
90. Bhatia S, Kishore S et al (2020) Stock price trend prediction. ISSN: 2321-9653; IC value: 45.98; SJ impact factor: 7.429, vol 8, issue VI

Chapter 11
A Comparative Analysis of ML Stratagems to Estimate Chronic Kidney Disease Predictions and Progression by Employing Electronic Health Records

Shruti Jain, Mayank Patel, and Konika Jain

Abstract Chronic Kidney Disease is a gradually increasing problem which can lead to reduced kidney function converting into complete Renal Failure. Chronic Kidney Disease is a public health matter all over the world, and it impacts the mortality rate drastically. Early diagnosis is a key factor to reduce the risk of kidney disease progression. This study is an effort toward analyzing the machine learning methods used for making predictions about chronic kidney disease along with identifying the most significant parameters supporting the early diagnosis. The analysis work has been done on the Weka software by implementing the J48 algorithm which shows that Serum Creatinine, diabetes mellitus, hemoglobin, pedal edema and specific gravity are the predictors of CKD. This study also identified various machine learning algorithms used so far for the early prediction of CKD.

Keywords Chronic Kidney Disease · Machine learning · Diagnosis · Biomarkers

1 Introduction

Chronic Kidney Disease or Renal disease is a continuing loss of kidney function over time. The main function of the kidney is to filter fluids and electrolytes present in the blood. CKD is a long-term condition in which Kidney function is reduced over time and leads to kidney failure [1]. Kidney Disease can lead to other health issues such as heart problems and stroke [2]. It is essentially required to assess the causes of reduced kidney function which helps in early diagnosis resulting in better patient care. Assessment of kidney function can be done by conducting urine test, blood test and medical imagining, i.e. renal ultrasound. Estimated Glomerular Filtration Rate (GFR) is the measure popularly used for Kidney function estimation. However, knowing about the biomarkers and the tests, CKD is usually identified when the

S. Jain (✉) · M. Patel
Geetanjali Institute of Technical Studies, Udaipur, RJ, India

K. Jain
Government Medical College, Ratlam, India

R. Mathur et al. (eds.), *Emerging Trends in Data Driven Computing and Communications*, Studies in Autonomic, Data-driven and Industrial Computing,
https://doi.org/10.1007/978-981-16-3915-9_11

patient is in the last stage causing dialysis, and transplantation is the only option remaining.

In the present scenario, all medical records are kept in digital form or in other words kept in Electronic Health Records (EHRs); therefore, it is easier to access and analyze these records and make conclusions [3]. Health Informatics and machine learning are the fields giving the opportunities to utilize the techniques in making diagnoses and predictions. The focus of this paper is to present a comparative study of the machine learning algorithms implemented for CKD predictions along with the assessment of key factors/biomarkers used in the study.

2 Systematic Outline for Literature Review

It is essential to understand the need and purpose of the literature review conducted, therefore, let's discuss the Significance of Early Diagnosis of CKD. Kidney Disease Improving Global Outcomes (KDIGO) and Nation Kidney Foundation are the non-profit organizations responsible for CKD guidelines. These guidelines have been used to analyze the CKD issues, eGFR Measurement and risk assessment. The population incidence of CKD exceeds 10%, and is > 50% of the high-risk subpopulations all over the world [4]. A recent survey in the United States shows that 37 million US adults suffer from CKD without their knowledge. Around 38% elders, and 13% middle-aged people from which 15% are women and 12% are men are also suffering from CKD. In the past decade, 125,000 people in the US got the treatment for ESKD, and 2 in every 1,000 people undergo dialysis or a kidney transplant [5]. In India, the prevalence of CKD is 800 per million populations (pmp), and the incidence of end-stage renal disease (ESRD) is 150–200 pmp [6]. The occurrence of CKD is 17.2% with stage 1 as 7%, 2 as 4.3%, 3 as 4.3%, 4 as 0.8% and 5 as 0.8% in India [7]. These figures clearly show that if CKD is diagnosed early in the stages, End-Stage Renal Disease can be prevented. The early diagnosis can drastically affect morbidity and mortality rates.

The present study focused on the utility of machine learning techniques for CKD predictions. This study will try to answer the following review questions:

RQ1-What are the necessary biomarkers/parameters that will help in CKD predictions?

RQ2-What are the machine learning techniques used for making early predictions about CKD?

Section 3 will try to answer RQ1 and Sect. 4 will try to answer RQ2. This Research will try to make the conclusions about

1. The effective set of parameters for making CKD predictions.
2. The Machine learning techniques with high precision and accuracy.
3. A future aspect of the research on CKD.

To perform this study, peer-reviewed papers, articles and journals have been done. The keywords used to find papers and articles are Chronic Kidney Disease, Machine

Table 1 Selection criteria of the literature

Type	Description
Included	Papers related to Chronic Kidney Disease
Included	Papers presenting Utility of machine learning techniques in disease predictions
Included	Full papers available
Excluded	Paper not directly related to Chronic Kidney Disease and machine learning
Excluded	Short or duplicate papers

Learning, diagnosis, prediction model, disease prediction and CKD Detection. The following Table 1 depicts the selection criteria of the literature:

3 Addressing RQ1 to Analyze Biomarkers Used for CKD Predictions

Biomarkers are the measures of the Biological state. For the diagnosis of any disease, it is required to perform physical examinations to measure these Biomarkers. According to the study in the research [8], there are many challenges in collecting Medical Data as the availability of electronic health records is difficult along with the presence of a unified record format and standards. It is very difficult to extract meaningful information from the data available. The analysis of authenticity and integrity of data is also very difficult. The most difficult work is to identify the significance of the parameters present in the records with the model's accuracy. To find the answer to RQ1, we will provide an insight into the biomarkers for CKD. To address RQ1, we need to dig deeper, therefore, RQ1 raises more issues which are discussed as follows:

RQ1.1 What are the biomarkers and diagnostic measures of CKD?
RQ1.2 What are the available datasets?
RQ1.3 What are the most significant Biomarkers affecting Kidney Function?

3.1 Addressing RQ1.1

To understand the measures of CKD, we need to understand individuals' genetic and phenotypic factors. According to Köttgen et al. [9], as a genetic component, a genome-wide study needs to identify vulnerability for glomerular filtration rate (GFR), estimated by serum creatinine (eGFRcrea) and cystatin C (eGFRcys). Another genetic component is AOPL1 and the participation of the renin–angiotensin system genes [10].

Apart from the Genetic components, there are several other phenotypic components causing CKD. According to R. Kazancioğlu [11], Family History, Gender,

Obesity, Nephrotoxins, Ethnicity, Age, Socioeconomic status, Smoking, Low Birth Weight, any prior Acute Kidney Injury, Diabetes Mellitus and Hypertension are the primary factors. KDIGO and Nation Kidney Foundation guidelines are the basis for the measurements [12, 13] so according to the guidelines provided, eGFR is the best measure to identify kidney function.

According to National Kidney Foundation 12, the stages of Chronic Kidney Disease are based on the eGFR value. The kidney function in percentage is equal to the GFR value. In Stage I, the Kidney works normally with a GFR value greater than and equal to 90 ml/min/1.73 m^2. In Stage II, the Kidney suffers from mild loss with a GFR value between 89 and 60 ml/min/1.73 m^2. Stage IIIa shows mild to moderate and Stage IIIb shows moderate to a severe loss of Kidney function with the GFR value of 59–45 ml/min/1.73 m^2 and 44–30 ml/min/1.73 m^2, respectively. In Stage VI, severe loss of kidney function will occur with the GFR value being reduced to 29–15 ml/min/1.73 m^2. Stage V will be considered as complete Kidney Failure with the GFR value below 15 ml/min/1.73 m^2.

eGFR below 60 for more than 3 months indicates a chronic kidney disease which can be identified by a urine test. The urine test reveals the albumin (a type of protein) and blood level. People with a high presence of albumin are at a higher risk.

Finally, to conclude RQ1.1, based on the formulas and observations hypothetically, the following are the main biomarkers of predicting CKD: Age, Gender, Weight, Blood Pressure, Sugar Level, SCr (serum creatinine), Blood Urea and Albumin are the primary factors to identify kidney disease.

3.2 *Addressing RQ1.2*

Dataset is a collection of data. To analyze any particular problem, a dataset is required which contains all the primary parameters required for the analysis. Open-source data libraries will help in getting the datasets for study. Some researchers will collect data by collaborating with some medical institutions to create their own dataset. Datasets "Chronic_Kidney_Disease.arff" on the UCI machine learning repository [14] and "kidney_disease.csv" on Kaggle [15] are available for open access.

3.3 *Addressing RQ1.3*

By answering RQ1.2, it is clear that most of the studies used the same parameters for generating predictions about chronic kidney disease. The next question that arises is, how to find the most significant parameters among the available list for CKD predictions. To answer that, we take the dataset available at the UCI machine learning repository as a base mark and try to find the relevance of the parameters. We conducted one experiment with the help of the Weka software to analyze the parameters of the base dataset considered. For that, we statistically examine each attribute present in the dataset. The following sub-sections show the steps involved in the experiment:

(a) **Experimental Setup**: Weka software version 3.8.4 has been used to perform the analysis work. Firstly, we need to analyze the parameters and their values. Tables 2 and 3 depict the description of the analysis. In Table 2, the results of the evaluation of all the numerical parameters have been shown which represents the statistical statistics of the parameters along with their distribution using a bar chart. In Table 3, statistics and distribution of nominal parameters have been shown. All these tables give an insight into the data. Some conclusions are as follows:

   i. Through Tables 2 and 3, we analyzed that there are 24 variables and 1 class variable. It also provides information about Value domains and baselines of each parameter.
   ii. In some cases, sod, pot, wc, rc, rbc and pvc have high missing percentage. Missing data is not necessarily a drawback but shows that feature engineering is required during data preprocessing.
   iii. Datasets have 250 class variables with values "CKD" and 150 as "NOTCKD", which means 62.5% data represent CKD confirmation and the remaining 37.5% represent No CKD. It shows Bias toward "CKD" which can affect the model performance. In general practice, 50–50% distribution will be considered; hence the dataset suffers from Bias which may affect the outcomes.
   iv. Table 3 shows data distribution of numerical parameters through a bar chart; it reflects that parameters like BP, bgr, bu, sc, sod and pot are highly skewed data. It is also required to do some preprocessing to reduce the skewness of data.

(b) The next step of the experiment is to estimate the most significant parameters among the 24 to predict CKD. This task has been accomplished by implementing the J48 algorithm on the dataset "Chronic_Kidney_Disease.arff" using the Weka software. The J48 algorithm also known as the C4.5 Decision Tree algorithm is used for classification problems. During the construction of a decision tree, the C4.5 algorithm selects the most effective attribute which splits the samples into subsets supplemented in one class or the other. By implementing this algorithm, with the help of a decision tree we can identify the most effective attributes to estimate CKD. The result of the J48 implementation is depicted in Fig. 1. The result shows that sc, pe, dm, hemo and sg are the most significant predictors of CKD. We can consider these results satisfactory as the algorithm works well with an accuracy of 99% and with a root relative squared error of 16.6603%.

Table 2 Analysis of numerical parameters

Parameter	Statistical values	Data distribution represented by bar chart
Age	Min: 2 Max: 90 Mean: 51.483 StdDev: 17.17 Missing percentage: 2% Distinct: 76 Unique percentage: 4%	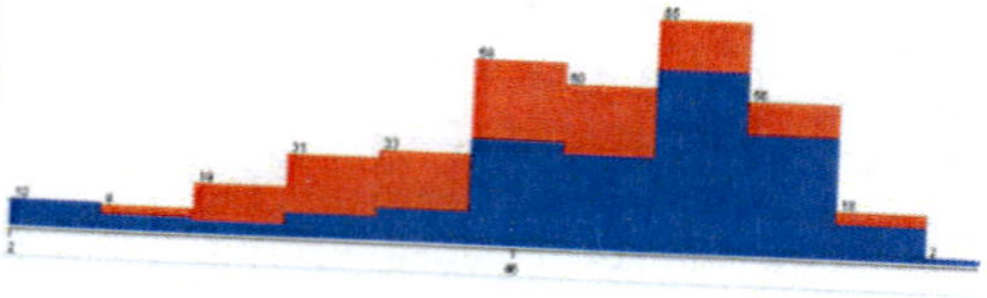
BP	Min: 50 Max: 180 Mean: 76.469 stdDev: 13.684 Missing percentage: 3% Distinct: 10 Unique percentage: 1%	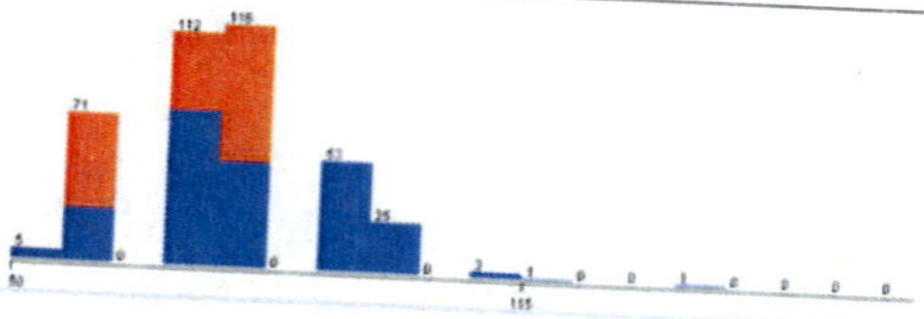
Bgr	Min: 22 Max: 490 Mean: 148.037 stdDev: 79.282 Missing percentage: 11% Distinct: 146 Unique percentage: 16%	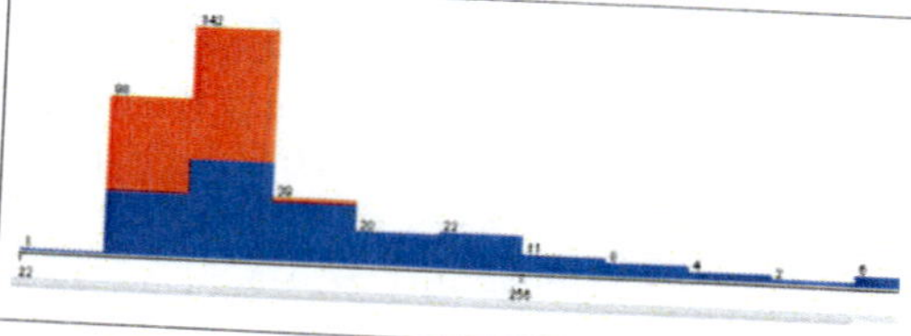
Bu	Min: 1.5 Max: 391 Mean: 57.426 stdDev: 50.503 Missing percentage: 5% Distinct: 118 Unique percentage: 14%	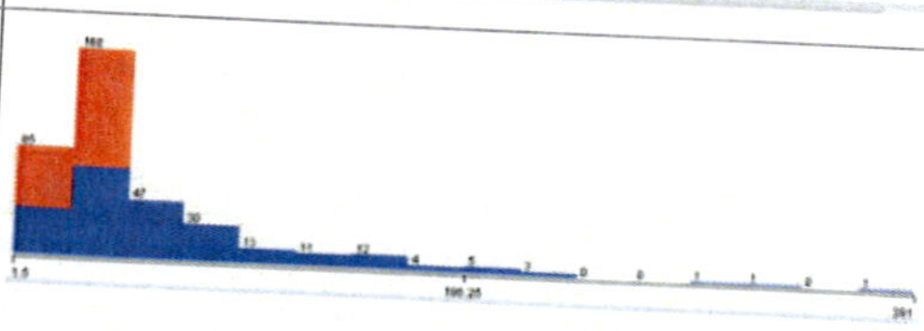
Sc	Min: 0.4 Max: 76 Mean: 3.072 stdDev: 5.741 Missing percentage: 4% Distinct: 84 Unique percentage: 10%	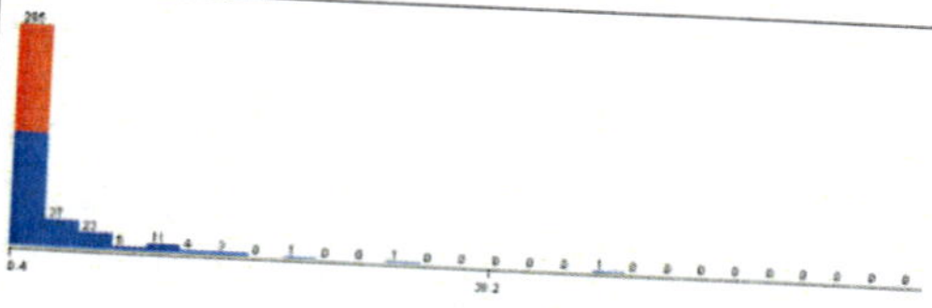
Sod	Min: 4.5 Max: 163 Mean: 137.529 stdDev: 10.409 Missing percentage: 22% Distinct: 34 Unique percentage: 2%	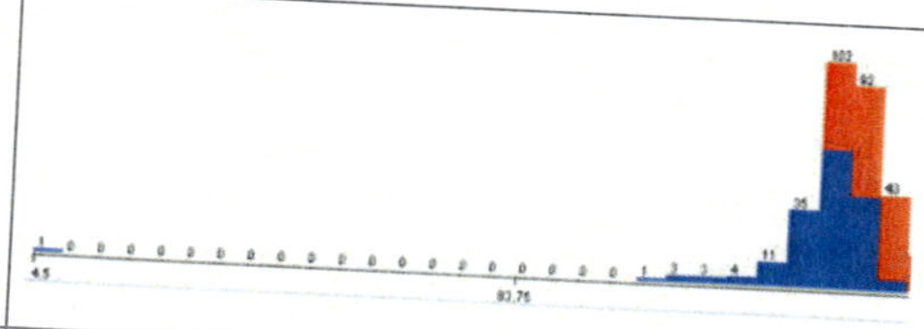
Pot	Min: 2.5 Max: 47 Mean: 4.627 stdDev: 3.194 Missing percentage: 22% Distinct: 40 Unique percentage: 2%	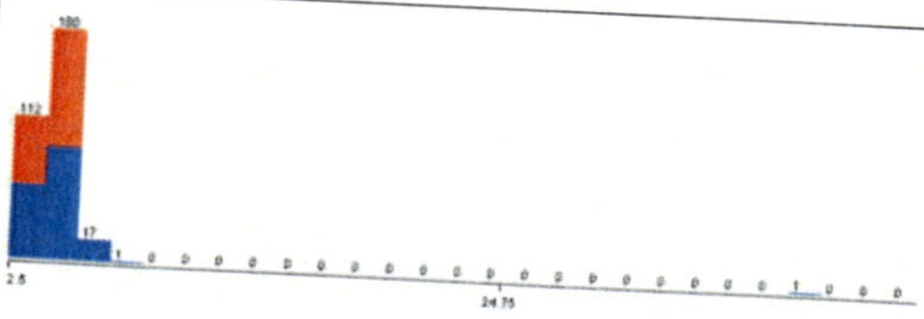

(continued)

Table 2 (continued)

Parameter	Statistical values	Data distribution represented by bar chart
hemo	Min: 3.1 Max: 17.8 Mean: 12.526 stdDev: 2.913 Missing percentage: 13% Distinct: 115 Unique percentage: 7%	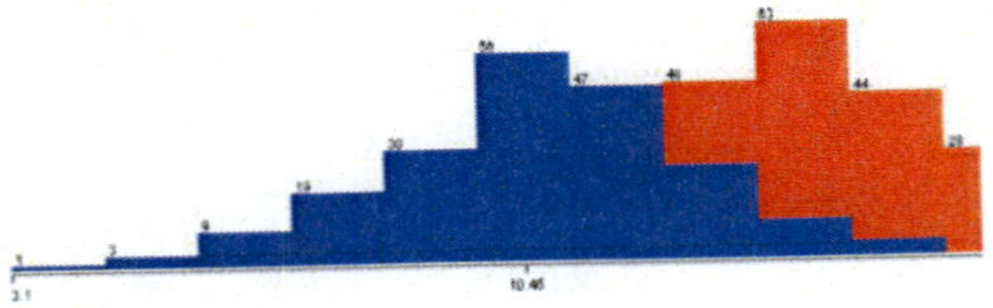
Pcv	Min: 9 Max: 54 Mean: 38.884 stdDev: 8.99 Missing percentage: 18% Distinct: 42 Unique percentage: 2%	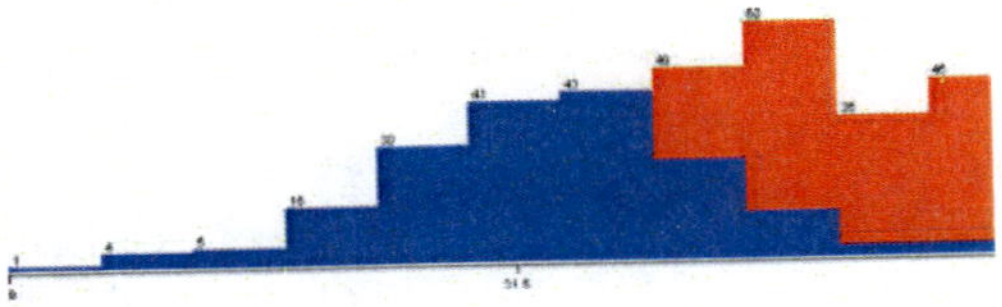
Wc	Min: 2200 Max: 2400 Mean: 8406.122 stdDev: 2944.474 Missing percentage: 27% Distinct: 89 Unique percentage: 8%	
Rc	Min: 2200 Max: 26,400 Mean: 8406.122 stdDev: 2944.474 Missing percentage: 27% Distinct: 89 Unique percentage: 8%	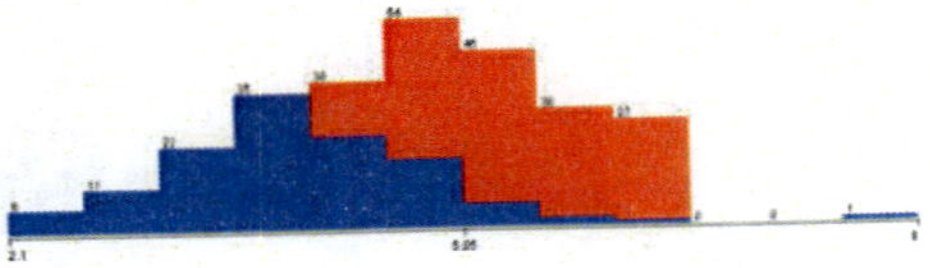

4 Addressing RQ2 to Identify Machine Learning Algorithms Implemented for CKD Predictions

Machine Learning models are efficient in the prediction tasks. The predictive modeling by implementing machine learning classification and regression algorithms plays a very beneficial part in disease predictions. Machine Learning models have been utilized for the early diagnosis of kidney failure for a long time now. In this section, various papers will be discussed to provide information regarding which algorithms have been used for CKD Detection.

Al-Hyari et al. [16] implemented three classification algorithms Naïve Bayes, Decision Tree algorithms and one ANN. Performance has been assessed by calculating accuracy, sensitivity and specificity. Here, the algorithms were implemented using the Weka software for comparison purposes. In this paper, the Decision Tree algorithm performs best with an accuracy of 92.2%.

Table 3 Analysis of nominal parameters

Parameter	Statistical values		Data distribution represented by bar chart
	Label	Count	
Sg	1.005	7	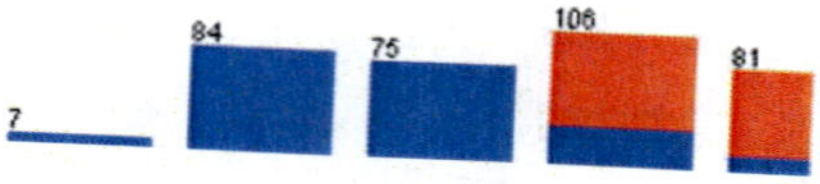
	1.010	84	
	1.015	75	
	1.020	106	
	1.025	81	
	Missing Percentage: 12% Distinct: 5 Unique Percentage: 0%		
Al	0	199	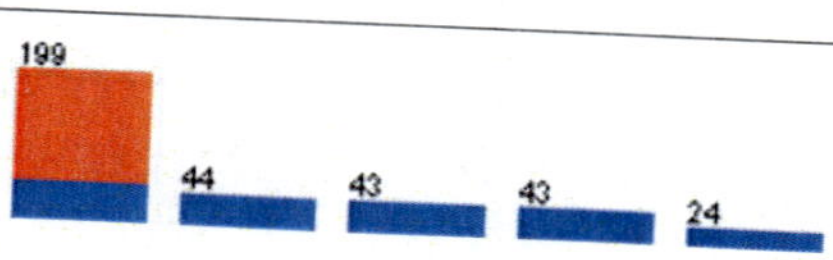
	1	44	
	2	43	
	3	43	
	4	24	
	5	1	
	Missing Percentage: 12% Distinct: 6 Unique Percentage:0%		
Su	0	290	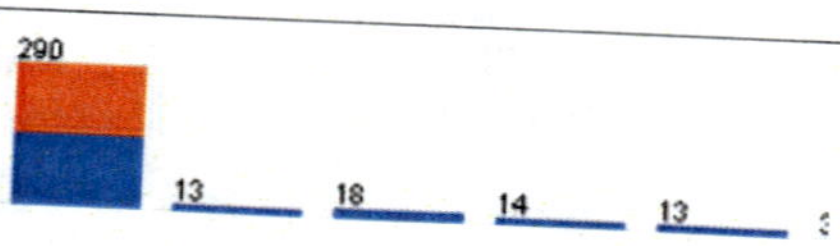
	1	13	
	2	18	
	3	14	
	4	13	
	5	3	
	Missing Percentage: 12% Distinct: 6 Unique Percentage: 0%		
Rbc	Normal	201	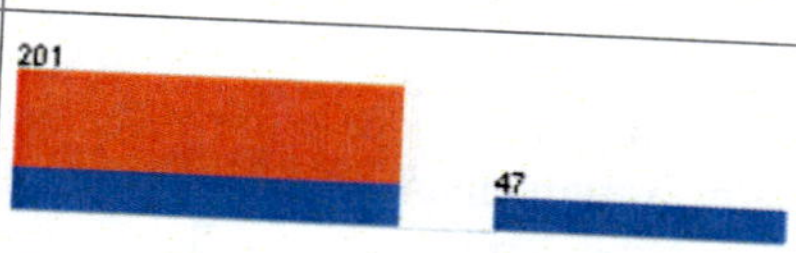
	Abnormal	47	
	Missing Percentage: 38% Distinct: 2 Unique Percentage: 0%		
Pc	Normal	259	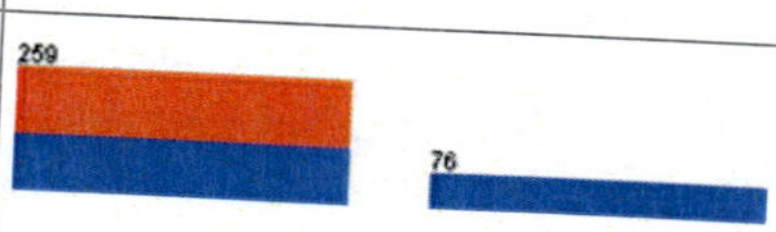
	Abnormal	76	
	Missing Percentage: 16% Distinct: 2 Unique Percentage: 0%		

(continued)

Table 3 (continued)

Parameter	Statistical values		Data distribution represented by bar chart
	Label	Count	
Pcc	Present	42	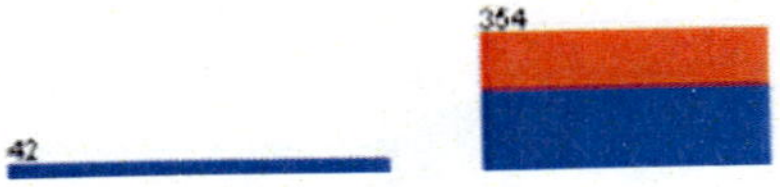
	Notpresent	354	
	Missing Percentage: 1% Distinct: 2 Unique Percentage: 0%		
Ba	Present	22	
	Notpresent	374	
	Missing Percentage: 1% Distinct: 2 Unique Percentage: 0%		
Htn	Yes	147	
	No	251	
	Missing Percentage: 1% Distinct: 2 Unique Percentage: 0%		
Dm	Yes	137	
	No	261	
	Missing Percentage: 1% Distinct: 2 Unique Percentage: 0%		
Cad	Yes	34	
	No	364	
	Missing Percentage: 1% Distinct: 2 Unique Percentage: 0%		
Appet	Good	317	
	Poor	82	
	Missing Percentage: 1% Distinct: 2 Unique Percentage: 0%		
Pe	Yes	76	
	No	323	
	Missing Percentage: 1% Distinct: 2 Unique Percentage: 0%		
Ane	Yes	60	
	no	339	
	Missing Percentage: 1% Distinct: 2 Unique Percentage: 0%		

```
J48 pruned tree
------------------

sc <= 1.2
|   pe = yes: ckd (15.57/0.08)
|   pe = no
|   |   dm = yes: ckd (12.63/0.14)
|   |   dm = no
|   |   |   hemo <= 12.9: ckd (16.02/0.56)
|   |   |   hemo > 12.9
|   |   |   |   sg = 1.005: notckd (0.0)
|   |   |   |   sg = 1.010: ckd (4.4/0.14)
|   |   |   |   sg = 1.015: ckd (2.91/0.09)
|   |   |   |   sg = 1.020: notckd (76.27/0.45)
|   |   |   |   sg = 1.025: notckd (70.64)
sc > 1.2: ckd (201.57/2.52)

Number of Leaves  :     9

Size of the tree :    14

=== Stratified cross-validation ===
=== Summary ===
Correctly Classified Instances        396          99     %
Incorrectly Classified Instances        4           1     %
Kappa statistic                         0.9786
Mean absolute error                     0.0225
Root mean squared error                 0.0807
Relative absolute error                 4.7995 %
Root relative squared error            16.6603 %
Total Number of Instances             400
```

Fig. 1 Result of J48 Algorithm Implementation

Research presented by Weng et al. [17] was completely focused on ensemble neural networks for disease prediction. In the paper, three types of neural networks have been implemented: Individual Classifier (IC), Solo Classifier (SC) and Ensemble Classifier (EC). For the evaluation of performance, accuracy, precision, F-score, TPR and TNR have been used. To test the significance of the performance, statistical testing has been done which confirms that the ensemble classifier (EC) performed best among the three.

Salekin, Stankovic [18] implemented K-NN, ANN and Random Forest. Root Mean Square Error (RSME) and F1-Measure have been implemented to calculating accuracy. Ten-Fold cross-validation technique is also used.

In the paper presented by Ani et al. [19], ANN, Naïve Bayes Classifier, K-NN, Decision Tree, LDA Classifier and Random Subspace Classification (RSC) Algorithms are compared. RSC algorithm is an ensemble learning model which is based on the attribute bagging method. It has multiple classifiers, where each uses a subspace of the originally created feature space. Linear Discriminant Analysis (LDA) Classifier extracts the most significant features/attributes from the original dataset and then generates the predictions. In this paper, Random Subspace Classification Algorithm performs best with an accuracy of 94%.

Decision Tree, SVM and Logistic Regression Classifiers are implemented in the paper by Charleonnanet et al. [20]. The experiment was conducted on MATLAB and the Weka software and both to compare the results. SVM gets the highest accuracy (98.3%) among the four implemented classifiers. The performance measures used are accuracy, sensitivity and specificity. It also performed 5-fold cross validation.

Gunarathne et al. [21] implemented an ensemble learning model for classification. It implemented Multiclass Decision Forest by building multiple decision trees and voting on the most popular output class. It also implemented Multiclass Decision Jungle which is used to predict multiple values and used Directed Acyclic Graphs (DAGs). The third model implemented was the Multiclass Logistic Regression Model and the fourth model is Multiclass Neural Networks. All these models are focused on predicting multiple values and Multiclass Decision Forest provided the highest accuracy as 99.1%. All four models were implemented using Microsoft Azure Machine Learning Studio.

A regression model has been implemented by Fki and ayed [22]. This paper suggests an integrated approach by collecting data using Internet of Things (IoT) devices to predict End-Stage Renal/Kidney Disease. IoT devices mainly collect data/biomarkers for dialysis. These biomarkers were combined with the CKD dataset to generate predictions.

Almansour et al. [23] implemented ANN and SVM for CKD prediction. The Weka software is used for data preprocessing and model implementation. ANN showed the best performance among the models implemented by providing an accuracy of 99.75%. The parameter tuning method has been used for both ANN and SVM as optimization strategies to improve results.

Xiao et al. [24] made the prediction of CKD using an increase of urinary protein in patients. Nine different predictive models have been established: logistic regression, support vector machine, random forest, XGBoost, neural network, Elastic Net, Ridge Regression Model, LASSO Regression Model and k-nearest neighbor. As model evaluation criteria, AU-ROC, sensitivity (recall), specificity, accuracy, log-loss and precision measures have been utilized for each model. The models including Elastic Net, lasso regression, ridge regression and logistic regression showed the highest accuracy and precision above 0.87 and 0.8, respectively.

Qin et al. [25] implemented six machine learning models that have been compared for CKD prediction. K-NN is used to remove missing values from the dataset. Support Vector Machine, Logistic Regression, Random Forest, K-NN, Naïve Bayes and feed-forward neural network are implemented. Among these algorithm, random forest performs best with an accuracy of 99.75%. They also suggested an ensemble method by combining logistic regression and random forest by using perceptron is the best predictors with an accuracy of 99.83% after 10 simulations.

Khan et al. [26] implemented seven ML techniques together with Support Vector Machine, NBTree, J48, LR, NB, Multilayer Perceptron and Composite Hypercube on Iterated Random Projection (CHIRP). J48 Decision Tree is an advanced technique of C4.5 which utilizes a pruning strategy for tree construction by calculating information gain and entropy measures. NBTree is an approach combining Naïve Bayes classifier and decision tree. Assessment of these techniques has been done using recall, precision, F-measure, accuracy, MAE, RMSE, RAE and RRSE. CHIRP performs best with an accuracy of 99.75% among other techniques.

Sobrinhoet et al. [27] compared K-NN, SVM, Random Forest (RF), Naïve Bayes, decision tree and MLP neural network, which were used to classify the CKD. Five iterations of 10-fold cross validation are used for model evaluation. The experiments

Table 4 Summary of literature reviewed

References	Machine learning models used	Task Accomplished	Features	Limitations
Al-Hyari et al. [16]	Naïve Bayes and Decision Tree algorithms and one ANN	Classification	Compared CRF system performance with Weka Implementation	Patients' quality-of-life and work environment not included
Weng et al. [17]	Individual Classifier (IC), Solo Classifier (SC) and Ensemble Classifier (EC)	Classification	IC and SC are single classifiers implemented with different size of the training dataset. EC is a multiple-type classifier that encompasses of many ICs	For IC, external validity has to be done in advance. EC and SC suffer from overfitting if the proportion of training data is not normalized
Salekin and Stankovic [18]	K-NN, ANN and Random Forest	Classification	Used Wrapper Approach and Embedded Approach for Feature Selection	Less data means less accuracy
Ani et al. [19]	ANN, Naïve Bayes Classifier, K-NN, Decision Tree, LDA Classifier and Random Subspace Classification Algorithms	Classification	Used Random subspace classification method employing K-NN as base classifier	kernel or neural-based classifiers were not used as base classifier

(continued)

are performed on the Weka software. The classifiers that show the best performances are the Decision Tree and RF with 95.00 and 93.33%.

Summary of papers reviewed to understand the role of machine learning algorithms for CKD prediction has been shown in Table 4.

5 Conclusion

Early diagnosis of CKD can increase the survival rate of human beings. It also reduces the number of renal transplantations by providing better medical care. This research review was an effect toward identifying the parameters or factors for early CKD diagnosis. A systematic literature review has been done which provides the answers

Table 4 (continued)

References	Machine learning models used	Task Accomplished	Features	Limitations
Charleonnan et al. [20]	Decision Tree, SVM, Logistic Regression	Classification	The experiment was conducted on MATLAB and the Weka software both to compare the results	Limited toward model's performance comparisons
Gunarathne et al. [21]	Multiclass Decision Forest, Decision Jungle, Logistic Regression Model, Neural Network	Classification	Ensemble Approach used	Strength of the data is low as the dataset is small and it contains missing values
Fki and ayed [22]	Regression Model	Classification	Integrated approach by collecting data using Internet of Things (IoT) devices to predict End-Stage Renal/Kidney Disease	There is no check for the consistency of attributes chosen for each biomarker
Almansour et al. [23]	ANN and SVM	Classification	Optimization technique is used	Deep Learning models with medical images can be used for further studies
Xiaoet al. [24]	LR, Support Vector Machine, Random Forest, XGBoost, ANN, Elastic Net, Ridge Regression Model, LASSO Regression Model and k-nearest neighbor	Classification	It used non-urine indicators as clinical predictors	The sample size used is small The parameters used fortuning which did not reduce to avoid overfitting
Qin et al. [25]	Support Vector Machine, Logistic Regression, Random Forest, K-NN, NB and feedforward neural network	Classification	K-NN is used to remove missing values from the dataset, also suggested an ensemble method by combining logistic regression and random forest by using perceptron	Performance of the models will be limited due to less-sampled data. Model cannot predict the severity of CKD

(continued)

Table 4 (continued)

References	Machine learning models used	Task Accomplished	Features	Limitations
Khan et al. [26]	Support Vector Machine, NBTree, J48, Logistic Regression, Naïve Bayes, Multi-layer Perceptron and Composite Hypercube on Iterated Random Projection (CHIRP)	Classification	Missing values in the dataset are replaced by medians. Composite Hypercube on Iterated Random Projection (CHIRP) is used which produces high accuracy	Works on limited sampled data
Sobrinho et al. [27]	K-NN, SVM, Random Forest (RF), Naïve Bayes, J48 Decision Tree and MLP neural network	Classification	Both Qualitative and quantitative analyses have been done using k-fold cross-validation methods	Dataset size considered is small Risk evaluations of individual subjects considering with the help of well-accepted medical guidelines and nephrologists needs to be done. No data augmentation is used to improve confidence

to two major review questions. Through this research, we identified the predictive biomarkers of CKD along with the machine learning models implemented for CKD predictions. Although it is an ongoing research area, one needs to focus on certain issues in future studies. The main objectives which can be fulfilled in future work are:

- Subject/patient-specific research needs to be done.
- Medical images need to be incorporated with the biomarkers to generate better implementations.
- Most of the papers work with limited data, so it is required to generate a dataset specifically covering CKD predictors and models will be implemented based on that data sets.
- CKD is a long-lasting, irreversible condition, so it is required to analyze it accordingly. Deep learning models can be utilized for image classification as well as time-series analysis.

References

1. Romagnani P, Remuzzi G, Glassock R et al (2017) Chronic kidney disease. Nat Rev Dis Prim Publ Online. https://doi.org/10.1038/nrdp.2017.88
2. Ene-Iordache B, Perico N, Bikbov B et al (2016) Chronic kidney disease and cardiovascular risk in six regions of the world (ISN-KDDC): a cross-sectional study. Lancet Glob Heal Publ online. https://doi.org/10.1016/S2214-109X(16)00071-1
3. Farfán F, Varadarajan R, Hristidis V (2009) Electronic health records. In: Information discovery on electronic health records. https://doi.org/10.4018/jhisi.2010070101
4. Eckardt KU, Coresh J, Devuyst O et al (2013) Evolving importance of kidney disease: from subspecialty to global health burden. Lancet 382(9887):158–169. https://doi.org/10.1016/S0140-6736(13)60439-0
5. CDC (2019) Chronic kidney disease in the United States, 2019 fact sheet. Cdc. 1:1–6. https://www.cdc.gov/kidneydisease/pdf/2019_National-Chronic-Kidney-Disease-Fact-Sheet.pdf
6. Agarwal SK, Srivastava RK (2009) Chronic kidney disease in India: challenges and solutions. Nephron—Clin Pract 111(3):197–203. https://doi.org/10.1159/000199460
7. Varma PP (2015) Prevalence of chronic kidney disease in India—where are we heading? Indian J Nephrol 25(3):133–135. https://www.ncbi.nlm.nih.gov/pmc/articles/PMC4446915/. Accessed 17 Oct 2020
8. Li Q, Fan QL, Han QX et al (2020) Machine learning in nephrology: scratching the surface. Chin Med J (Engl) 0(6):687–698. https://doi.org/10.1097/CM9.0000000000000694
9. Köttgen A, Glazer NL, Dehghan A et al (2009) Multiple loci associated with indices of renal function and chronic kidney disease. Nat Genet Publ online. https://doi.org/10.1038/ng.377
10. Su SL, Lu KC, Lin YF et al (2012) Gene polymorphisms of angiotensin-converting enzyme and angiotensin II Type I receptor among chronic kidney disease patients in a Chinese population. JRAAS—J Renin-Angiotensin-Aldosterone Systems. Published online. https://doi.org/10.1177/1470320311430989
11. Kazancioğlu R (2013) Risk factors for chronic kidney disease: An update. Kidney Int Suppl 3(4):368–371. https://doi.org/10.1038/kisup.2013.79
12. Estimated Glomerular Filtration Rate (eGFR) | National Kidney Foundation. Accessed 20 Oct 2020. https://www.kidney.org/atoz/content/gfr
13. KDIGO—Kidney disease | Improving global outcomes. https://kdigo.org/. Accessed 20 Oct 2020
14. UCI Machine Learning Repository: Chronic_Kidney_Disease Data Set. https://archive.ics.uci.edu/ml/datasets/Chronic_Kidney_Disease. Accessed 20 Oct 2020
15. Chronic KIdney Disease dataset | Kaggle. https://www.kaggle.com/mansoordaku/ckdisease/metadata. Accessed 20 Oct 2020
16. Al-Hyari AY, Al-Taee AM, Al-Taee MA (2013) Clinical decision support system for diagnosis and management of Chronic Renal Failure. In: 2013 IEEE conference on applied electrical engineering and computing technologies AEECT 2013. Published online 2013. https://doi.org/10.1109/AEECT.2013.6716440
17. Weng CH, Huang TCK, Han RP (2016) Disease prediction with different types of neural network classifiers. Telemat Inf 33(2):277–292. https://doi.org/10.1016/j.tele.2015.08.006
18. Salekin A, Stankovic J (2016) Detection of chronic kidney disease and selecting important predictive attributes. In: Proceedings—2016 IEEE international conference on healthcare informatics, ICHI 2016.https://doi.org/10.1109/ICHI.2016.36
19. Ani R, Sasi G, Sankar UR, Deepa OS (2016) Decision support system for diagnosis and prediction of chronic renal failure using random subspace classification. In: 2016 International Conference Advances in computing, communications and Informatics, ICACCI 2016. Published online, pp 1287–1292. https://doi.org/10.1109/ICACCI.2016.7732224
20. Charleonnan A, Fufaung T, Niyomwong T, Chokchueypattanakit W, Suwannawach S, Ninchawee N (2016) Predictive analytics for chronic kidney disease using machine learning techniques. In: 2016 management and innovation technology international conference

MITiCON 2016. Published online, pp MIT80–MIT83. https://doi.org/10.1109/MITICON.2016.8025242
21. Gunarathne WHSD, Perera KDM, Kahandawaarachchi KADCP (2017) Performance evaluation on machine learning classification techniques for disease classification and forecasting through data analytics for Chronic Kidney Disease (CKD). In: Proceedings—2017 IEEE 17th international conference on bioinformatics and bioengineering BIBE 2017. 2017;2018-Janua:291–296. https://doi.org/10.1109/BIBE.2017.00-39
22. Fki Z, Ammar B, Ben AM (2018) Machine learning with internet of things data for risk prediction: application in ESRD. In: Proceedings of the international conference on research challenges in information science, 1–6 May 2018. https://doi.org/10.1109/RCIS.2018.8406669
23. Almansour NA, Syed HF, Khayat NR et al (2018) Neural network and support vector machine for the prediction of chronic kidney disease: a comparative study. Comput Biol Med 2019(109):101–111. https://doi.org/10.1016/j.compbiomed.2019.04.017
24. Xiao J, Ding R, Xu X et al (2019) Comparison and development of machine learning tools in the prediction of chronic kidney disease progression. J Transl Med 17(1):1–13. https://doi.org/10.1186/s12967-019-1860-0
25. Qin J, Chen L, Liu Y, Liu C, Feng C, Chen B (2020) A machine learning methodology for diagnosing chronic kidney disease. IEEE Access 8:20991–21002. https://doi.org/10.1109/ACCESS.2019.2963053
26. Khan B, Naseem R, Muhammad F, Abbas G, Kim S (2020) An empirical evaluation of machine learning techniques for chronic kidney disease prophecy. IEEE Access. 8:55012–55022. https://doi.org/10.1109/ACCESS.2020.2981689
27. Sobrinho A, Queiroz ACMDS, Dias Da Silva L, De Barros Costa E, Eliete Pinheiro M, Perkusich A (2020) Computer-Aided diagnosis of chronic kidney disease in developing countries: a comparative analysis of machine learning techniques. IEEE Access 8:25407–25419 (2020). https://doi.org/10.1109/ACCESS.2020.2971208

Chapter 12
Application of Machine Learning Algorithms in Agriculture: An Analysis

Kalpana Jain and Naveen Choudhary

Abstract Machine learning is being rapidly adopted in various industries: According to Research and Markets, the machine learning market is projected to grow to $8.81 billion by 2022, at a compound annual growth rate of 44.1%. One of the main reasons for its increasing use is that companies are collecting big data from which they need to obtain valuable information. Machine learning is an efficient way to make sense of that data. In the current situation, we are talking about the emerging concept of smart farming that makes farming more efficient and effective with the help of high-precision algorithms. The mechanism that drives it is machine learning, the scientific field that gives machines the ability to learn without being strictly scheduled. It has emerged alongside big data technologies and high-performance computing to create new opportunities to unravel, quantify, and understand data-intensive processes in agricultural operational environments. This paper reviews the exiting techniques and methods of machine learning applicable in the agriculture sector.

Keywords Agriculture · Machine learning · Training · Data · Algorithms

1 Introduction

Agriculture is essential to the world's sustainability. Humans profit in one direction or the other from agriculture, which makes agricultural practices a crucial field of research. Farmers often need details, particularly when growing crops that are not popular in their land or crop. The ordinary farmer has links to crude data outlets such as television, radio and journals, fellow growers, government departments, farmers, and merchants. Therefore, a framework is required, which enables farmers to access relevant details. Machine learning is one of the trends; thus, different strategies and applications function within the context of machine learning. In recent years, many machine learning systems have been validated and established in agriculture.

K. Jain (✉) · N. Choudhary
Department of Computer Science and Engineering, College of Technology & Engineering, MPUAT, Udaipur, India

R. Mathur et al. (eds.), *Emerging Trends in Data Driven Computing and Communications*, Studies in Autonomic, Data-driven and Industrial Computing,
https://doi.org/10.1007/978-981-16-3915-9_12

Research has also been carried out on the efficacy of multiple machine learning algorithms in agriculture and other applications since machine learning is an extremely effective method for efficient utilization of tools, forecasts, and management that is required for agriculture. Machine learning is the skill and the implementation of information through an electrical processing device.

The planet has to be more conscious of sustainable agriculture for some time now. According to the UN, by 2050, the population of the planet could reach 9.7 billion. Experts from the World Resource Institute predict that food demand could grow from 50 to 70% to feed these people. Agriculture can, however, have disastrous environmental consequences because of its wastewater use, transport of carbon dioxide, and the improper use of fertilizers that cause pollution. India has a long history as one of the world's biggest farmers. The nation has a comprehensive network of commercial farmers and small-scale growers; it is one of the major rice, cotton, sugar cane producer and exports large amounts of wheat, maize and produce other crops. India is therefore the ideal testing ground to transform its farming sector using revolutionary new technology. Intelligent farming supports farmers (large and small) for the value chain by the provision of materials, consultancy services and storage services, transport, and international exchange aid [3].

Now several companies are developing a software solution that guarantees sustainable agriculture both for the world and for farmers. They plan to create an open digital network for farmers and others to help grow more at less cost and with less environmental effects. This new technology uses innovative machine learning software, geospatial data mining, and cloud computing to provide farmers with advice and feedback in real time [33]. The challenge was to design a framework that could process data in real time from many sources, achieve visibility in all farming phases, and automatically provide farmers with recommendations. With satellite and drone geo-spatial data, the company may track ground conditions, which imply possible field productivity [19]. This data can be mixed with other outlets, including weather and market data, in order to consider farmers' real-time conditions and to make intelligent, real-time recommendations. The machine also gives a simpler way of identifying and avoiding accidents such as plague or water shortage and helps to avoid crop destruction. As the industry expands and evolves, different forms of machine learning are developing that can be explored in new applications. Many examples of machine learning implementations today, therefore, fall into two categories: supervised and unregulated learning.

1.1 Supervised Learning

A common method of machine learning is supervised learning that is often used to build training patterns to predict future happenings such as fraudulent credit card purchases in applications that use historical data. It is a method of machine learning that recognizes inputs and outcomes and uses tagged examples to train algorithms.

Supervised learning uses techniques for pattern analysis such as grouping, regression, estimation, and gradient change. These trends are then used to predict mark values of unlabeled results [10]. In drug research and development, this form of machine learning has already been used, with applications including target validation, biomarker identification, and automatic clinical pathology analysis. This means that machine learning promotes data-driven decision-making and facilitates the discovery and development process and improves success rates [12].

1.2 Unsupervised Learning

Unsupervised learning works without previous validation for data sets compared with controlled learning. It instead analyzes the collected data to describe the framework and trends. Unregulated machine learning is being used in factories for predictive maintenance. Machines can analyze and use data and algorithms that cause device failures to anticipate problems before they occur. This allows less needless downtime as factories order parts from the retailer of automation equipment to be replaced before a failure takes place. Research by Deloitte found that the use of machine learning technologies [18] in development decreased the unexpected downtimes by 15–30% and hence the costs of maintenance by 30%. Humans are not the only people able to think about themselves: machines like Google Duplex can also pass the Turing test today. Manufacturers may use machine learning to boost maintenance processes and make informed decisions in real time based on findings.

1.3 Problem Statement

It is well known that the prevention and timely diagnosis of any disease will bring us the strategic advantage over said disease and in agriculture it is no exception, because knowing what ails a crop or plant increases the chances of success in treatment [24]. In the developing world, more than 80% of agricultural production is generated by small farmers [1], and reports of yield losses of more than 50% due to pests and diseases are common [9]. Besides, the greater proportion of people with problems of poverty and famine (50%) live in these productive areas, which makes the small farmers particularly vulnerable group to disruptions in the food supply caused by pathogens. There are methods to determine the diseases of any plant, such as taking samples of vegetative tissue to a specialized laboratory or taking an expert agronomist to the cultivation site; in either of the two methods, the disadvantage lies in the time necessary to obtain the results. That is why they have been considered the use of artificial vision techniques and pattern recognition, as well as some classification algorithms that automatically determine the possible disease, facilitating the task of specialists to develop their work and that they can find a timely diagnosis for treatment. And as [7] says, the Tools for the automatic recognition of plant diseases have the potential to become a valuable source of information to aid decision-making in agriculture.

In the next section, the techniques used are shown, as well as the most significant research works that address the challenge of detecting a disease, based on the analysis of the characteristics present in the leaves of the crops.

2 Techniques and Algorithms

To arrive at the classification of diseases, the authors rely on proven methodologies, like data acquisition, pre-processing, feature extraction, and recognition, which are the steps or procedures followed to obtain results. For the acquisition of the images, digital cameras are used to capture the leaves or the parts where the damage caused by the disease is visible, as well as set so that the images available on the Web are made available to everyone. Public serves as the basis for training the model. For the images acquired by the camera, the shots were in controlled environments with acceptable resolutions, although it is important to mention that [27] used a mobile phone, to acquire the images and do an alternate experiment to see how much it affected the quality of the images.

Once the images are acquired, they go to pre-processing, where they will be treated such as scaling, noise elimination, color space transformation, histogram equalization, and everything else that can be done to maximize the characteristics. By the time you move on to the next process, the image is processing and noise removed. Therefore, when applying segmentation techniques, they will separate the points of interest with better precision, obtaining valuable data that will be more descriptive of the disease. Once you have the characteristics, then next thing is to do the classification, make use of the algorithms, obtain the results, and make a description of them. It is in this step when the effectiveness of the procedure is shown, if it has been classified according to expectations.

The next section shows the significant works related to the detection of diseases in various types of plants , using various machine learning algorithms as listed in Table 1.

3 Literature Review

Using different machine learning algorithms such as mathematical and statistical approaches, crop prediction can be carried out. Any of the approaches that are currently being tested are discussed here.

Table 1 Machine Learning Algorithms and their characteristics

Algorithm	Properties
Fuzzy logic	It is based on heuristic rules; it is used for processes highly not linear [15]. Easy implementation
SVM	Look for a hyperplane that works as a separator. In typical training and problems, it is very efficient [32]
Bayes	It is very efficient where this type of environment is used supervised learning. Large amounts of data are not required for their training [4]
KNN	Search for observations closest to the one you are trying to predict and classify the point of interest based on most training data [34]
ANN	Qualifying is very efficient, at the cost of training computationally expensive [37]
CNN	The performance of convolutional neural networks in the recognition object and image classification has made great progress in recent years. They tend to be more accurate at the cost of high computational cost. In training, it may require a considerable number of images to produce reliable results [5]

3.1 Artificial Neural Network

The Network of Artificial Neurons is the Artificial Neural Network (ANN). It is founded upon the biochemical functions of the human brain. This is one of the markers of controlled learning. Neural network has to be trained once, for example, after equivalent trends can be expected in future data, practical solutions to problems can be generated even though the input data is incorrect/incomplete. The accuracy of ANN continues to improve by including more and more details. ANNs are often willing to embrace their ambiguity without understanding the values behind them. For any method, ANN may extract the correlation between input and output, and compare the SVM and ANN algorithms to classify diseases in various crops that have been attacked by fungi, bacteria, nematodes, and nutrient deficiencies. They mention that the symptoms of plant diseases exhibit different properties such as color, shape, and texture, and based on this, the characteristics are obtained. They consider color as an important dimension, but applying dimensionality reduction they discover by experimentation that, out of 24, only 8 characteristics are significant for the classification of diseases. In the end, they put the two algorithms to the test and found that with SVM they obtained 92.17% precision and that with ANN only 87.48%, which is why they show that, for this case, the SVM is a better classifier. Another paper was presented by [19]. In Ireland, over 80% of farmland is grassland, a source of feed for the dairy and livestock industries. There have been very few studies worldwide that estimate small weeds like Ireland using remote sensing data. Certain computational models have been developed to estimate the amount of grassland biomass available in two intensively controlled grassland farming companies in Ireland such as the multilinear regression (MLR), the artificial neural network (ANN), and the adaptive neuro-fuzzy inference system (ANFIS). For 12 years (2001–2012), in situ weekly biomass measurements were used for model creation on the first test site (Moorepark) (2001–2012). In all three versions, 5 vegetation indices plus two spectral (red and close infrarouge) bands extracted from an 8-day MODIS product were fed. ANFIS has provided better biomass figures (RM2 Moorpark < 0.85, RMSE Moorepark = 11.07; RG range 2 > = 0.76, RMse grange = 15.35) than ANN and MLR. This project will provide a blueprint for the analysis and measurement of spatial data for the recovery of various biophysical variables. Another paper introducing the method [25] network integration is very useful for many vector-based algorithms and is now a hot subject of study in network analysis. A lot of studies have been conducted on network integration, but much of the material that is already present in the network was overlooked. In this article, we suggest a new method of network embedding using the semi-supervised SSKNMF kernel, from which previous knowledge can be implemented and more practical network features learned. By using the objective function based on the L2,1 standard, it can become more resilient against noise. These evolutionary principles are very useful in managing the following agent-based paradigm. Our proposed algorithm performs with thorough testing considerably better than current members. The author of [7] recommends that a deep-rooted network architecture be extended through the use of VGGNet [17] and AlexNetOWTBn [9] to automatically identify diseases arising in tomato leaves. Early plague, powdery mildew, and mildew were the pathogens to be classified. The photographs captured are preprocessed using techniques of image processing such as noise reduction, regression, and the improvement of the image processing to minimize expense and time. Later on, the data set characteristics were derived using convolution charts, in which the input data were imaged by healthy and contaminated leaves. While the architecture showed very accurate results; for this analysis, only AlexNetOWTBn architecture was used to achieve 32.23% and VGG 33.27%.

3.2 Information on Fuzzy Network

The remote sensing and other parameters for crop yield prediction were evaluated through aggressive neural network processing. They employed the versatile method of Neuro-fuzzy Inference (ANFIS). The inputs to ANFIS are the soil moisture content, field biomass, and the repository organ. It has only a single number, or another output node, i.e. yield, that is pursued. In that remote sensing data, the other problem in forecasting yield does not go far behind in time. In order to construct a design to estimate future values, every forecasting attempt is therefore compelled to add a very limited number of previous years. By leaving one year out and utilizing all the other material, the agreement is disciplined. In relation to the return of the year that is left out, they measure the variance of our calculation. The system is used for all years and the average efficacy of prediction has been given. A paper presented by [17] seeking a suitable motor imaging machine is a huge task. For positive discrimination, the collection of discriminatory characteristics is important. This research will include a controlled method to classify biased traits in the EEG signal MI classification. The function selection approach eliminates dimensionality. Any EMG measure is mapped to a continuum. The spatial characteristics are derived from each sub-band. A high-dimensional function vector is paired with several sub-bands. In order to allow an effective definition, the neighborhood component analysis-based filtering approach is used to select appropriate features. Least Squares is a supervised mathematical learning technique used in categorized regression data to improve accuracy. The chosen features are used to identify the SVM. In order to reduce the function factor, the irrelevant features are discarded. Evaluation of the strategy proposed is performed with the data sets 4a and IV 2b of the BCI Competition III. Both data sets can be used as benchmark data types to help test BCI machine learning. Simulations demonstrate the supremacy of the time series prediction system proposed a study that works to predict disease [36]. All the major factors influencing agricultural development are agricultural diseases and insect pests. Early detection and pestilence prediction can mitigate economic harm from plagues. This paper uses a neural network to automatically classify crop diseases. The statistics are taken from the 2018 public data sets of the AI competitor with 30 illnesses of 10 crops. In this post, we use the training model Inception-ResNet-v2. These are approaches widely used in prediction networks dependent on convolution. After the integration is complete, the ReLU function is enabled. The findings reveal that the average outcome rate of the model is 86.1%. After being educated in this approach, our students developed and introduced a smartphone application for the iPhone. And we carried out an actual examination to validate the theory. The findings revealed that the device would correctly predict crop diseases. Another concept is given by [6] that used a system of fuzzy logic and decision trees and with the help of a human expert, they were able to make a recognition of coffee diseases, where they obtained the characteristics of the symptoms that occur in the plant and thus be able to do your decision tree. The results they obtained are within 85% accuracy. Although this research does not deal with artificial vision techniques, it can be noted that a very important step has been taken for the detection of diseases, using expert systems applying decision trees with fuzzy logic.

3.3 Support Vector Machine

Cluster analysis or clustering is a method in which artifacts that are identical but distinct from individuals in other classes are defined. It is used mostly for data processing. In several areas, clustering is used, such as computer education, pattern detection, image processing, knowledge collection, and agriculture. There are different clustering algorithms like k-means and k-medoid set, but the popular and important clustering SVM is the algorithm. A paper presented the concept of the cluster [23]. The objective of this paper was to develop an effective method for harvesting automated cucumber harvesting. The proposed algorithm comprises several processing and data mining techniques aimed at classifying the image of carrots properly. Computer Vision project uses an SVM, a Euclidean distance transform, a bag-of-visual-words (BoW) classifier, and a watershed transform in order to classify images into seven broad categories. Several experiments were carried out to generate the data sets which will be used for training and validation of the classifiers. Detection was tested at both levels of pixel and cucumber by evaluating its result to the ground truth data. The high percentage accuracy at the pixel level and at the tissue level proves that the proposed algorithm is highly reliable in cucumber harvesting applications. Comparative analysis is done by [7]. According to the Food and Agriculture Organization, world production of date fruits is forecast at about 8.5 million tons and 1.3 million tons in 2018 in Saudi Arabia. The most popular varieties of dates are Badri, Khalas, Muneer, NabootSaif, and Suleh in Saudi Arabia. The co-dependencies in maturity are immature, Khalal, Khalal, Rutab, Pre-Tamar, and Tamar. The way fruit is processed has a big influence on earnings. This paper proposes a smart harvest method by means of computer vision and DL to decide the form, maturity, and weight of dates. The method contains three sub-systems: maturity estimate, form estimate, and date weight estimate (DWES). We used four architectures of Deep Neuro Networks, ResNet, VGG-19, Inception-V3, and NASNet for DWES as well as the SVM (regression and linear) support for DMES. We also focused on the Smart Robotics Research Center to test the device proposed. DTES attained a median performance of 99.175%, a 99.225% F1 score, average accuracy of 99.8%, and an average recall of 99.05% using further performance metrics. The DMES has achieved a high-quality precision, which averages 99,058, 99,34, 99,64, and 99,08%. DWES reached a maximum score of 84.27% with SVM Linear.

3.4 *Convocational Neural Network*

A CNN network was developed to determine the impact of climate change on potato development. They have also seen a network of beliefs integrating the volatility of potential climate change, taking into account the fluctuations of current weather parameters including temperature, radiation, rainfall, and potato production. They claimed their network would help politics Agricultural makers. They validate their model with synthetic weather conditions and equate the findings with the traditional mathematical model and conclude that the performance of the faith network is greater. A paper was presented by [17]. AI has been recently extended to a variety of sensing activities to anticipate, monitor, and/or understand. However, it is limited to embedded devices. We propose a low-power sensor with AI onboard to improve the application in agriculture. For this reason, we used a CNN to come up with a system which achieves 83% of average IoU and 97% of seeds recognition accuracy on the test set. The proposed solution would conduct seed identification and seed germination detection via spectral domain image processing. For the training of CNN, we collect data of the image of the seed germination process over the span of many stages. The whole framework would be measured by industrial regulations. The studies have demonstrated that the device can open up a massive amount of possibilities for smart applications. [21] write another paper: The argument over finding an optimal sample set for signals on graphs is an excellent one. The proposed sighter collection is based on a localization operator that uses both vertex and spectral domain positions. We show the relationships between the new procedure, selecting sensors in machine learning methods, and using graph frequency methods to select sensors. The method suggested does not have to measure the vectors of the variable operator while also considering (graph) information. By measuring performance metrics such as quality and memory, we evaluate the effectiveness of our approach. Another definition is the same [2]. Accurate data will greatly reinforce spatial growth and sustainable natural resource and farm management in Singapore. The research focuses on the detection of greenhouse photographs from SPOT-7, and MSI photographs from Sentinel-2. In chosen test areas, multiple classifiers are employed, such as k-nearest neighbor (KNN), random forest (RF), and support vector machine (SVM). The image is created by multi-resolution scanning. Secondly, for each image unit, spectral, texture, and remote sensing indications were obtained. The greenhouses were classified by classifiers. The classification accuracy of this task was evaluated using an ambiguity matrix. The study of the KNN and RF classification revealed a slightly higher mean accuracy (OA) and Kappa (β) of 0.69 and 0.85. The image of Sentinel-2 was rated according to KNN techniques. Another article has been published by [1]. Recently, deep learning methods are becoming extremely relevant to solve image processing problems. Among various algorithms, Convolutional Neural Network (CNN)–based and Recurrent Neural Network (RNN) based systems have achieved state-of-the-art results on satellite and aerial imagery in several applications. While these methods are part of research are for some, there is currently no functional or generic version at the consumer level for remote sensing culture. In this letter, we introduce a method to use distance learning strategies with remote sensing imagery and geographic details. This project is focused on the libraries Orfeo toolbox for image processing and TensorFlow for numerical computation. It is capable of processing a large amount of information without any limitation on image size or hardware configuration. Another paper is written by [3]. There is also a need for more refinement and specificity in the study of human behaviors. Deep education will soon have a significant effect as a study is carried out on security applications in the area of elderly tracking and the detection of individuals and objects left in the public room. Wearable devices for human activity have been developed but these technologies can cause users unwanted physical and mental distress. Researchers have concentrated on using image-based HAR and have placed it at the forefront of electronic consumer design. This essay explores an intelligent method to classify human behaviors through image analysis and profound learning methods. A skeleton-based approach can also produce detailed findings in a number of conditions and domain structures. This paper explores the creation of an efficient HAR embedded knowledge approach. Two public databases of people's routine behaviors, such as experimental outcomes, are used for tests to render the proposed method more reliable and successful on both data sets than most equivalent schemes [13]. As robotics and computer technologies evolve, the agriculture industry is increasingly evolving. Farmers shift to agriculture in which selective productivity is significant. This is called agricultural precision. To prevent weed growth in crops, it is important to correctly identify seed rows. We are proposing a new approach called CRowNet which uses CNN and Hough to detect crop rows in a drone image (UAV). It consists of a "SegNet" model made up of "SegNet" and a

"CNN-based Hough transform" (HoughCNet). The approach suggested could yield better results than traditional techniques. An accuracy rate of 93.58% was obtained for the identification of crop rows with IoUs over 70%. Furthermore, the trained model can detect crops on a given field of crops.

3.5 Deep Learning

Deep learning is an artificial intelligence (AI) feature that imitates the operation of the human brain in data processing and generates templates for decision-making use. Deep learning is a branch of artificial intelligence machine learning, which has networks that can learn unattended from unstructured or unscheduled results, often named profound neural learning or deep neural network. A concept is given by [14]. The Internet of Things would revolutionize health systems, agriculture, banking, electricity, and transportation. With the incorporation of software-defined networks (SDNs) and network feature virtualization (NFV) in the edge-cloud interplay, IoT platforms are significantly improving with developments. Deep learning has been getting its existence more influential due to enormous evidence of Internet of Things (IoT). DL algorithms may have privacy issues when applying to confidential data. These algorithms depend on conventional server-based approaches, that is, high server processing capacity is needed. We would like to suggest a new learning method by differentially private training algorithm called LATENT. The functionality of LATENT allows a data owner to introduce randomization until the data are exposed to possibly untrusted machine learning service. It divides the neural design into three layers: (1) convolutional module; (2) randomization module; and (3) completely linked module. Therefore, the Randomization module will effectively perform a privacy guard service for SDN-enabled NFV. There would be substantial variation in government policy-making owing to this aspect. Our empirical review on classification using convolutional deep neural networks indicates fine results in accuracy (e.g., 92–96%). Another concept is presented by [20] where they conducted 60 experiments using deep convolutional neural networks to identify 14 crop species and 26 diseases, using trained models such as AlexNet [16] and GoogleNet. They used the approach of [12] that demonstrates for the first time that end-to-end supervised training with a CNN architecture is a potential option for a large number of classes, going beyond the traditional approach of using hand-built features. Within the Plant Village data set convolutional with RelU activation function, each followed Max Pooling layer and the second part contain two dense layers to focus color and gray scale. In the end, they showed that working with color characteristics gives them 99.84% better results compared to grayscale with 95.54%. Another article [22] shows a reinforcement learning (RL) method of maximizing charging strategies in a public electric vehicle (EV) charging station. "Online" in the sense that this price mechanism takes choices on a real-time basis, and "model-free" in the sense that it does not use simulations of unknown occurrences. A challenge with RL is we need to optimize the total charging rates to satisfy customers' demands before departure time.

This feature-based method can further improve the performance of the proposed algorithm. Compared to representative benchmark algorithms; our algorithm achieved the highest charging-station profit. This paper presents [31] a novel Computational Intelligence vision sensing approach to analyze the content of nutrients in plant leaves. We propose developing the Deep Sparse Extreme Learning Machines (DSELM) framework for fusion and GA for image normalization as well as to reduce variation of images due to differing degrees of sunlight. Of course, we do relevant working image segmentation with DSELM. In this paper, four moments of color distribution of the leaves images (mean, variance, skewness, and kurtosis) are used as predictors in the estimation of nutrient profiles. Our machine learning model consists of a variety of DSELMs and combines them to predict nitrogen content in wheat leaves. Studies also show that the proposed approach outperforms other approaches in terms of efficiency and processing time. Another method is proposed by [24] that proposes algorithms to identify multiple plant diseases, based on color analysis and using a pair-wise classification algorithm. According to them, their methodology allows them to operate in uncontrolled conditions and thus be able to cover a large number of diseases. This method was tested with a large, unrestricted set of leaf images containing symptoms belonging to 74 diseases, 4 pests, and 4 abiotic disorders, affecting 12 different plant species. The results obtained were between 40 and 80% precision. [29] used an ANN, KNN, Naive Bayes, a hybrid of self-organizing maps (SOM), and radial base function (RBF); all this in order to determine the best algorithm to classify diseases of rust, coffee wilt (CWD), and the CFD that affects the coffee fruit. In their work, they state that they obtained 58.16% for KNN, for NaiveBayes they obtained 53.47%, for ANN 79.04%, and for the combination of RBF and SOM they obtain 90.07% accuracy, which shows a great improvement with respect to the previous algorithms, although they note that the latter takes them longer in training. [26] tell us that they found a feasible solution for the diagnosis and identification of four alfalfa diseases. They extracted 129 characteristics of texture, color, and shape from the 1651 images using the Relif F, 1R, and CFS methods. To classify the diseases, they used SVM, KNN, and Random Forest. The best classifier was SVM and the Relif method for obtaining the characteristic, since they achieved 97.64% in precision for the training set and 94.74% for the test set. [28], experiments have been done to recognize diseases or burns presented in the leaves. The crops examined were bananas, rice, citrus fruit, and roses. They suggest segmentation using genetic algorithms after analyzing the photos and clustering. They used the color rivalry technique to remove attributes and they think it easier to use the color picture than the standard grayscale. To identify them, they used MDCs achieving K-Mean 86.54%, MDCs with an algorithm they proposed, enhanced 93.63% and SVMs with a proposed algorithm, which received substantial improvements of 95.71%. Many of these percentages represent the total average of 54.306 photos from the four crops of 58 groups with 14 crop types and 26 diseases (or their absence), as shown by the highest precision of 99.35%. This target was accomplished. Therefore, without any feature engineering, the model correctly classifies crops and diseases from 38 possible classes in 993 out of 1000 images. The authors noted that training requires a lot of computational work but for classification it is less than a second; they believe it could be implemented on a cell phone [30].

Propose a web tool, where the farmers who grow the pomegranate will upload an image of the fruit to be analyzed, with the trained model to check whether or not the fruit is infected. The technique used is based on extracting the characteristics of the pomegranate images, such as color, morphology, and the color coherence vector, to later perform the classification using the SVM algorithm, giving them an accuracy of 85% with a 10-megapixel camera. Knowing that farmers will not always have the optimal capture means to upload the images to the system, the authors carried out significant tests with the cameras of mobile devices with resolutions of 5 and 3 megapixels, and obtained results of 82 and 79%, respectively. Tang et al. [30] carried out a study that consisted of analyzing 9,000 images of tomato leaves, to produce a model that could be used on smartphones, with the purpose of identifying 5 types of diseases. Your model would be based on a deep convolutional network, but it would be made up of two parts, the first part of the model (the extraction functions) was the same for full-color focus and grayscale focus, consisting of 4 layers. It is used by [34] and has been working with convolutional neural networks of various crops, looking for disease levels in plants, being able to classify healthy crops with 89% accuracy, slightly ill with 31%, moderately ill 87% and seriously ill with 94%. Although the results are satisfactory, he concludes that the classification of plant diseases from digital images is very difficult. On the other hand, the limitations of the data set in terms of quantity and variety of samples continue to prevent the emergence of truly comprehensive systems to perform disease classification functions [35]. Many deep learning models have recently been used to identify various forms of plant diseases, but relatively little work has been done in predicting whether or not these illnesses are dangerous. In addition, it is necessary to thoroughly monitor the severity of plant diseases because it facilitates plant protection decisions. We create a data set of images with the aid of Plant Village and crowd AI of contaminated citrus fruit leaves. Six separate algorithms for the magnitude of citrus HLB were then tested. The experimental findings show that, when dealing with severity detection, the model Inception v3 with epochs = 60 will obtain a higher precision rate of 74.38% than other models. We also decided to explore whether GAN data increase would help substantially enhance the learning output of the model. A new set of 14,056 leaves was eventually used in the V3 network preparation. We have provided more than 85% precision and very high performance.

Table 2 shows the summary of the results obtained by the exposed investigations and the comparative analysis of crop production prediction accuracy with different machine learning classification methods. It shows that crop prediction classification depends on the method, crop and it varies on different parameters, result shows.

4 Discussion and Results

One of the advantages that we can observe when [6] used the fuzzy logic classifier algorithm is how easy and fast it could be implemented, since it does not require a large collection of data for its training, and it presented more or less favorable

Table 2 Summary of the algorithms and their accuracy results

References	Culture	Method	Accuracy in percentage
[4]	Coffee	Fuzzy logic	85.00
[5]	Pomegranate	SVM	82.00
[6]	Coffee	KNN ANN N. BAYES RBF and SOM	58.16 79.04 53.47 90.07
[7]	Banana, Bean, Lemon, Rose	MDC + K-Means MDC + SVM	86.54 93.63
[8]	Bean Fruit Cassava Citrus Fruit Coconut tree Coffee Corn Cotton Grape Passion fruit Soy Wheat Sugar Cane Passion fruit Soy Wheat Sugar Cane	Proposal Classification by pairs	50.00 46.00 56.00 71.00 53.00 40.00 76.00 58.00 56.00 58.00 59.00 70.00
[9]	Alfalfa	SVM	94.74
[10]	Wheat, sunflower, grape, corn, cucumber, cotton, cabbage, tomato	ANN SVM	87.48 92.17
[11]	Strawberry	Fuzzy logic	97.00
[12]	Various	CNN	99.35
[13]	Tomato	CNN	99.84
[14]	Bean Cassava Naranjo Coconut Corn Coffee Cotton Cashew Grape Kale Passion Fruit Soybean Sugar cane Wheat		95.00 83.00 62.00 97.00 66.00 77.00 100 61.00 82.00
[15]	Tomato	CNN	33.27

results when wanting to classify. Now, if the target class is decreased, then it would have better results, as shown by the work of [26], which got better accuracy just by predicting just one disease. The proposal made by [37], by using RBF and SOM, seems quite good, since they are the ones that gave them the best results, compared to KNN, ANN, and Bayes, and this is because their approach to obtain the characteristics by their texture and color made it have better results, since the choice of the latter is the one that works best for these cases.

From the SVM algorithms, it can be concluded that it is another good alternative for solving these cases, although at first, they were not intended to solve this type of problem; with the passage of time, they have been adapted and presenting very good results, such as those of [3, 4].

When the classification is for few diseases, it could be said that it is relatively easy, but when you really want to cover more diseases or more crops, that is when the complexity increases, and a proposed solution is the one presented [24]. Classification by pairs it is based on the theory where plants present similar symptoms when attacked by the same disease and that the algorithm can be retrained for new diseases. Another proposal for the same plants and the same diseases was presented by [5] when using convolutional neural networks; he declares that it is the best way to do this type of study, and that the only limitation today is the availability of large amounts of data to do a good training of the net.

5 Conclusion

Nowadays, an increasing range of computer application Agricultural learning strategies are required for which a lot of accessible data from several resources can be examined to find the secret information. This is an advanced area of study and is anticipated to develop in the near future. The convergence of computer science with agriculture allows predicting crops. It must draw on objective methods for forecasting pre-harvest crops. The creation of an effective model would have some merits over the conventional approach of forecasting as it could be seen in this review of articles, the works that obtained the best results are convolutional neural networks; their use is increasingly biased for this type of problem, since diagnoses are obtained closer to what a human expert would determine. The only problem would be that the training is too computationally expensive and requires a lot of data to do it. If the study data were not enough, the recommendation would be decision trees with fuzzy logic, since it is the algorithm with the best results after convolutional networks. Although it is not very precise, we can trust that as technologies advance; more optimal values can be reached at a lower computational cost. Another important point that we can note is that with the use of new technologies such as machine learning and pattern recognition, diseases in crops can be detected and a timely diagnosis made, reducing the risk of agricultural and economic losses, which would bring a direct benefit to the farmers who implemented it.

References

1. Ai Y, Sun C, Tie J, Cai X (2020) Research on recognition model of crop diseases and insect pests based on deep learning in harsh environments. IEEE Access 14
2. Anand P, Singh Y, Selwal A, Alazab M, Tanwar S, Kumar N (2020) IoT Vulnerability assessment for sustainable computing: threats, current solutions, and open challenges. IEEE Access 8:168825–168853
3. Arachchige PCM, Bertok P, Khalil I, Liu D, Camtepe S, Atiquzzaman M (2020) A trustworthy privacy preserving framework for machine learning in industrial IoT systems. IEEE Trans Industr Inf 16(9):6092–6102
4. Chaudhury A (2009a) Machine vision system for 3D plant phenotyping. In: IEEE/ACM transactions on computational biology and bioinformatics, vol 16, No 6
5. Chen J, Huang YY, Li YS, Chang CY, Huang YM (2020) An AIoT based smart agricultural system for pests detection. IEEE Access 8:180750–180761
6. Du G, Wang Z, Li Z (2020) A new cloud robots training method using cooperative learning. IEEE Access 8:20838–20848
7. Elavarasan D, Vincent PMD (2020) Crop yield prediction using deep reinforcement learning model for sustainable agrarian applications. IEEE Access 8:86886–86901
8. Faisal M, Alsulaiman M, Arafah M, Mekhtiche MA (2020) IHDS: intelligent harvesting decision system for date fruit based on maturity stage using deep learning and computer vision. IEEE Access 8:167985–167997
9. Feng S, Zhao J, Liu T, Zhang H, Zhang Z, Guo X (2019) Crop type identification and mapping using machine learning algorithms and Sentinel-2 time series data. IEEE J Sel Top Appl Earth Obser Remote Sens 12(9):3295–3306
10. Fleming SW, Good body AG (2019) A machine learning metasystem for robust probabilistic nonlinear regression-based forecasting of seasonal water availability in the US West. IEEE Access 7:119943–119964
11. He Q, Zhang Y, Tang S, Liu H, Liu, (2019) Network embedding using semi-supervised kernel nonnegative matrix factorization. IEEE Access 7:92732–92744
12. Horng G-J, Liu M-X, Chen C-C (2020) The smart image recognition mechanism for crop harvesting system in intelligent agriculture. IEEE Sens J 20(5):2766–2781
13. Jin (2020) 'Clustering Life course to understand the heterogeneous effects of life events, gender, and generation on habitual travel modes. IEEE Access 8
14. Josef S, Degani A (2020) Deep reinforcement learning for safe local planning of a ground vehicle in unknown Rough Terrain. IEEE Robot Autom Lett 5(4):6748–6755
15. Khan TM, Robles-Kelly A (2020) Machine learning: quantum versus classical. IEEE Access 8:219275–219294
16. Lee W, Ham Y, Ban T-W, Jo O (2019) Analysis of growth performance in swine based on machine learning. IEEE Access 7:161716–161724
17. Lee W, Kim M, Cho D (2019) Deep learning based transmit power control in underlaid device-to-device communication. IEEE Syst J 13(3):2551–2554
18. LeVoir SJ, Farley PA, Sun T, andXu, C. (2020) High-Accuracy adaptive low-cost location sensing subsystems for autonomous rover in precision agriculture. IEEE Open J Ind Appl 1:74–94
19. Li W, Ni L, Li Z-L, Duan S-B, Wu H (2019) Evaluation of machine learning algorithms in spatial downscaling of MODISL and surface temperature. IEEE J Sel Top Appl Earth Obser Remote Sens 12(7):2299–2307
20. Meng B (2020) Modeling alpine grassl and above ground biomass based on remote sensing data and machine learning algorithm: a case study in east of the Tibetan Plateau, China. IEEE J Sel Top Appl Earth Obser Remote Sens 13:986–2995
21. Minh DL, Sadeghi-Niaraki A, Huy HD, Min K, Moon H (2018) Deep learningapproachforshort-termstocktrendspredictionbasedontwo-streamgated recurrent unit network. IEEE Access 6:55392–55404

22. Molla MKI, Shiam AA, Islam MR, Tanaka T (2020) Discriminative feature selection-based motor imagery classification using EEG signal. IEEE Access 8:98255–98265
23. Phyo CN, Zin TT, Tin P (2019) Deep learning for recognizing human activities using motions of skeletal joints. IEEE Trans Consum Electron 65(2):243–252
24. Ren A (2020) Machine learning driven approach towards the quality assessment of fresh fruits using non-invasive sensing. IEEE Sens J 20(4):2075–2083
25. Sakiyama A, Tanaka Y, Tanaka T, Ortega A (2019) Eigen decomposition-free sampling set selection for graph signals. IEEE Trans Signal Process 67(10):2679–2692
26. Shafi U (2020) AMulti-Modal approach for crop health mapping using low altitude remote sensing, internet of things (IoT) and machine learning. IEEE Access 8:112708–112724
27. Sharma A, Jain A, Gupta P, Chowdary V (2021) Machine learning applications for precision agriculture: a comprehensive review. IEEE Access 9:4843–4873
28. Sott MK, Furstenau LB, Kipper LM, Giraldo FD, Lopez-Robles JR, Cobo MJ, Zahid A, Abbasi QH, Imran MA (2020) Precision techniques and agriculture 4.0 technologies to promote sustainability in the coffee sector: state of the art, challenges and future trends. IEEE Access 8:149854–149867
29. Sun L (2020) Application of machine learning to stomatology: a comprehensive review. IEEE Access 8:184360–184374
30. Tang Z, Wang H, Li X, Li X, Cai W, Han C (2020) An object-based approach for mapping crop coverage using multiscale weighted and machine learning methods. IEEE J Sel Top Appl Earth Obser Remote Sens 13:1700–1713
31. Wang S, Bi S, Zhang YA (2021) Reinforcement learning for real-time pricing and scheduling control in EV charging stations. IEEE Trans Industr Inf 17:849–859
32. Wu T, Luo J, Dong W, Sun Y, Xia L, Zhang X (2019) Geo-Object-Based soil organic matter mapping using machine learning algorithms with multi-source geo-spatial data. IEEE J Sel Top Appl Earth Observ Remote Sens 12(4):1091–1106
33. Yu J (2020) A deep learning approach for multi-depth soil water content prediction in summer maize growth period. IEEE Access 8
34. Z (2020) Smart farming becomes even smarter with deep learning-a bibliographical analysis. IEEE Access 8:105587–105609
35. Zeng Q, Ma X, Cheng B, Zhou E, Pang W (2020) GANs-Based data augmentation for citrus disease severity detection using deep learning. IEEE Access 8:172882–172891
36. Zerrouki N, Harrou F, Sun Y, Hocini L (2019) A machine learning-based approach for land cover change detection using remote sensing and radiometric measurements. IEEE Sens J 19(14):5843–5850
37. Zhu J, Zhao X, Li H, Chen H, Wu G (2018) An effective machine learning approach for identifying the glyphosate poisoning status in rats using blood routine test. IEEE Access 6:15653–15662

Chapter 13
Combined DCT-DWT Color Image Digital Watermarking Technique

Devender Kumar Dhaked

Abstract In present time, information transmission is becoming increasingly different through digitization. But security is a measure concern between information transmissions. Therefore, the digital watermarking technique gives a better method to prevent illegal access, unlike exploitation and allocation and information about the copyright protection of digitized data. It must be strong; therefore, if one part of this is changed then it will be recoverable. Several different methods are provided for embedding and extracting watermarks from the original image. In proposed paper, watermarking is completed by applying the combined approach of Discrete Cosine Transformation (DCT) and Discrete Wavelet Transformation (DWT). DWT technique is applied on the Red color (or R-channel) component of the color image for embedding the watermark. Once completing the embedding process, DCT technique is applied to 8X8 block of the color image. The results of proposed algorithm strongly maintain image quality and also it is sufficient for general image processing tasks.

Keywords Digital image watermarking · Discrete cosine transformation (DCT) · Discrete wavelet transforms (DWT) · Color image · RGB components

1 Introduction

The security of digital data over the Internet from unwanted access is a main problem in the current scenario. For this problem digital watermarking becomes a fastest growing solution. In digital watermarking process, some parts of information are known as watermarks are attached to social networking data like pictures, videos & audio, and text documents too. To avoid repetition of data and provide copyright security digital watermarking approach is used wherein to confirm the authenticity of individuals. Whenever it is a requirement to confirm the authenticity then the user can extract the watermark and validate it [1]. In the last few years, range of the digital technology has been speedily increasing day by day. Therefore, unauthorized

D. K. Dhaked (✉)
Rajasthan College of Engineering for Women, Jaipur, India

R. Mathur et al. (eds.), *Emerging Trends in Data Driven Computing and Communications*, Studies in Autonomic, Data-driven and Industrial Computing,
https://doi.org/10.1007/978-981-16-3915-9_13

access to watermarking and the possibility of attacks is too rising, but different safety techniques are available to protect data from illegal uses.

The watermarking techniques are classified into two separate parts on the basis of operational domains, first part is a spatial domain and the second part is a frequency domain. In the spatial domain watermark object directly embedded in the original image by updating the pixel value of an original image. In this type of technique very small changes occurred in the pixel intensity value of the original image. Frequency domain is a useful technique for embedding a watermark which provides high imperceptibility and robustness. In these techniques for embedding the watermark, transform coefficients are modified. The original picture splits by using transform domain and after this a watermark is embedded in it. In this research we have study about discrete wavelet transform (DWT)[2] and discrete cosine transform (DCT) technique[3]. Also discusses in what way these techniques are followed to embedding and extracting the watermark from the image [4].

2 Literature Survey

According to the research provided by Dimple Bansal and Manish Mathuria the article titled- "Color Image Dual Watermarking using DCT and DWT Combine Approach", the article published in ICEI-2017. In this research paper proposed color image watermarking by combining the feature of DCT and DWT. For embedding the watermark red color component is selected. Then completing the embedded process, applied DCT on 8×8 blocks [5].

Another research on the article titled "A Robust Watermarking Approach using DCT-DWT", provided by S. Pratap, R. Paresh and A. Sudhir, the article published in IJETAE-2012. Author presents a robust watermarking approach for color and grayscale images. The complete study of this paper shows that DCT and DWT watermarking is better for color images comparing to gray images. This research work also shows full watermark embedding is possible on color image which is not possible in earlier proposed techniques [6].

The research on the article titled "Comparison of Hybrid Watermarking Technique on Different color spaces" provided by N. Fatima and D. J. Tuptewar, the article published in Conference on Advances in Signal Processing. This research paper proposed an efficient watermarking method used for color image by combining the feature of DWT and SVD technique. DWT technique is used to find the areas in the image to insert the watermark and SVD is used to build the watermark more robust [7].

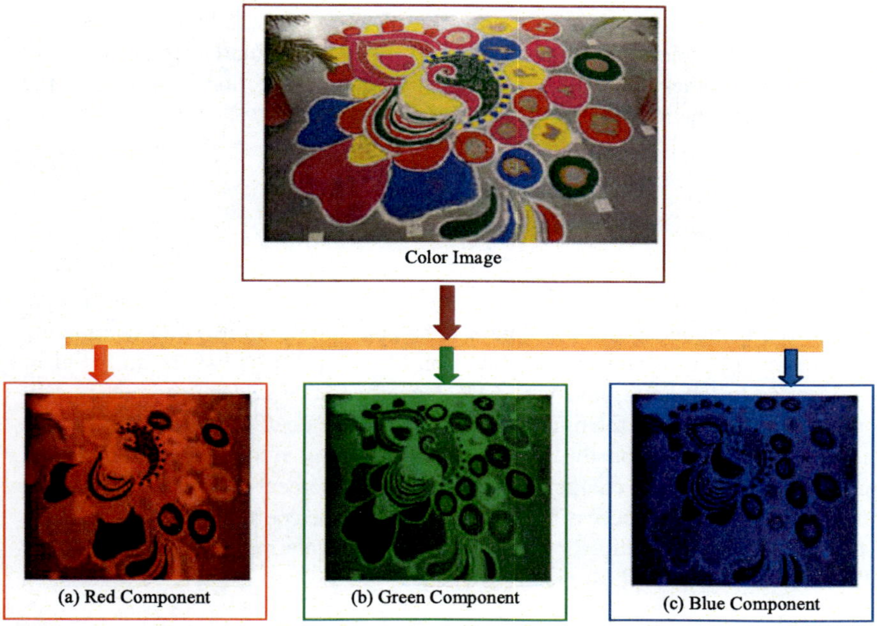

Fig. 1 Original image decomposition into RGB component

3 Proposed Work

3.1 Color Image Component

A color image holds different color items that gives the information regarding the situation for the reason that human eyes are more sensitive or responsive to color deviations. Therefore, each and every one is familiar with regarding the fundamental concept of color and these colors are developed with the combination of three basic colors components like red color (R), green color (G), and blue color (B) [10]. In many experiments it is found that human beings attraction regarding to the color, therefore red (r) color component is more significant compare to others reason to their wavelength. Proposed research paper used red (r) color component to embed the watermark on original image and after this DWT was applied to it (Fig. 1) [8, 9].

3.2 Discrete Wavelet Transform

The 2D Cosine Transform is addressed their forward discrete change capacity of the spatial frequency u or v as F (u,v). No immediate data about pixel and spatial variables is there. For the most part Discrete Wavelet Transform utilizes other kinds of capacities, for example, Daubechies, Haar, and so on. These fundamental capacities are also called Mother Wavelet. Here in this work, Haar function is utilized as the Mother Wavelet function [8].

3.3 *Discrete Cosine Transform*

Since due to high wavelength red color is selected therefore DCT is relegated to complete red component and watermark is be appropriate on it. After completion of transformation, the energy block focus on low frequencies.Watermark's binary bit is inserted in the low frequency of DCT coefficient of selected frequency sub-band of DWT [8]. During the time spent checking DCT coefficients are arranged by their whole extent. After that the beneficiary builds a level of the all-out energy P and characterizes the biggest n coefficients that frame the P percent of the absolute energy. At that point watermark succession is added to the whole coefficient in the rundown. At the point, when the estimation of P is increased, watermarked components are further to be added to the host picture [11, 12].

3.4 *Block Diagram of Proposed Algorithm for Watermarking*

Every algorithms having their security objectives, but here combined approach of DWT-DCT is applied. Through using this technique, the watermark provides more robustness against different attacks (Fig. 2).

A. **Embedding of Watermark**:

- Input Images

(i) Original Image
(ii) Watermark Image

Step 1: Decompose the colored image into RGB color component.
Step 2: After RGB decomposition, select the red component Matrix of Size (512 $\times$ 512) image for watermark embedding process.
Step 3: Then selecting red component, applying DWT on the selected matrix of the red Component and find at A (Approximated), H (Horizontal), V (Vertical), and D (Diagonal) value of the image.
Step 4: Choose the approximated value of the red component, then after declaring the block size of the A value, here in this algorithm block size = 8 used.
Step 5: Apply the DCT on image specified Block Area and after this, embed the Watermark to the specified Block Area.

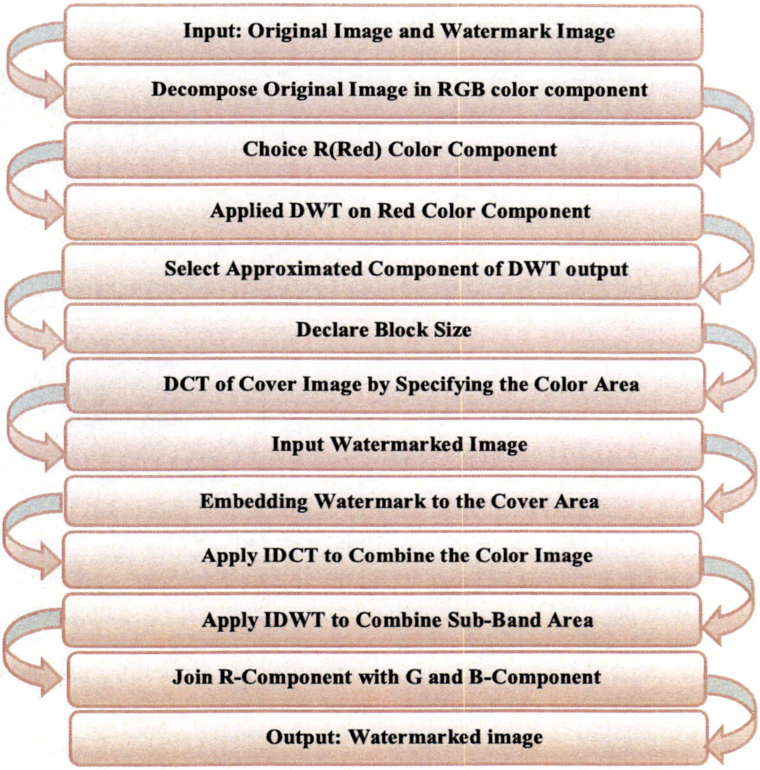

Fig.2 Proposed algorithm

Step 6: Apply IDCT on this image to Association the main object and apply to IDWT to Combine Sub-Bands Once more.
Step 7: After applying IDWT, Join R-Component with G and B-component.
Step 8: Output: Watermarked Image.

B. **Extraction of Watermark:**

- Input Images:

(i) Watermarked Images
(ii) Original Watermark Image

Step 1: After pass the watermarked image, Divides the Watermarked color image into RGB component.
Step 2: After RGB decomposition, select single color (red) component which contains Watermark information.
Step 3: Apply DWT on the Red component and then after gives A, H, V, D values of the red component, select the approximated coefficient.

Fig. 3 Original input images

Fig. 4 Original watermark image: fingerprint.jpg (author's own finger)

Step 4: Select block size which used for embedding here block size = 8.
Step 5: Use original watermark and obtain the size of Watermark image.
Step 6: Now apply the DCT on it and transform the block for getting message vector then reshape the message vector to obtain recovered image
Step 7: Now, write the recovered image on the disk in required format and then finally, read the recovered image and display.

4 Experimental Results

4.1 Selected Original Input Images

In the direction of analyzing the effectiveness of proposed algorithm the experiment work is implemented on four different color images (Fig. 3).

4.2 Selected Original Watermark Image

See Fig. 4.

4.3 Output of Invisible Watermarking

See Fig. 5.

Fig. 5 Four output of watermarked images

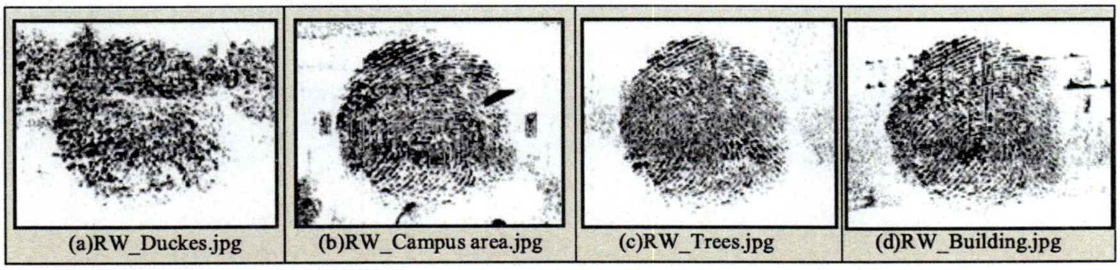

Fig. 6 Recovered watermark image from the original image

4.4 Recovered Watermarks

See Fig. 6.

4.5 Image Quality and Similarity Analysis

Generally PSNR value is used to make a decision regarding quality of image. Usually when the PSNR value varies between 30 and 50 db then good image quality is creating. Apart from the PSNR, the correlation coefficient is also a factor which is used for calculating image quality. The correlation coefficient of image discovers the similarity between the images. The similarity between the images are calculated in the range of 0–1. If the value of correlation coefficient is 1, then both the images (original image and watermark image) are strongly similar to each other. And other then, when the value is 0, then both pictures are not similar and when value comes in –1, then both the images are completely anti-correlated. Here Table 1 presented comparative results of the measurement between the original image and watermark image.

Table 1 presents the PSNR (Peak signal-to-noise ratio) values for each color component of color image. The maximum value of PSNR is 38.9774 for Bulding.jpg at GREEN component. Table 2 below presents the result of Image Quality and Similarity Measurement for Original and Extracted Watermark image.

It is analyzed by the result obtained, reflects the changes in Watermark Image after extracting from the original image. The result state that the Extracted Watermark is

Table 1 Between original and watermarked image

Color images			PSNR value		
Sr. No	Original image	Watermarked image	Red	Green	Blue
1	Duckes.jpg	W_ Duckes.jpg	35.5712	37.8593	37.9373
2	Campus area.jpg	W_ Campus area.jpg	35.7083	37.0624	37.0599
3	Trees.jpg	W_ Trees.jpg	34.0272	35.5638	35.4456
4	Bulding.jpg	W_ Bulding.jpg	37.9662	38.9774	36.8792

Table 2 Between original and recover watermarked image

Color images			Correlation coefficient
Sr. No	Original image	Watermarked image	
1	Duckes.jpg	W_ Duckes.jpg	0.9961
2	Campus area.jpg	W_ Campus area.jpg	0.9914
3	Trees.jpg	W_ Trees.jpg	0.9671
4	Bulding.jpg	W_ Bulding.jpg	0.9704

acceptable in and it can easily be recognized. The Correlation coefficient of all original images and watermarked images is less than 1 and near to the 1, it shows the similarity between the original image and its watermarked images. So, this value verifies the efficiency of the algorithm.

5 Conclusion

In this research work, color image watermarking is completed by performing a combined approach of DCT-DWT to providing image security. Since Color image watermarking is now vital for secure transformation of digital images, which makes a user to freely distribute images. Using a combination of DWT and DCT, it strengthens the image against attacks, which shows the fake user on the image. The research reviewed regarding DCT-DWT based watermarking is especially significant; therefore the research work is done based on multiple research writings on this topic. The DWT approach is popular for robust watermarking nature while DCT is popular for its feature of image security. The complete research work depends on combined DWT-DCT color image digital watermarking technique which gives the high quality of digital watermarking. Proposed algorithms for watermarking can be watermarked color images while red, green, and blue channel watermarks can be used separately to embed. This is a colorful image watermarking but in this research work, the RED component was used for embedding watermarks. The conclusion of this research work shows that watermark image is the same and later recovery we can simply recognize watermarks. This experimental work is done in MATLAB simulation software, implemented algorithm is tested for different images to prove the algorithm efficiency.

Main features of the proposed algorithm are as follows:

(a) It can perform Invisible watermarking.
(b) It can watermark color images.
(c) It takes very less time in computation.
(d) This algorithm has good capacity of hiding watermark information (Invisible watermarking).
(e) The input image to the watermarking can be of any Size and Resolution.
(f) It completely supports the Verifiability property of watermarking, i.e., extracted watermark can prove the ownership of the author of image.

References

1. Saied Amirgholipour Kasmani, Ahmadreza Naghsh-Nilchi (2008) A new robust digital image watermarking technique based on joint DWT DCT transformation. In: Third international conference on convergence and hybrid information technology IEEE
2. Angshumi Sharma, Amrita Ganguly (2012) A combined DCT/DWT based image algorithm. IRNet transactions on electrical and electronics engineering (ITEEE) ISSN 2319–2577
3. Kaushik Deb, Md Sajib Al-Seraj, Mir Md Saki Kowsar, Iqbal Hasan Sarkar (2013) A joint DWT-DCT based watermarking technique for avoiding unauthorized replication IEEE
4. Mingwei Zhao, Yanzhong Dang (2008) Color image copyright protection digital watermarking algorithm based on DWT & DCT. 978–1–4244–2108–4/08 IEEE
5. Dimple Bansal, Manish Mathuria (2017) Color image dual watermarking using DCT and DWT combine approach. IEEE conference publications of ICEI, ISBN: 978–1–5090–4257–9
6. Surya Pratap Singh (2012) Paresh Rawat and Sudhir Agrawal A robust watermarking approach using DCT-DWT Int J Emerg Technol Adv Eng (IJETAE). ISSN 2(8):2250–2459
7. Nazish Fatima, Tuptewar DJ (2016) Comparison of hybrid watermarking technique on different color spaces. IEEE conference publications of CASP, DOI: https://doi.org/10.1109/CASP.2016.7746129
8. Nuruzzaman M (2005) Digital image fundamentals in MATLAB. Book, ISBN 1–4208–6965–5 (sc)
9. Unjal BL (2011) Wavelet based color image watermarking scheme giving high robustness and exact correlation. Int J Emerg Trands Eng Technol I(1). ISSN 2248–9592
10. Li X, Fengyuan Z (2020) Double-color digital image watermarking technology based on double transform domain. Int Conf Artif Intell Comput Appl (ICAICA). https://doi.org/10.1109/ICAICA50127.2020.9182708IEEE
11. Chandra M, Pandey S, Chaudhary R (2010) Digital watermarking technique for protecting digital images. IEEE, 978–1–424–5540–9/10
12. Bhargava N, Sharma MM, Garhwal AS, Mathuria M (2012) Digital image authentication system based on digital watermarking. IEEE conference publications of ICRCC, pp 185–189

Chapter 14
A Performance Analysis of Face and Speech Recognition in the Video and Audio Stream Using Machine Learning Classification Techniques

Roshni Wadhwani, Akhil Pandey, and Vishal Shrivastav

Abstract Object detection and tracking is usually the first step in applications such as video surveillance. The static camera face recognition and tracking system's main purpose is to estimate the speed and distance parameters. We propose a general detection and tracking method for motion based on the visual system and using the image difference algorithm. Then recognize the person's voice to get feedback from the corresponding person. The process focuses on detecting people on stage and then completes the voice signal processing. We propose a new person recognition technology that uses face and voice fusion compared to a single biometric recognition, and this technology can greatly improve the recognition speed. The development of security systems uses the Viola–Jones face recognition algorithm. The proposed method uses the Local Binary Pattern (LBP) as a function extraction technique to calculate local functions. Our project uses Mel-Frequency Divergence Coefficient (MFCC) extraction technology for speech recognition. The extracted functions are used as input to the multi-SVM classifier to provide a gender to identify individuals and display the results. The new system can be used in various areas, such as identity verification and other potential commercial applications.

Keywords SVM · KNN · LBP · Machine learning · Viola–Jones

R. Wadhwani (✉) · A. Pandey · V. Shrivastav
Arya College of Engineering and IT, Jaipur, Rajasthan, India

Rajasthan Technical University, Kota, India

A. Pandey
e-mail: akhil@aryacollege.in

V. Shrivastav
e-mail: vishalshrivastava.cs@aryacollege.in

R. Mathur et al. (eds.), *Emerging Trends in Data Driven Computing and Communications*, Studies in Autonomic, Data-driven and Industrial Computing,
https://doi.org/10.1007/978-981-16-3915-9_14

1 Introduction

Biometrics is a technique of using unique features of a person to determine his identity. When a single biometric feature is used, the chance of compromise results in the growth of multimodal biometrics. With the development of science and technology, the transition from unimodal biometric (one single trait at a given an example) to multimodal biometric techniques (combos of two or more qualities) has been observed to increase safety. The most favorable biometric system is personality, durability, acceptability, collectability, and safety. However, no unique biometric identifier has all these properties. As an explanation, several biometric identifiers are used in a single organization, which is usually a known multimodal biometric system. As support, a multimodal system can use both face recognition and iris gratitude to validate people. Due to reliable or effective security explanation in security-critical submission, multimodal biometric recognition association has recently emerged in biometric society to replace established unethical systems. The system works by first taking feature samples or taking digital color images for face recognition [1]. The sample is then converted to a biometric template using a specific mathematical function. The biometric template provides a standardized, efficient, and highly differentiated format of functions, which can then be compared objectively with other templates to determine identity. Most biometric systems agree on two modes of operation. The registration mode is used to insert the template into the database. The verification mode where the template generates for the device, and then the matching element of the database of pre-registered templates searches for overview of biometric identification systems can be used for personal verification or personal identification. Personal verification answers the following questions: "Do I claim to be me?" It determines the validity of the alleged identity by comparing the verification template with the registration template. Verification requires a declaration of identity or find the individual's registration template and compare it with the verification model. Therefore, the comparison required for verification is called a one-to-one comparison. During the verification, some knowledge of the system's identity usually provides along with the biometric identifier. These additional factors present the unique registered identity or remove biometric functions to the classification database thus providing the relevant biometric machine presentation. In everyday life, most people who trade with us or trade with us use verification to confirm our identity. Therefore, in the recognition system, matching is mentioned in one-to-many ways. These are two types of recognition systems: positive recognition and negative recognition. The closest neighbor is one of simplest machine learning algorithms support on control knowledge. The K-NN algorithm considers the new case/data similarity to the existing case and places the new case in the same category as the existing category. The K-NN algorithm stores all existing data or organize new data points based on connection. This means that it can easily be categorized into drill kit categories using the K-NN algorithm when new data appears [2, 3]. The K-NN algorithm can be used for organization and classification, but in most cases, it is used for organization problems. K-NN is a non-parametric algorithm, which means that it does not predict basic data. It is also called non-lazy learning algorithm because it does not immediately learn from the education response, but stores the data and performs the work on the data stored during the classification [4, 5].

Objective: The process's overall goal is to produce a classification that can pre-preview feeds in video examination or help users recognize reaction actions. To attain this goal, we have residential an automatic resolution that can perform three key functions: identify contact, monitor relationships, and characterize statement behaviors.

2 Existing Method

The first method is to use the video structure to gather end-to-end candidates from the shooting range among the existing methods. Then, by using the MCMC technique to choose real-time boundaries from these candidates to be selected, it is possible to divide space-bound. It should be celebrated that when given the previous probability of number of intention video shows, the MCMC method can provide more accurate distribution results. Thus, in the second method of planned method, the first probability parameters were set at most select charge by Multiple Regression Analysis (MRA).

3 Related Work

With the proper implementation of the information transmission system, based on human speech processing, there are major technical problems related to the complexity of the automatic speech detector. When the sound information is distorted due to noise and others' interference in nature, even the most reliable information will be compromised. These include noise, speech, external noise, etc. Therefore, to recover performance of the audio input data, it is recommended to examine the audio and the video stream, i.e., identify the audio in the video. To solve this difficulty, organization of objects, that is, organization of common phonetic elements is the most important.

Efim V. Zatonskikh et al. This article describes the results attain when creating a complex software prototype that implements speech recognition through the lips through neural networks. This speech appreciation based on lip movement is measured a two-step biometric verification process. This article also suggests developing a neural network model based on the LSTM layer, which is the basis for speech appreciation. To train the model, we collected and organized a record of words with words from a class. To rotate reproduce words in video, a model of lips develop and update, in which the geometric organize of main points in the lip image defined by the change of time. As a result, we get a model that can identify words in a category and has a score of 73.1% (Fig. 1).

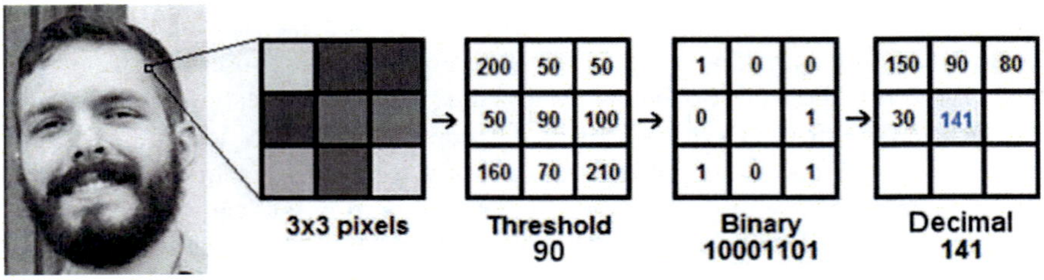

Fig. 1 LBP operation

M.A. Anusuya et al. present a brief review of the acceptance of automated speech and discusses key topics and.

4 Basic Concept

Applying the LBP operation: The first step in LBPH is to produce a transitional reflection that describe original image better by highlighting facial features. For this reason, algorithm uses thought of sliding window support on restriction radius or neighbors.

Pedestal on image above, let's fracture it down into numerous small ladder to appreciate: assume that we have a grayscale facial picture easily. We can display division of this picture as a 3 × 3 pixel window. It can also be expressed as a 3 × 3 matrix, enclose the power of each pixel (0~255). Then we need to use a center value of a matrix as a threshold. This worth is used to describe new values from eight neighbors. For each neighbor to center charge (threshold), we set a new binary. We set the value identical to or higher than threshold to 1, either values less than or equal to level 0. Now, the matrix is only a binary value (regardless of the center value). You need to connect the two binary values of each position in the matrix to the new binary values (e.g., 10,001,101) alternately. Note: some playwrights use other methods to link binary values (e.g., to a clock), but results will be same [6].

Then, we convert this binary value as a decimal worth or place it in the center value of the matrix, which is actually first image's pixel. At the end of the process (LBP process), we have a new image that can define the first image's properties. The LBP program is extended using a specific radius and no port, called a cyclic LBP (Fig. 2).

This can be done using bilinear exclamation. If a data point is among pixels, it uses morals of the last 4 pixels (2 × 2) to estimation value of the new data point [7]. Extract histogram: Now, using the image produce in the last step, we can divide the image into several grids using parameters Grid X or Grid Y, as shown in the following figure (Fig. 3).

MULTI SVM—The following formula presents the optimization problem solved by SVM:

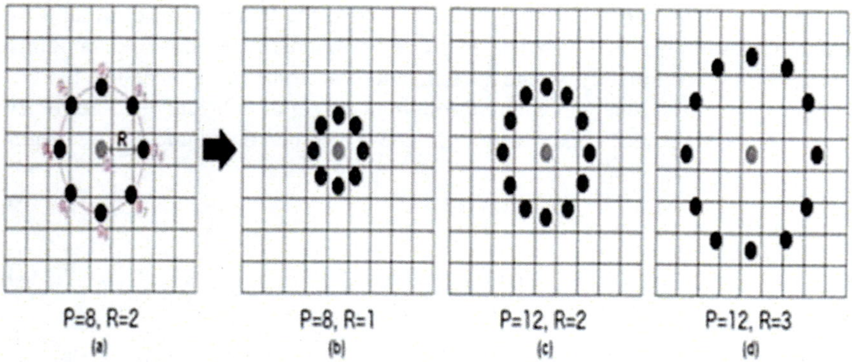

Fig. 2 LBP operation cyclic LBP

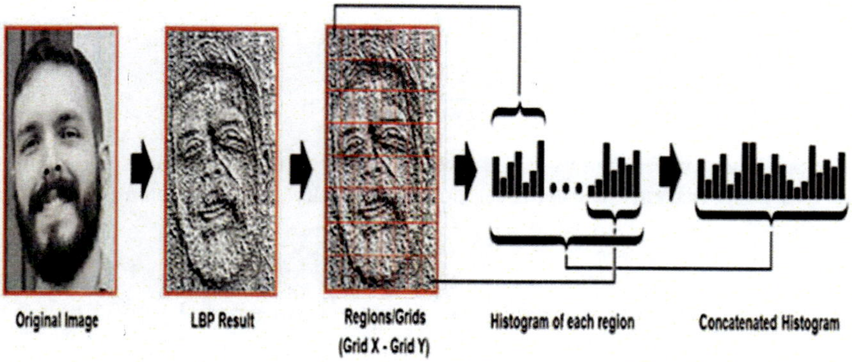

Fig. 3 LBP operation Extract histogram

$$\min \tfrac{1}{2}w^T w + C\sum_{I=1}^{N}\xi_I \tag{1}$$

$$Y_I(W^T\Phi x_i + b \geq 1 - \xi_1 \tag{2}$$

where ξ_I denotes the distance to the correct margin with $\xi_I \geq 0$, i = 1, … n,
C denotes a regularization parameter,
$W^T w = \|w\|^2$ denotes the normal vector,
(Φx_i) denotes the transformed input space vector,
b denotes a bias parameters, and
y_i denotes the i-th target value.

The goal is to organize as many data points as probable correctly by capitalizing on margin from sustain vector to hyperplane while minimizing expression, W^TW.

In other words, the target can also be interpreted as the sentence the optimal w or b, so that most samples can be envisaged properly. In most cases, not all data points can be distributed completely, so the detachment to accurate margin can use [8].

The normal vector generates a line throughout origin of coordinates. The hyper plane intersects this straight line in a straight line at a certain detachment from origin

$$b/(\|w\|2) \tag{3}$$

$$Yi(\omega T\emptyset(xi) + b \tag{4}$$

would be ≥ 1 and following predict completely. Having now data points with a detachment to their ideal position lets us an accurate ideal case of being ≥ 1 to $\geq 1 - \tau$ At the same time, a sanction period is established in the minimization method. C acts as a control parameter or pedals how strong sanction is for how several data points distribute incorrectly and the total distance is.

$$\sum_{i=1}^{n} \tau i$$

The optimization task can be called a dual problem that tries to reduce parameters while exploiting margin. To solving the double difficulty, a Lagrangian multiplier ($\alpha \geq 0$) is used.

$$L(w, b, a) = \frac{1}{2}\|w\|2 - \sum_{n=1}^{N} an\{yn(wt\emptyset(xn) + b) - 1\} \tag{5}$$

With a = (a1,….,an)t representing the Lagrange multipliers,
With b being the bias parameter,
With w being the normal vector,
Where $\varnothing(xn)$ denotes the transformed feature space, and
Where yi denotes the i-th target value.
Utilizing the following two situation:
$W = \sum_{i=1}^{N} aiyi\varnothing(xi)$
$0 = \sum_{i=1}^{N} aiyi$
w and b can be removed from L (w, b, a). This results in subsequent Lagrangian purpose being maximized to (Bishop, p. 325 ff., 2006):
$L(a) = \sum_{N=1}^{N} an - \frac{1}{2}\sum_{n=1}^{N} anamynymk(xn, xm)$
With the constraints an $\geq$ 0and $\sum_{n=1}^{N} antn = 0$,
k(xn,xm) denoting the kernel function.
Solving optimization difficulty, new data points can be confidential by using

$$\sum_{N=1}^{N} \text{anyn}\, k(kn, xm) + b. \tag{6}$$

For the core purpose k (x_n, x_m), core function is described earlier (Sigmoid, linear, polynomial, rbf).

Multi-classification using support vector machine—In its simplest type, SVM is used for binary categorization, which divides data points into 1 or 0. For an organization into several classes, the same standard use. Multi-class problems can break down into several binary categorization cases, also called one-to-one cases. In scikit learning, "vs-one" is not the default value and must be elected overtly (as you can see in the code). The default setting is "a few rests." It shares basic data points in x-class or rest class. Continuously, separate a category from all other categories.

$$\frac{n * (n - 1)}{2}$$

The number of classifiers required for the one-to-many organization can be recovered using the following formula (n is a number of classes): In one-to-one method, each classifier separates the points into two diverse includes all one-to-many classes [9, 10].

K-NN CLASSIFICATION—The principle of K-NN operation can be clarified by the following algorithm: In K-NN algorithm, select each support vector as the representation point and compare the distance between the samples. Test and vector are supporting each [11, 12].

- Step 1: Select neighborhood number K.
- Step 2: calculate the Euclidean distance of neighbor K.
- Step 3: Take your nearest K-neighbor depending on the Euclidean reserve.
- Step 4: Number of your k-neighbors, count number of data points per class.
- Step 5: Assign new data points to the largest number of neighbors in this category.

5 Proposed Methodology

In this process, a facial descriptor proposed uses a local binary pattern algorithm to extract feature information from an emotion-related facial appearance by using directional information and a ternary pattern to take fine edges in the face area while the face has a smooth area. By extracting, this technique performs better than other methods. Then, while sampling the information related to expression at different scales, the grid used to construct the face descriptor is classified. Reducing the number of dimensions by extracting distinctive features is based on the idea of maximizing total data distribution and minimizing differences within the class. It can be seen that the characteristic values of the six categories are highly merged, which can lead to a higher misclassification rate. Note that the actual number of functions can exceed

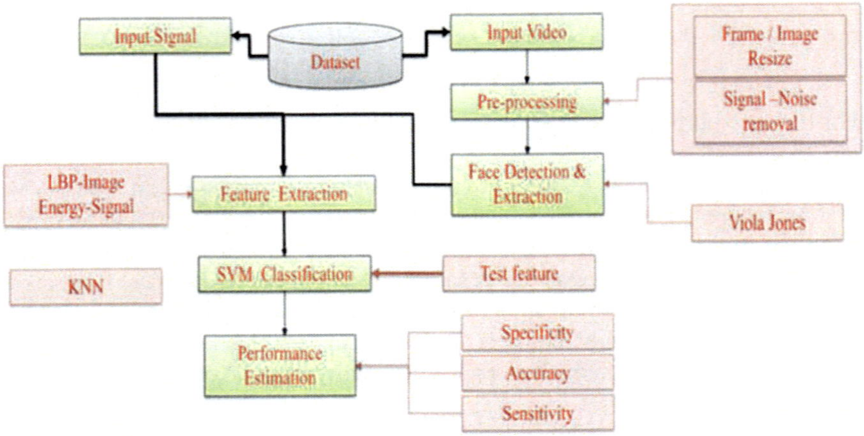

Fig. 4 Proposed flow chart

three, but the first three functions were selected for creation for the sake of intuition. Therefore, this work uses powerful features (Fig. 4).

In speech recognition, feature-by-feature's major goal is to calculate a reduced chain of quality vectors to provide a compacted illustration of a given input indicator. Extraction of functions is regularly achieve in three phases. The first phase is describe speech investigation or sound front end. It performs spectral time examination of signal or produce original functions that describe the power spectrum envelope for short speech distance. The second phase assembles the extensive function vector collected of static or dynamic functions to finish, last phase (not always present) converts these extended function vectors into more compacted or robust vectors and then provides them to recognizer. Multi-mode systems use multiple biometric recognition systems at the same time. In this face detection process, Viola–Jones perceives the facial area and extracts the detected area. In the function extraction procedure, we can implement LBP for outline removal in images as well as MFCC used for extraction in speech signals. The multi-SVM classifier and K-NN we implemented used to identify individuals and then display the results.

Face Detection using Viola–Jones Algorithm—The Viola–Jones algorithm is an extensively used instrument for article recognition. The major function of this algorithm is slow education speed but fast discovery speed. The algorithm uses Haar basic function filters, so multiplication not use. By first produce, an integrated image, the competence of Viola–Jones algorithm can appreciably improve [4, 13].

$$II(y, x) = \sum_{p=0}^{y} \sum_{q=0}^{x} Y(p, q)$$

The integrated image makes it possible to calculate the integral of the Haar by adding only four information. For instance, picture essential of the ABCD region

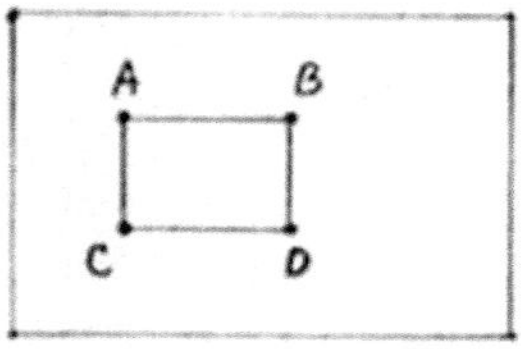

Fig. 5 Image area integration using integral image

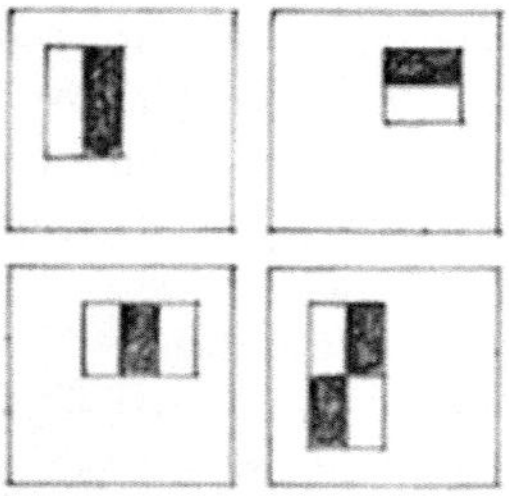

Fig. 6 Example rectangle features shown qualified to the enclosing discovery window

(Fig. 1) is calculated as II (Fig. 5).

$$(yA,\ xA) - \mathrm{II}(yB,\ xB) - \mathrm{II}(yC,\ xC) + \mathrm{II}(yD,\ xD).$$

Registration is done in the search window. Choose the minimum or maximum window size, or then select the amount of flying bread for each size. Then move the detection window to the image as shown below (Fig. 6):

- Set window size to a minimum and adjust the step size depending on the size.
- For selected window size, window slides straight or horizontally in the same steps. At each step, an N-type verification filter use. If the censorship gives a right answer, there is a form in the present widow.
- If window size is maximum size, stop the process, or else amplify the window's size and move the part according to next selected size, or then proceed to step 2.
- The filter (from filter N) contains various categories of filters. Each category looks at the rectangular window of recognition window or conclude if it is a human shape. If so, the following classifier use. If all the racists give good answers, the censors provide the right solutions and accept the face. Otherwise, run the next filter on the smoke N filter.
- Each category contains Haar miners (weak classification). Each Haar function is a combination of two-dimensional integrals in a rectangular area. The value may require a value of ± 1 Fig. 2 shows an example of the Haar function associated with a closed window. The weight of the gray area is good, and the weight of the white area is bad. Hair expansion are a great way to improve your lifestyle, so have fun or treat yourself [14].

fm, i is weighted sum of two-dimensional integrals. Is decision threshold for i-th feature extractor? αm, i, and βm, i are the constant principles connected with i-pull extractor. θm is the decision threshold for the m classification. Input a dataset that collects facial images and audio signals. Face image and language data-sets implement as input. The input image record in.jpg or.png format, and the signal recorded in ".wav" format [15–17].

Binary image: This is the characteristic of the simplest image with 2 Gy values (0 and 1 or black and white). Each pixel is replaced by a tiny bit. These types of imagery are useful for computer vision that do not just need image information or plans. You can create an image from a grayscale image that uses 0 to indicate pixels whose grayscale values are below the threshold and 1 to indicate other pixels. Still, this manufacturing method is useless because most of the information is lost or the image result is small.

Grayscale images: These pictures contain information about the brightness. The number of bits used to signify each pixel depends on the number of clarity levels available. The standard image is 8 frames per second. Pixels, which have 256 grayscale (Ng) or high values, range from 0 to 255.

Color image: Normally image is symbolized by the RGB model (red, green, blue), and each pixel has 24 bits. In numerous function, brightness information and color information are coupled and represented. The two pieces of information are estranged by transferring the RGB information to a numerical function.

Preprocessing: Noise diminution is a procedure of eliminating noise from a signal. All audio diplomacy, include digital or digital recordings, have a function that makes them vulnerable. Noise can be a sudden noise or a white noise with no direct or indirect noise introduced by a device mechanism or processing algorithm. In electronic recording devices, the main form of noise is the noise, which is caused by random electrons. With the force of high heat, random electrons will deviate from the direction indicated. These moving electrons will affect the signal strength of the signal, leading to detectable noise.

Face recognition: Face recognition is a computer technology that can be used in a variety of applications that recognize face in digital images. Facial recognition also refers to the person's psychological process seeing and caring for the face in the visual field. Paul Viola and Michael Jones introduced the Viola–Jones object detection system in 2001, which was the first object detection system to supply real-time article detection rates. Although it can be trained to identify dissimilar class of objects, it is due to form appreciation.

Feature extraction: Pattern acceptance is a division of machine learning that focuses on pattern recognition and data inequality. In some cases, it is almost considered a duplicate similar to machine learning. The tri-local state (LTP) is an extension of the local self-government (LBP). Unlike LBP, it does not set pixel entrance to 0 and 1, it uses a constant gate to set pixel level to three values. Using k as the threshold size, c as rate of center pixel, or use of adjacent pixel p as threshold.

MULTI SVM: In machine learning, a support vector machine (SVM, also known as a support vector network) is a standardized learning model with related learning algorithms that can classify and analyze data. Giving a series of education, each training instance is marked for one or both of two categories. The SVM training algorithm will build a model and provide new examples for one or another category to become -P binary linear classifier (although this method has problems like Platt extraction to use SVM in the field of probabilistic classification). The SVM models represent situations such as points in space and show them the distance between each category as much as possible. Then put the new examples in the same room and guess that they fall into the class based on which side of hole they fell.

In addition to the theater random classifications, SVMs can also use the so-called core technology to successfully perform seamless classifications and seamlessly integrate their access to high-end model locations. It requires a learning approach that attempts to classify nature into groups and then maps new data maps for those groups formed. The clustering algorithm for vector support machines is called clustering vector support, which is often used in industrial applications. If no data is validated, or only a few data are marked as preprocessing the classification cards, they will be used in industrial applications [18].

In accepting the speech, the system of classification of controlled patterns gave an example. In other words, each input mode has an additional class mark. The order arranger can also be trained without supervision. For example, in a technique called vector qualitative, some reactions to input data are grouped by searching for a solid group of data. The customized cluster center table is called a codebook, and the new vector can be identified by locating the cluster center closer to the new vector. For voting, see Fig. 4a. The case of the F1 child vowels is often and F2 is shown. Representative vowels are replaced by bot (/ a /) and boot (/ u /). Note that they are in a good group (Fig. 7).

Local Binary Pattern Functions (LBPs) work well in a diversity of applications, including texture cataloging or segmentation, image recovery, or exterior inspection. The task is to find the position and size of all substances belonging to a given class in the image. Examples include the upper body. The face detection algorithm focuses

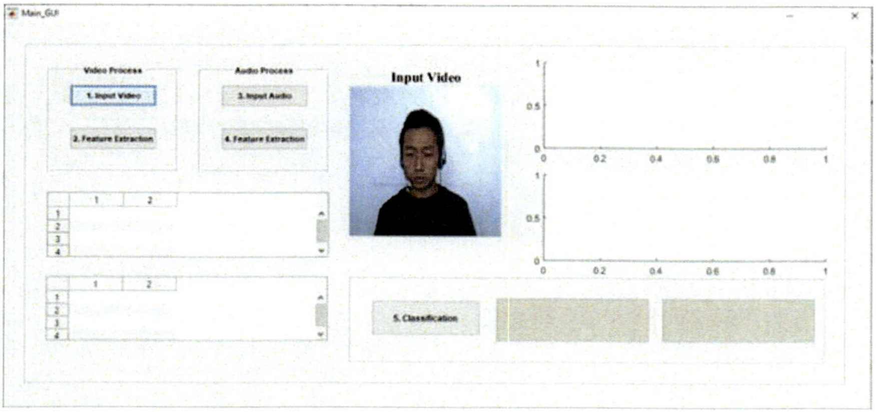

Fig. 7 Input video dataset

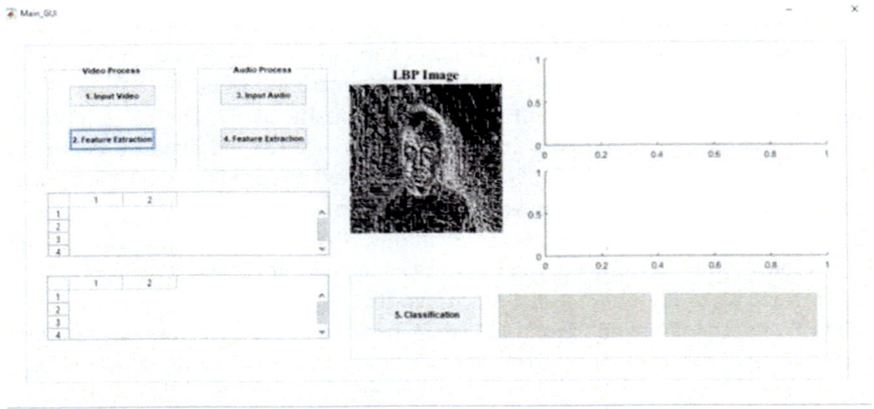

Fig. 8 LBP feature extractor

on perceive frontal faces. This is comparable to image detection, where imagery of people is matched little by little. The image matches image accumulates in the database, as shown in Fig. 8.

Predictably, the selected function will contain applicable information from input data so that this simplified representation can be use instead of the complete initial data to perform the required tasks shown in Fig. 9 (Fig. 10).

The LBP function works very well in different submission counting texture organization or segmentation, image retrieval, or exterior examination shown in Fig. 11.

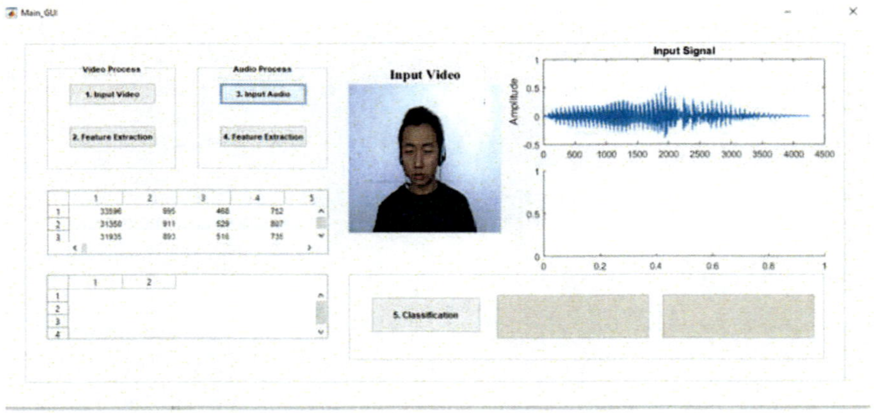

Fig. 9 Audio input signal feature extractor

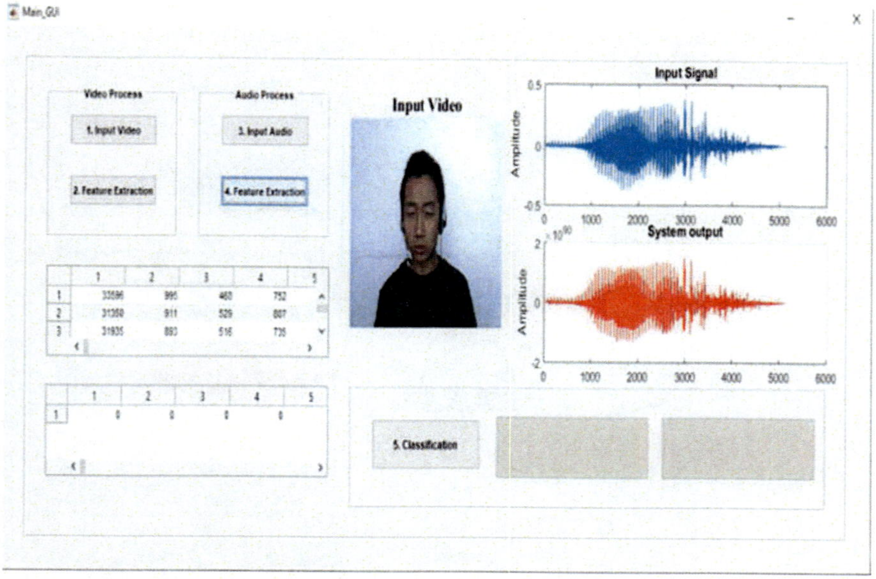

Fig. 10 Audio input feature extractor

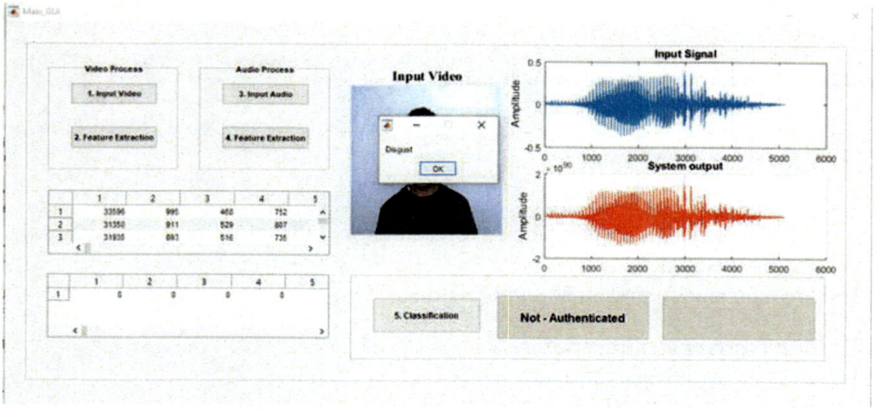

Fig. 11 Classification Result

Multi-SVM performs the mapping from input space to function space to support nonlinear classification problems. Kern tricks can help achieve this by tolerating the lack of accurate representations of mapping skin texture that can cause the curse of dimensionality (Fig. 12).

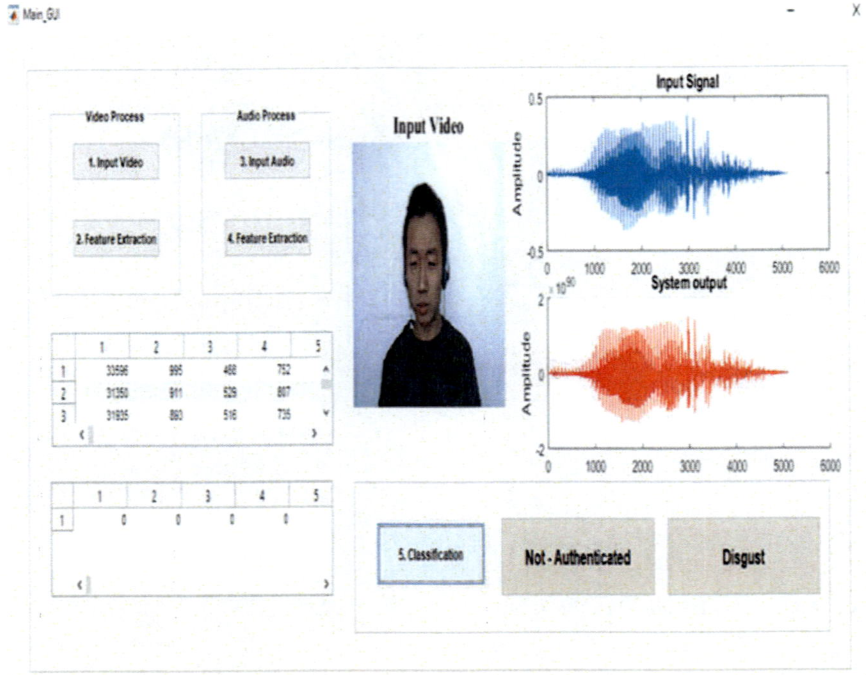

Fig. 12 Classification result person identification

6 Performance Measure

Process performance measure against presentation indicators such as accuracy, sensitivity, or specificity.

Terms associated to presentation indicators:

- TP-True (correct identification).
- TN true negative (reject correctly).
- FP false positive (identification error).
- UN false negative (wrong rejection).

Accuracy: Accuracy in a categorization difficulty is the number of correct predictions from the model in dissimilar calculations.

$$\text{Accuracy} = (\text{TP} + \text{TN}) / (\text{TN} + \text{TP} + \text{FN} + \text{FP})$$

Sensitivity: The aptitude of test to suitably identify people with the disease (true positive rate). Compute authentic positive rate correctly identified.

Table 1 Comparison of the results obtained with existing solutions

	Classification	Accuracy	Specificity	Sensitivity
Proposed work	Multi-SVM	99.68	99.83	99.63
	K-NN	99.36	99.90	99.60
Previous work	Neural network model	73.1	–	–

$$\text{Sensitivity} = \text{TP} / (\text{TP} + \text{FN})$$

Specificity: Ability of the test to correctly identify people without the disease (true negative speed). Measure proportion of original negatives properly identified.

$$\text{Specificity} = \text{TN} / (\text{TN} + \text{FP})$$

F-measure: F-measure (F1-score or F-score) calculates test exactness or is defined as the weighted harmonic mean of exactitude or recall.

f_measure = 2 ∗ ((precision ∗ recall rate) / (precision + recall rate));

The data in Table 1 show the comparison work with previous work.

7 Conclusion

Biometric recognition is related to human recognition through physiological individuality, fingerprints, iris, voice, face, etc. The biometric system can use for classification or verification of people. This paper detects the human face and voice signals. Therefore, the LBP function descriptor extracts the exact functions in pixels in the image. MFCC technology is used for speech recognition. We use multi-SVM and K-NN classifier for classification. Therefore, compared to other classifiers, the mutli-SVM and K-NN classifier require only a few positive examples. Therefore, the performance of the classifier also assess. The selected functions are introduced into the organization to achieve best results. Since the procedure is iterative, results attain filter through iteration. Therefore, future enhancements will make using PSO (Particle Swarm Optimization) knowledge. This technique will achieve a better degree of recognition. The result is better in terms of recognition speed, correctness, or best numeral of functions produce.

References

1. Zatonskikh V, Borzunov GI, Kogos K, Development of elements of two-level biometric protection based on face and speech recognition in the video stream efim department of cryptology and cybersecurity. National Research Nuclear University MEPhI (Moscow Engineering Physics Institute) Moscow

2. Anusuya MA, Katti SK (2009) Department of computer science and engineering, Sri Jaya chamarajendra College of Engineering, Mysore, India. (IJCSIS) Int J Comput Sci Inf Secur
3. Gaikwad SK, Marathwada BA, Gawali BW (2011), A review on speech recognition technique, pp 1561–1569
4. Therese S, Lingam C (2013) Review of feature extraction techniques in automatic speech recognition. Int J Sci Eng Technol
5. Speech Recognition Technique (2012) A Review Sanjib Das Department of Computer Science, Sukanta Mahavidyalaya, (University of North Bengal), India. Int J Eng Res Appl (IJERA) May, Jun
6. Deng L, Li J, Huang J-T, Yao K, Yu D, Seide F, Seltzer ML, Zweig G, He X, Williams J, Gong Y, Acero A (2009) Microsoft corporation, one microsoft way, Redmond, WA 98052, USA (2009)
7. Desai1 N, Dhameliya2 K, Desai3 V (2013) Feature extraction and classification techniques for speech recognition: a review. Int J Emerg Technol Adv Eng
8. Deng L, Platt JC (2010) Microsoft research, one microsoft way, Redmond, WA, USA, Nov 2010, ensemble deep learning for speech recognition
9. Bengio S, Heigold G (2007) Google Inc, Mountain View, CA, USA, Feb 2007, Word embeddings for speech recognition. Rubi, Int J Comput Sci Mob Comput 4(5), May 2015, pp 1017–1024 © 2015, IJCSMC All Rights Reserved 1024
10. Neti C (2006) Member, IEEE, Guillaume Gravier, Ashutosh Garg, audio-visual speech Gerasimos potamianos, member, IEEE, student member, IEEE, and Andrew W Senior, member, IEEE 2006, recent advances in the automatic recognition
11. Dandan M (2014) A survey on deep learning: one small step toward AI. 11. In: Aalto University publication series. Foundations and Advances in Deep Learning, Kyunghyun Cho, 4 Dec 2012
12. Abboud AJ, Sellahewa H, Jassim SA (2009) Quality approach for adaptive face recognition. In: Proceedings of mobile multimedia/image processing security, and applications, SPIE, vol. 7351, 73510 N
13. Aloysius G (2013) Efficient high dimension data clustering using constraint partitioning K means algorithm. The Int Arab J Inf Technol 10(5):467–476
14. Amoli G, Thapliyal N, Sethi N (2012) Iris preprocessing. Int J Adv Res Comput Sci Soft Eng 2(6), 301–304
15. Ang R, Safavi-Naini R, McAven L (2005) Cancelable key-based fingerprint templates. In: Boyd C, Gonzalez Nieto J (eds) Australasian conference on information security and privacy, pp 242–252
16. Anil K, Arun R, Sharath P (2006) Biometrics: A Tool for Information Security. IEEE Trans Inf Forensics Secur 1(2):125–143
17. Anil J, Karthik N, Arun R (2005) Score normalization in multimodal biometric systems. Pattern Recogn 38(12):2270–2285
18. Anil K, Sharath P (2007) Fingerprint-Based fuzzy vault: implementation and performance. IEEE Trans Inf Forensics Secur 2(4):744–747

Chapter 15
Comparative Analysis of AFE Cardiac Amplifiers Used in IOT Application

Tina Bhati, Mohammad Sabir, Vijendra K. Maurya, and Latif Khan

Abstract In this paper, we present the architecture of the current reuse technique-based inverter stacked-4 amplifier which provide low power consumption and low noise, so as to be used as an analog front end instrumentation cardiac amplifier which decide the overall efficiency of ECG recording system as well as the performance of current reuse technique-based inverter stacked-4 amplifier is also compared with the existing architecture of stacked 2 and stacked 3 amplifier. Based on the comparison result we can conclude that the presented low power architecture of stacked-4 amplifier can be used to gather sensory signal from sensor nodes (electrodes) in IOT-based ECG application. As sensor nodes are generally operated by battery so, the key challenge in the design of IOT-based application is low power consumption.

Keywords AFE (analog frontend) · Stacked inverter · IOT · Sensor nodes · Current reuse · ECG

1 Introduction

The concept behind this is to boost the overall trans-conductance of the amplifier without rising the bias current I_D. The technique involves in biasing the MOS in such a way to drive MOS in a weak inversion region to maximize the g_m/I_D [2, 3].

In order to increase the trans-conductance g_m, the input pair of PMOS is stacked on the NMOS input pair which result in inverter-based input stage, as a result of the setup the amplifier overall g_m get doubled, without causing the need of extra biasing current, since the current is shared among both the pairs.

With vertical stacking of N number of inverters, 2 N time current can be reused by single input channel. By combining output current from N number of branches, power of the circuit is reduced which is responsible to boost the overall efficiency of the

T. Bhati (✉)
Rajasthan Technical University, Kota, RJ, India

M. Sabir · V. K. Maurya · L. Khan
Geetanjali Institute of Technical Studies, Udaipur, RJ, India

R. Mathur et al. (eds.), *Emerging Trends in Data Driven Computing and Communications*, Studies in Autonomic, Data-driven and Industrial Computing,
https://doi.org/10.1007/978-981-16-3915-9_15

circuit [1]. In order for minimizing the requirement of voltage associated with power supply, the sources of tail current between stacked inverters are basically eliminated meanwhile not compromising with the PSRR and CMRR. So that, Stacked-4 low power amplifier is used for gathering sensory signal from sensor nodes in IOT based ECG application [9–11].

2 Core Schematic Explanation of Current Reuse Technique-Based Inverter Stacked-4 Amplifier

In this design of amplifier, four stages based upon stacking of inverters, such that the same bias current is shared. The common gate braches originating from the stacked amplifiers tend to combine the current signal coming from all the eight pairs. Thus the overall trans-conductance of the circuit obtained is 8 g_m. The approach behind designing of this amplifier is to embed it in between the capacitive feedback loop [6–8].

The input pairs are separated such that they can be biased at different voltage levels. There are total of eight input pairs, as result this is known as stacked-4 amplifier. The circuit shown in Fig. 1 employs eight common gate transistors which are PMOS_ (8/10), NMOS_ (8/12), PMOS_ (9/11), and NMOS_ (9/13) in order to combine the signal obtained from the input pairs.

Small signal behavior can be analyzed, such that the total trans-conductance g_{mt} of the amplifier can be given by Eq. (1).

$$g_{mt} = \frac{g_m}{I_D}(4I_D + 4I_D) \sim 8g_m \tag{1}$$

I_D is the input bias current of the transistor; it is assumed that all the transistors have identical g_m/I_D and the intrinsic gain is shown by $g_m r_o$ for all the MOS in circuit is greater than 1. The open-loop gain DM, A_{DM} is given by Eq. (2).

$$A_{DM} = g_{mt} r_{ot} \tag{2}$$

The differential mode gain (DM) is the intrinsic gain square and can be compared to a telescopic or folded cascode amplifier. The amplifier is the combination of folded cascode amplifier and telescopic amplifier [4, 5]. If lower end of amplifier is considered NMOS_10 and NMOS_11 and rest of the inputs are dc biased then amplifier overall structure behaves like same as telescopic amplifier and if lower end of the PMOS circuit is considered as input pair PMOS_12 and PMOS_13, such that the relationship between input and output is similar to the folded cascode amplifier topology. The same consideration will be considered for the upper pair of the MOS transistors. The amplifier input referred 1/f noise can be given by the Eq. (3).

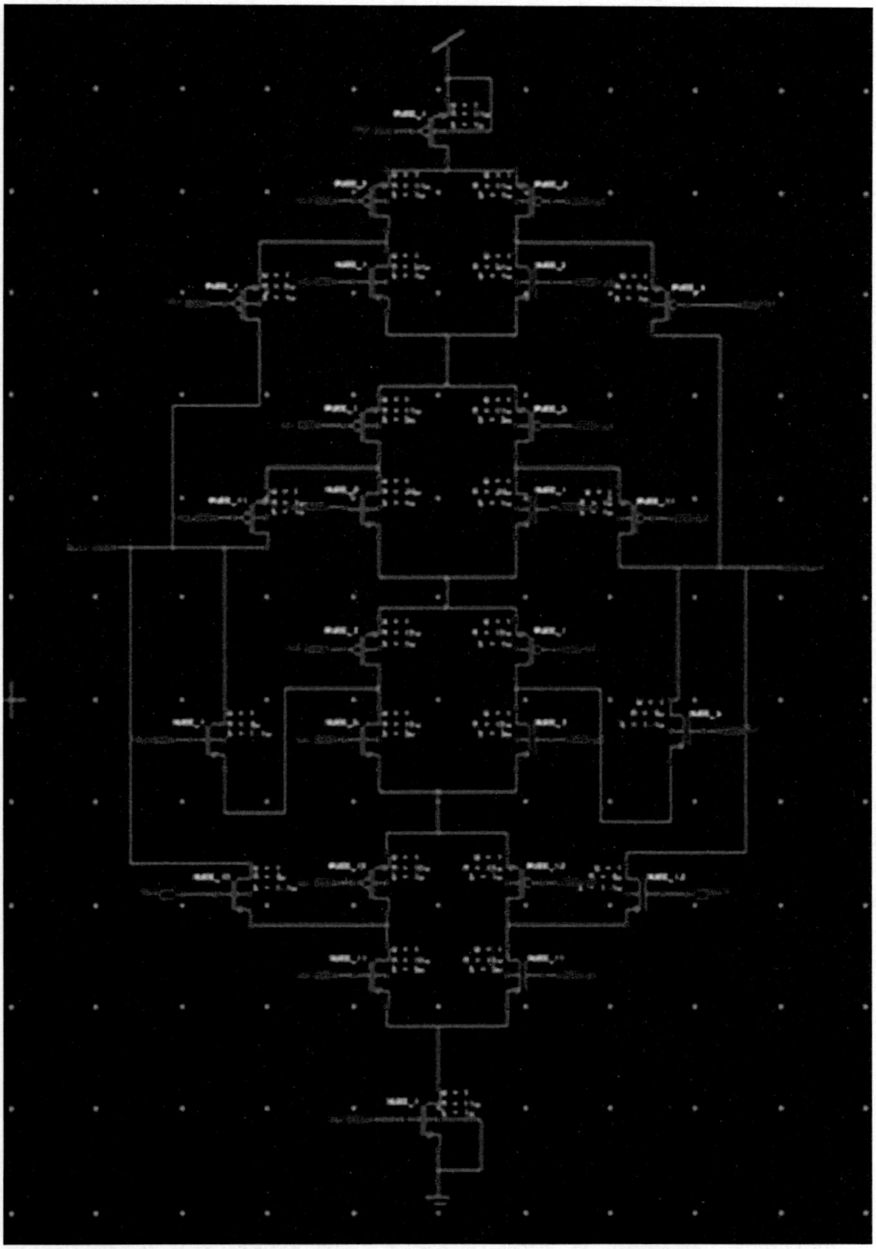

Fig. 1 Schematic implementation using current reuse technique-based inverter stacked-4 amplifier

$$N_{PSD} = \frac{K_f}{4C_{ox}WL} \times \frac{1}{f} \quad (3)$$

K_f is considered as process dependent parameter, width and length of the transistor is denoted by W and L, 1/f noise associated with the circuit can be minimized by increasing the size of the input transistor, as result thermal noise is responsible to dominate the band noise [12].

The CMRR of the circuit topology can be calculated, as CM input is applied and output DM is derived in the presence of mismatch can be given by the Eq. (4).

$$CMRR = \frac{A_{DM}}{A_{CM-DM}} \quad (4)$$

In above mentioned Eq. (4) $A_{CM\text{-}DM}$ expresses the CM to DM gain. The PSRR of the circuit can be illustrated by Eq. (5).

$$PSRR = \frac{A_{DM}}{A_{VDD-DM}} \quad (5)$$

A_{VDD-DM} symbolize the voltage gain from V_{DD} to V_{DM}. In order to reduce the requirement of supply voltage the DC bias voltage tends to be different. The lower end of the NMOS and PMOS pairs of circuit is quite simple to bias; in case if small deviation is existed causes minimal influence on operation of amplifier.

PMOS (8/10), NMOS (8/12), PMOS (9/11), and NMOS (9/13) is responsible to form cascode transistors which is responsible to provide low impedance node at the drain of MOS pairs can be given as NMOS_1, PMOS_2; NMOS_2, PMOS_3; NMOS_3, PMOS_4; NMOS_4, PMOS_5; NMOS_5, PMOS_6; NMOS_6, PMOS_7; NMOS_10, PMOS_12; and NMOS_11, PMOS_13. The current induced due to the mismatches between the bias and input transistors tends to flow in the CG transistors.

3 Simulation Waveform

The circuit of current reuse-based inverter stacked-4 amplifier is implemented using 0.18 μm technology, from the obtained simulation it can be observed that only 2 mV of weak input cardiac signal is applied to the input stage of amplifier which result in large output voltage swing which is 603 mV at the output node, the resulting amplified signal will further be processed by the signal conditioning circuitry. The circuit operates on 1 V supply voltage. The DC response is shown in Fig. 2, the AC response plot is illustrated in Fig. 3 and Noise analysis is shown in Fig. 4, in which magnitude of input noise (innoise), output noise (onnoise), and total noise (totnoise) is represented.

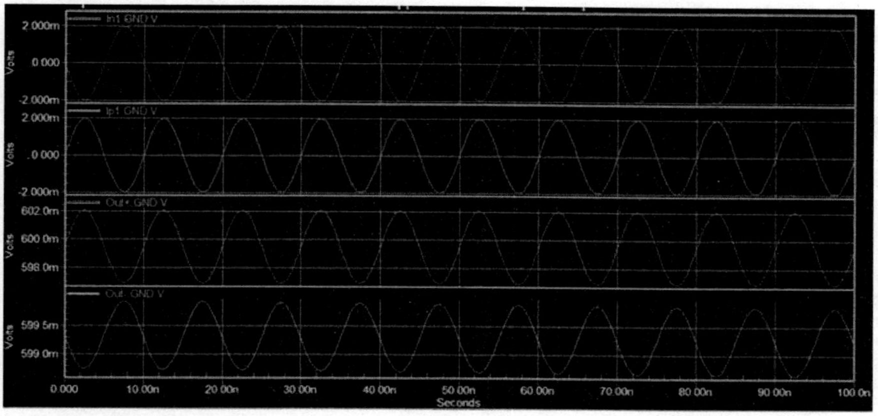

Fig. 2 Simulation waveform current reuse technique-based inverter stacked-4 amplifier

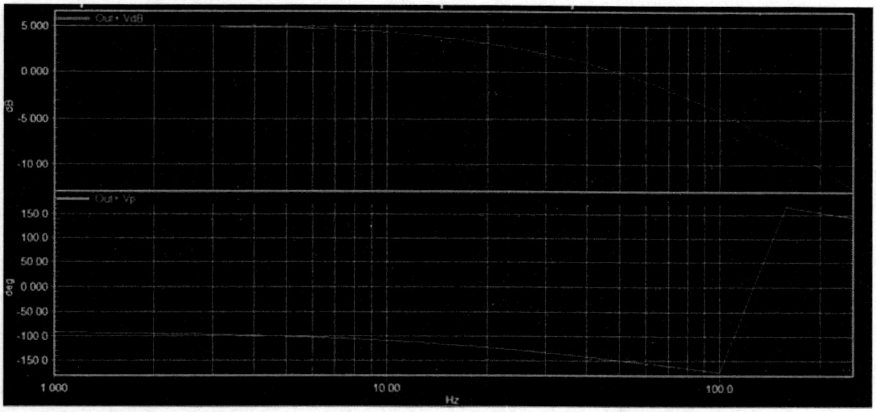

Fig. 3 AC response of inverter stacked-4 amplifier **a** Magnitude Response (dB) **b** Phase response (Deg)

The performance of stacked-4 amplifier is compared with the existing architecture of stacked 2 and stacked 3 amplifiers; the result of the comparison is tabulated in Table 1.

4 Conclusion

A technique for reducing power consumption and comparison with existing design is presented here. With the stacking of four inverters, the stacked-4 cardiac amplifier for ECG applications consuming 18.9 pW from a 1 V supply and attain NEF of 1.5.

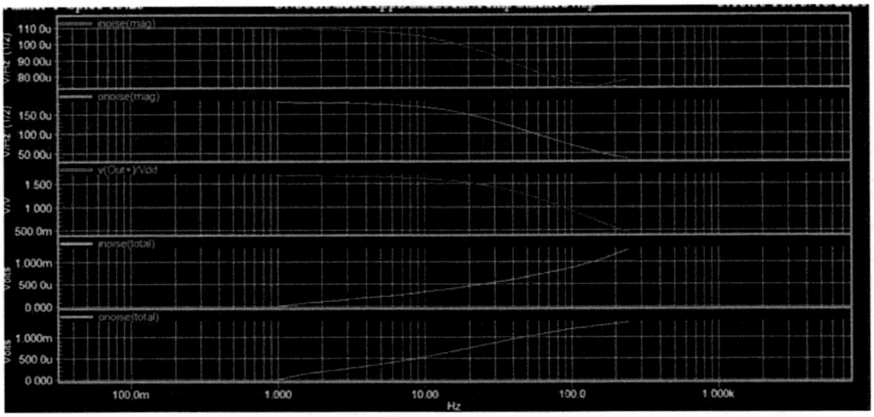

Fig. 4 Noise analysis response of inverter stacked-4 amplifier

Table 1 Comparison between presented and existing design

S. no	Parameter	Unit	Stacked-4 amplifier	[1]	
				Stacked-3 amplifier	Stacked-2 amplifier
1.	MOS technology	Nm	0.18	0.18	0.18
2.	Supply voltage	V	1	1	0.9
3.	Gain	dB	49.57	25.6	25.4
4.	NEF	NA	1.5	1.07	1.26
5.	CMRR	dB	85.29	84	82
6.	PSRR	dB	45.22	76	81
7.	**Power dissipation**	**W**	**1.89×10^{-11}**	**0.25×10^{-06}**	**0.23×10^{-06}**

It is best to use this as an AFE for IoT-based ECG applications, which meet the low noise as well as power requirement.

References

1. Shen L, Lu N, Sun N (2018) A 1-V 0.25-μW Inverter stacking amplifier with 1.07 noise efficiency factor. IEEE J Solid-State Circ 53(3):896–905
2. Revanth Kumar G, Naga Sunanda K, Durga Prakash M (2019) An 86 dB gain 18.06 mVrms input-referred noise LNA for bio-medical applications. Int J Innov Technol Explor Eng (IJITEE) 8(8)
3. Kosari A, Breiholz J, Liu NX, Calhoun BH, Wentzloff DD (2018) A 0.5 V 68 nW ECG monitoring analog front-end for arrhythmia diagnosis. J Low Power Electron Appl
4. Weng OT, Isaak S, Yusof Y (2018) Low power CMOS electrocardiogram amplifier design for wearable cardiac screening. Int J Electr Comput Eng (IJECE) 8(3)

5. Sangeetha M, Geetha S (2017) An Efficient VLSI architecture implementation based on ECG classification system. Int J Recent Eng Res Dev (IJRERD) 02(05)
6. Mubin MA, Marzuki A (2019) A low-noise amplifier utilizing current-reuse technique and active shunt feedback for medradio band applications. Int J Electr Electron Eng Telecommun
7. Pradhan UU, Naveen Kumar SK (2015) Design of a three lead ECG amplifier system. Int J Adv Res Comput Sci Softw Eng 5(12)
8. Sathwara H, Shah K (2015) Design of low power CMOS low noise amplifier using current reuse method-a review. Int J Electr Electron Eng 07(01)
9. Khanchuea K, Siripokarpirom R (2018) An IoT-enabled platform for power consumption measurement of low-power wireless sensor devices. In: 2nd international conference on engineering innovation (ICEI)
10. Fialho V, Fortes F (2019) Low power IoT network sensors optimization for smart cities applications. In: International conference on smart energy systems and technologies (SEST)
11. Pronath M (2017) Low power circuit optimization for IoT. In: 30th IEEE international system-on-chip conference (SOCC)
12. Wagaj SC, Chavan YV (2016) Smart trends in information technology and computer communications. Springer Science and Business Media LLC

Chapter 16
Design and Analysis of Multi-band Fractal Antennas for Wireless Communications

Gaurav Kumar Soni, Sonam Gour, and Abha Sharma

Abstract In highly performance structure like aircraft, satellite, missile application, and spacecraft, where the mostly used design parameters weight, cost, size, ease of installation, and profile of aerodynamics parameters be required in low profile antenna. In the present scenario radio and wireless communications have been used in many other government and non-government applications, which has similar requirements as per in the mobile. Microstrip antenna is used in this entire requirement. It is used for low profile, planer, and non-planer effects. It is in high demand due to the simple and inexpensive design used by manufacture in support of printed-circuit technology. It is converted into mechanically robust when it is mounted on rigid surfaces; they are very highly efficient in the polarization, pattern, impedance, and resonant frequency. Since few decades the microstrip antennas are almost used in every small or big devices due to their smaller size, very light weight, low profile, and simple circuitry. These antennas consume less power so this property of microstrip antenna can be used in low power transmitting devices receiving applications. Due to these properties microstrip fractal antennas will be in demand in coming future corresponding framework. In this paper work new designs of fractal geometries are simulated using software like CST Microwave studio software. The proposed fractal antennas that are square shape multiband fractal patch antenna and elliptical shape multiband fractal patch antenna. Following parameters like return loss, voltage standing wave ratio (VSWR), gain, and directivity and radiation pattern are taken to test the design of the antennas. As the iterations increases, it is examined that return loss and VSWR are minimized and gain and directivity are increased. It is also examined that the radiation pattern is quite stable.

Keywords Fractal patch antenna · VSWR · Microstrip · Wireless communication

---

G. K. Soni (✉) · S. Gour · A. Sharma
Arya College of Engineering & Research Centre, Kookas, Jaipur, RJ, India

R. Mathur et al. (eds.), *Emerging Trends in Data Driven Computing and Communications*, Studies in Autonomic, Data-driven and Industrial Computing,
https://doi.org/10.1007/978-981-16-3915-9_16

1 Introduction

In wireless or current communication systems antennas is required with enhancing bandwidth and smaller dimension. In UMB (ultra wideband) communication, UMB antenna is used as the main components. Wideband antennas are changed as per the available frequency bands. It will not work as efficiently if the size of the antenna is used as less than a quarter of the wavelength. To achieve the efficient work in multi-band antenna and antenna used in low profile antennas, fractal geometry provides a very good option. It allows changing the antenna element like reduced size of the antenna, multiband frequency, broadband signal properties, and many more.

In wireless communication, antennas are used widely and are irreplaceable by any other parts. The devices and free space are connected by the use of antennas. Beside many advantages it also has many disadvantages like it can operate only in single frequency, provides very low gain, and limits radiation pattern directivity. The radiation pattern affected the size and the efficiency of the antenna [1].

To overcome these remedies designer used different types of techniques. Fractal shape is one type of technique which uses different types of shapes. All shapes have different fractal dimensions, it is not affected by Euclidean geometry. It shows some interesting properties, for example, space-filling and self-similarity. Fractal antenna used all the structures to enhancing the characteristics. Designer used multiple bands for designing it. They also enhance the characteristic of the antenna by reducing the size and increasing the bandwidth [2–4]. It was developed by the Nathan Cohen in 1995. In his design he explained different fractal antennas and their characteristics [5]. It was the starting of many research works in this field.

Fractals structure defines various shapes and it can also be divided into different sub-parts such that the every sub-part has improved the characteristic of the previous one. It used self-similarity in design of fractals idea and it is the main reason for broadband and multiband properties. The antenna size can be reduced by using their complicated shapes. It used jagged and convoluted shapes so that these structures provide an increment in the desire bandwidth and provides radiation pattern. The structured used space-filling property of fractals design. It leads to a design which provides high electrical length with reduced admirable volume [6–10].

Some designer utilized MIMO OFDM for reducing the errors in multiple channels. The full form of the OFDM is an orthogonal frequency division multiplexing that uses multiple inputs with multiple outputs. It uses multipath channels which provide in transferring lot of data. Some designers used passive channel technique for weakening patterns in desired wireless system. It can be reduced by using cooperative communications [11]. The Ka-band defined by IEEE is a frequency range from 27 to 40 GHz. The Ka-band is mostly used for satellite communications. Ka-band has numerous advantages like broad bandwidth, smaller wavelength, high-resolution, and higher focus [12–14]. In weakening situation, the reliable transmission can be achieved by the utilization of multiple relays input from one position to a different position. The choices of the relay methodology provide the effect in the adverse channel [15].

2 Proposed Antenna Design

We simulate our design by using Fractal antennas. Microstrip Patch Antenna is used for fractal antennas design. Design is simulated by CST Microwave Studio simulation software. Fractal antenna is designed by using Fr4 glass epoxy substrate. The height of the used substrate is 1.6 mm and the dielectric constant is 4.3. We used two types of feeding techniques. Microstrip feeding technique is used for first antenna and the second antenna is designed by using coaxial Probe feeding technique.

3 Design and Analysis of Square Fractal Antenna

The proposed antenna includes a FR4 (lossy) glass epoxy dielectric substrate is used. It has 1.6 mm thickness and 4.3 dielectric constant. Microstrip feed line is used for feeding. In design we used a square-shaped fractal antenna on a microstrip patch. The proposed antenna work in X-band (8–12 GHz) and in lower Ku-band (12–14 GHz). As we know there are endless fractals present in nature. Every fractal has its unique geometry and mathematical analysis. The designer uses a square fractal antenna which has a symmetrical geometry. Our motive is to design an antenna which radiates on multiple frequencies. After reviewing many research papers a new design is created. The design of the proposed antenna has three iterations.

3.1 Base Shape of Antenna

Base shape is a simple patch antenna with microstrip feeding at (−2.5, −4.5). The length and width of patch are 14×15 mm, therefore the area of the patch is 225sq.mm and perimeter of the patch is 60 mm; another square is removed from the patch structure. The length and width parameters of the removed square are approximately10×10 mm, having area of about 85sq.mm and perimeter 40 mm. After carrying out the iterations the area of the patch decreases. As a result the antenna radiates on multiple bands.

In the first iteration geometry In this iteration frequency on which the antenna radiates are 8.7 and 10 GHz as shown in Fig. 1. Table 1 shows the results obtained in iteration 1 with various parameters.

Figure 2 shows the VSWR plot of iteration 1 which is less than 2 and radiation pattern.

In the second iteration another square of area 25sq.mm is removed from the patch. After second iteration the antenna radiates at 8.5, 10.2 GHz as shown in Fig. 3. Table 2 shows the results with various comparative parameters. Figure 4 shows the radiation pattern which is omnidirectional.

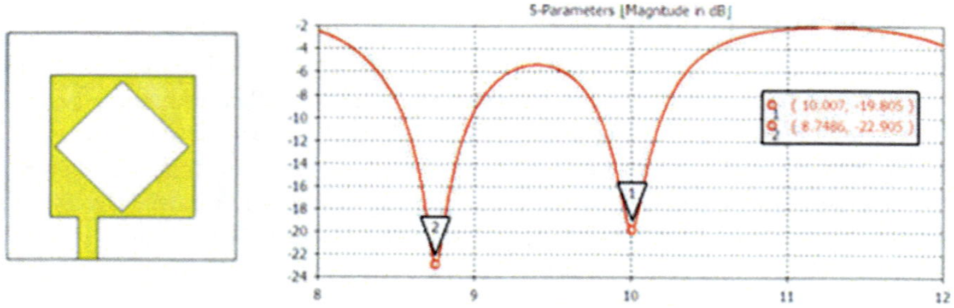

Fig. 1 Structure and return loss plot of iteration 1

Table 1 Result for iteration 1

Band number	Frequency (GHz)	Return loss (dB)	VSWR	Directivity (dBi)
I	8.7	−22.9	1.1	7.74
II	10	−19.8	1.2	7.07

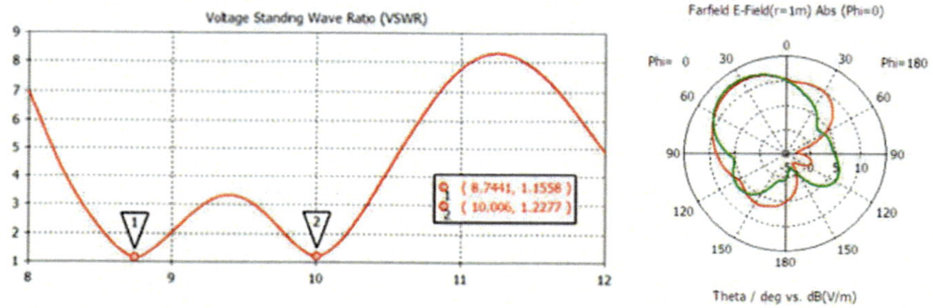

Fig. 2 VSWR and radiation pattern plot iteration 1

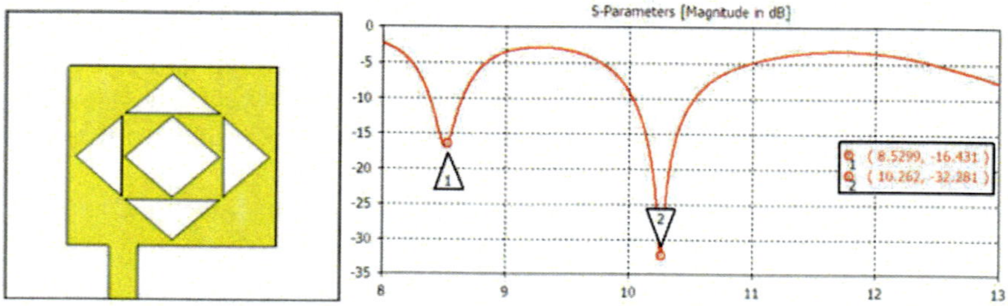

Fig. 3 Structure and return loss plot for iteration 2

Table 2 Result for Iteration 2

Band number	Frequency (GHz)	Return loss (dB)	VSWR	Directivity (dBi)
I	8.5	−16.4	1.3	5.12
II	10.2	−32.2	1.05	6.37

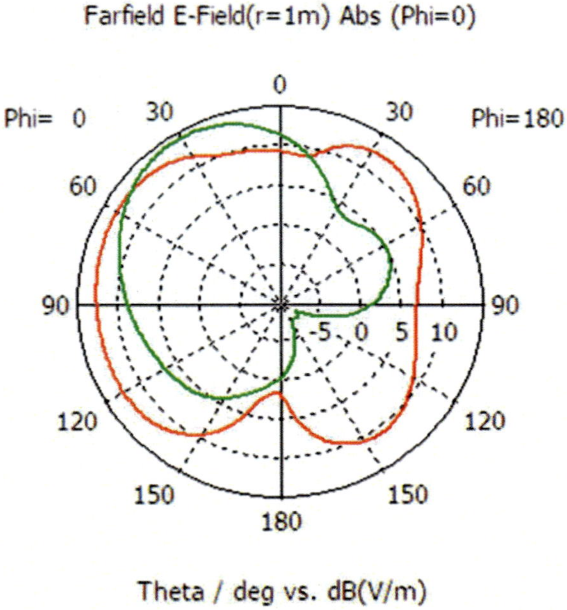

Fig. 4 Radiation pattern plot iteration 2

Similarly in the final iteration the square of area 5sq.mm is a crop from the patch. After last iteration the antenna radiates at 8.5, 10.1, 12, 13.4 GHz as shown in Fig. 5. The radiation pattern is shown in Fig. 6 at 8.5, 10.1, 12, 13.4 GHz. Table 3 shows the results with parameters of VSWR,frequency, and directivity.

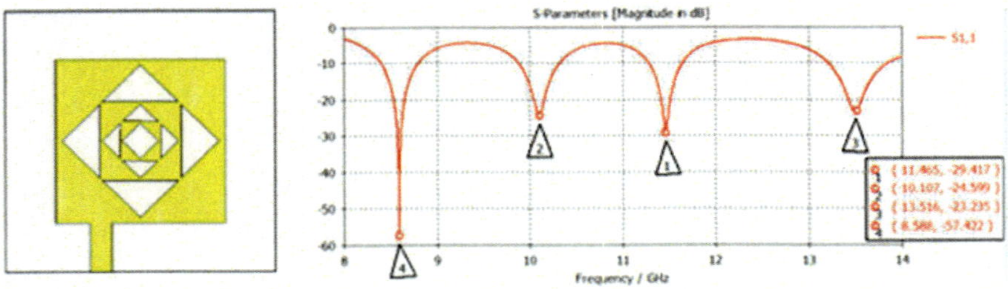

Fig. 5 Structure and return loss of final iteration

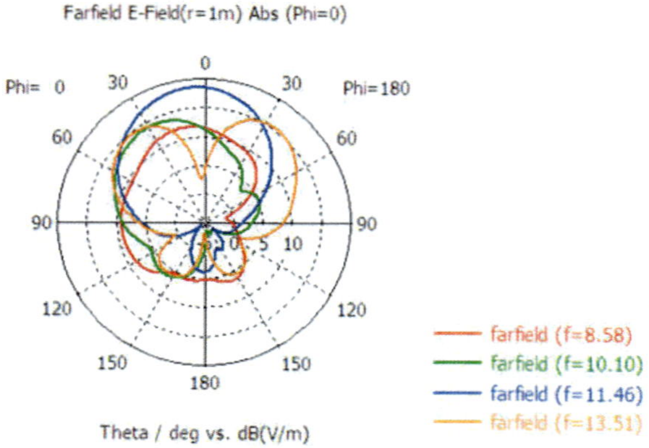

Fig. 6 Radiation pattern plot final iteration 3

Table 3 Result for iteration 3

Band number	Frequency (GHz)	Return loss (dB)	VSWR	Directivity
I	8.5	−57.4	1.02	7.73
II	10.1	−24.5	1.1	7.52
III	11.4	−29.4	1.07	9.19
IV	13.5	−23.2	1.1	7.1

4 Measured Results

The design of the square fractal slot and three iteratives were studied and simulated by using CST microwave studio 2017(student edition). The results are measured by Microwave Analyzer. In this part we discuss some illustrative results like Return loss, VSWR, Radiation pattern, and gain.

As shown in Fig. 7 the measured return loss values are close to simulated values. In measured result the radiating frequencies are 8.5, 10, 11.3, 13.7 GHz the return loss values at respective frequencies are −19.45, −23.37, −23, −30.96 shown in Fig. 8.

As shown in Fig. 9 the measured VSWR values are very close to simulated values. The values at various frequencies are as follows at 8.5 GHz it is 1.2, at 10 GHz It is 1.3, at 11.3 GHz it is 1.1, and at 13.8 GHz it is 1.06.

5 Results and Conclusion

The proposed square shape multiband fractal antenna is compact sized 24 mm × 24 mm × 1.6 mm, which finds possibilities to be used in hand-held devices like mobile. Table 4 shows the comparative analysis of simulated and measured results.

Fig. 7 Fabricated structure of final iteration 3

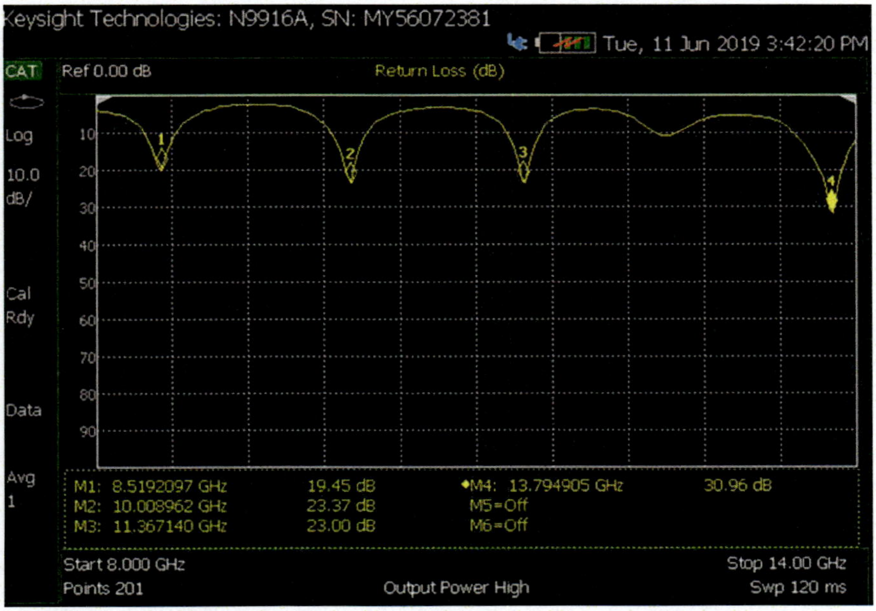

Fig. 8 Measured return loss fabricated structure of final iteration 3

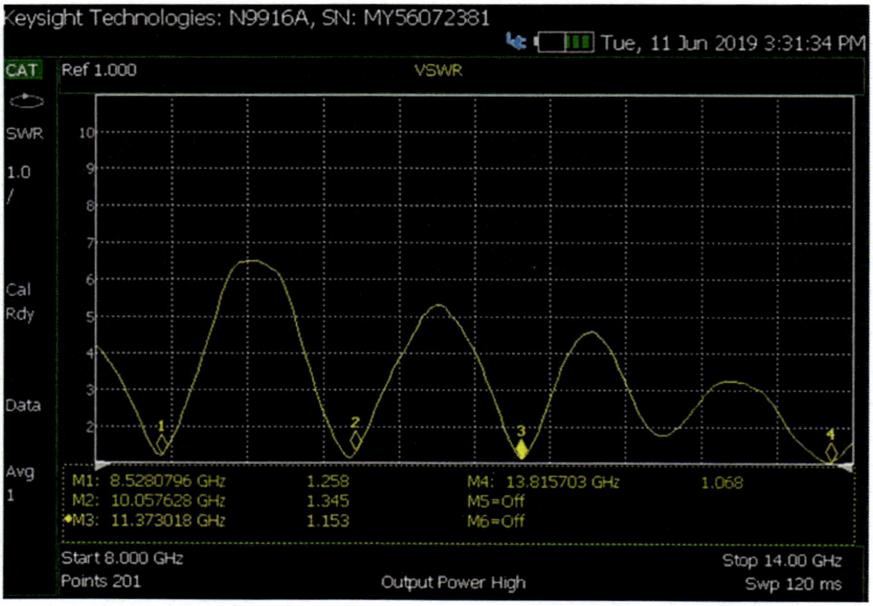

Fig. 9 Measured VSWR fabricated structure of final iteration 3

Table 4 Comparison of simulated and measured results

Parameters	Simulated results	Measured results
Frequency bands (GHz)	8.5, 10.1, 11.4, 13.5	8.5, 10.0, 11.3, 13.8
VSWR	1.02,1.1,1.07,1.1	1.2, 1.3, 1.1, 1.06
Return loss (dB)	−57.4, −24.5, −29.4, −23.2	−19.47, −23.37, −23, −30.96

Great outcomes have been found at X-band (8–12 GHz) and lower Ku-band (12–14 GHz) which can be used for radar applications, wireless communication. The designed square shape multiband fractal antenna can be used in long distance, space, satellite, radar, communication applications.

References

1. Balanis CA (2005) Antenna theory: analysis and design, 3rd edn. Wiley, Inc, Hoboken, NJ
2. Puente C, Romeu J, Pous R, Garcia X, Benitez F (1996) Fractal multiband antenna based on the Sierpinski gasket. IEEE Electron Lett 31(1):1–2
3. Puente C, Romeu J, Pous R, Ramis J, Hijazo (1998) A small but long koch fractal antenna. Electron Lett 34(1): 9–10

4. Best SR (2002) On the significance of self-similar fractal geometry in determining the multiband behavior of the Sierpinski gasket antenna. Antennas Wirel Propag Lett IEEE 1(1):22–25
5. Cohen NL (1995) Fractal antennas: Part 1. Commun Q, summer, 5–23
6. Yoshimura Y (1972) A microstrip slot antenna. IEEE Trans Microw Theory Tech MTT-20:760–762
7. Adams T (1967) Flush mounted rectangular cavity slot antennas–theory and design. IEEE Trans Antennas Propag 15:342–351
8. Cockrell CR (1976) The input admittance of the rectangular cavity-backed slot antenna. IEEE Trans Antennas Propag 24(3): 288–294
9. Hadidi A, Hamid M (1989) Aperture field and circuit parameters of cavity-backed slot radiator. IEE Proc 136(2):139–146
10. Akhavan HG, Mirshekar-Syahkal D (1997) Study of coupled slot antennas feed by microstrip lines. In: Proceedings of IEEE 10th international conference on antenna and propagation, pp 290–293
11. Jangir D, Shankar G, Jain BB, Soni GK (2020) Performance analysis of LTE system for 2 × 2 Rayleigh and Rician fading channel. In: Proceedings of the international conference on smart electronics and communication (ICOSEC 2020), pp 961–966
12. Sharma A, Saini Y, Singh AK, Rathi A (2020) A recent advancements and technological challenges in flexible electronics: mm wave array for 5G networks. submitted and presented in two-day conference on flexible electronics for electric vehicles (FlexEV -) Mar 5–6 2020, Manipal University, Jaipur
13. Sharma A, Rathi A (2020) Analyzing reconfigurable slot antenna for cognitive radio applications: submitted and presented. In: IEEE 8th international conference on reliability, infocom technologies and optimization, (ICRITO'2020)
14. Sharma A, Suvalka R, Singh AK, Agrahari S, Rathi AA (2019) Rectangular annular slotted frequency reconfigurable patch antenna. submitted and presented in international conference on communication, devices and networking (ICCDN), Dec 2019
15. Singh H, Soni GK, Suryawanshi P, Shankar G (2020) Performance analysis and BER comparison of OFDM system for 4 × 4 MIMO fading channel in different modulation scheme. In: Proceedings of the third international conference on smart systems and inventive technology (ICSSIT 2020), pp 255–259

Chapter 17
A Summary on Neurodegenerative Disease Detection

Kavita Mundra and Manju Mandot

Abstract Due to the need for proprietary equipment and expert workers, digital medical imaging is extremely costly and complicated. The most popular imagery technology faces graphic assessments, manual reorientations, and other challenges with limited resolution and a low signal-to-noise ratio. Several algorithms have been developed over the years, which are effective enough to recognize biomarkers that show neurodegenerative diseases of interest to clinicians. This is a detailed theoretical paper on various unsupervised strategies for better detection. This paper discusses the best algorithm among the unattended mapping methods available, based on evaluation criteria such as MSE, PSNR, performance, and sensitivity. This gives an overview of the unsupervised techniques used for neurodegenerative diseases.

Keywords Alzheimer's disease · SOM · GHSOM · AMSOM · CNN

1 Introduction

A major danger to human health is disorders that are neurodegenerative. These age-dependent diseases are becoming more common, partly because the aging population has grown in recent years. There are also neurodegenerative diseases that are very difficult to cope with, such as Alzheimer's disease brain dysfunction, Parkinson's disease, Huntington's disease [1], amyotrophic lateral sclerosis, frontotemporal dementia, and spinocerebellar ataxia. These disorders are diverse in their pathophysiology, with some inducing memory and cognitive impairments and others affecting a person's ability to walk, speak, and breathe. Effective treatments are urgently necessary, but they can come with only a detailed knowledge of the causes and mechanisms of each disease.

One way to learn about how a disease works is to construct a model system that recapitulates the signature characteristics of the disease. Strong laboratory model organisms such as the mouse, fruit fly, nematode worm, and even baker's yeast have

K. Mundra (✉) · M. Mandot
Department of Computer Science and Informtion Technology, Janardhan Rai Nagar Rajasthan Vidhyapeeth University, Udaipur 313001, RJ, India

R. Mathur et al. (eds.), *Emerging Trends in Data Driven Computing and Communications*, Studies in Autonomic, Data-driven and Industrial Computing,
https://doi.org/10.1007/978-981-16-3915-9_17

been used to study neurodegenerative diseases for many years and provided useful insights into the mechanisms of disease [2].

Owing to the need for proprietary software and expertise, digital medical imaging is extremely expensive and complicated.

A simulation program called MATLAB is used to simulate brain images. This software would include a comprehensive interpretation of Magnetic Resonance Imaging (MRI) brain scans that may be useful for more clinical analysis [3].

With the introduction of various imaging modalities, i.e., in Functional Magnetic Resonance Imaging (fMRI), Magnetic Resonance Imaging (MRI), and Positron Emission Tomography (PET), the complexities and emerging challenges come into being. When it comes to the recognition point of view that suffers from noise, resolution problems, and other artifacts due to patient activity at the time of acquisition and also internal triggers that create host images in the direction of phase encode, the MRI scan image must be examined. Thus, this will lead to false ALZ and mild cognitive recognition, so an appropriate approach must be developed that can function on MRI images since it can often not be subjected to such a high frequency and is therefore expensive [4–8].

ALZ is a heterogeneous brain transition that contributes to the elimination of dead brain nerve cells and tissues, contributing to an extreme decrease in hippocampal volume [1], a significant decrease in hippocampal cortical width, and a serious enlargement of the inner ventricles. Essential associative behavioral data includes neuropsychological screenings such as the Mini-Mental State Appraisal Score Card (MMSESC) [9, 10], the Depression Scale Short Form Geriatric (GDS-SF) [10], the Hachinski Ischemic Score (HIS) [11], the Functional Evaluation Questionnaire (FAQ) [12], and Trail Making Measures of TMT-A and TMT-B [1, 13]. For cohort research, comprehensive neuropsychological measures such as the forward and backward digit interval test (Wechsler memory scale, WMS-III), instant and delayed recall (WMS-III), and the Boston name test (BNT) [14] are used [15]. Brain information provided by MRI plays a role in identifying robust biomarkers for early diagnosis of ALZ [1].

The neurochemical knowledge would be in terms of the individual's metabolic peak position, concentration, height, and line-broadening, and in terms of concentration, height, and line-broadening. The MMSE, CDR, TMT-A, and TMT-B neuropsychological scores given as functions define the characteristics of the input material supplied and the findings extracted. A broad variety of various classifiers, such as help vector machines, analysis of main elements, and linear discrimination analysis, can be used to differentiate against ALZ patients from healthy control groups and to model migration from MCI to ALZ. In addition, techniques of feature reduction are used to remove only the distinctive and most significant features of high-dimensional outcomes in order to estimate the exact prediction model. Such a predictive study of multi-model biomarkers would cause stable and pathological cohorts, both at and beyond the single-subject stage, to be diagnosed differently. In order to develop neural network intelligence, especially deep networks with multiple learning parameters, artificial intelligence needs enormous amounts of training data in clinical diagnosis [1].

One of the top ten most common diseases is the cause of Alzheimer's disease (ALZ). Alzheimer's is an incurable disease that affects the elderly and mostly attacks memory and thinking skills. By the time a person has Alzheimer's, they can no longer successfully execute even the simplest of tasks. ALZ (also known as Lou Gehrig's disease) is currently the sixth leading cause of death in the United States; however, in recent unreported cases, it may be read as having the same health risk as 7th or even 8th. Early diagnosis has proven to be of significant value for improving the lives of the patient, but no cure for the illness has ever been identified. As well, one can incorrectly diagnose schizophrenia and administer medications with worse side effects that can impair neurological or behavioral function [1] in patients.

Parkinson's disease is yet another neurological condition that mainly affects nerve cells that produce dopamine, a compound that operates as a neurotransmitter across the brain (PD). Parkinson's Disease (PD) is a degenerative condition in which dopamine systems in the brain deteriorate gradually and leave the patient with decreased motor system function. It has been said that three out of every one thousand people suffer from this disease. Due to the lack of proven PD drugs, new drugs are focusing on different areas of people's lives, such as quality of life. Although the liquid might not be ideal for long-term use, it could help to get an impression or diagnosis of the disease early on. Detections can be performed by sophisticated advanced neuroimaging techniques [16]. Moreover, but these symptoms appear idiopathic, or occur without a known cause, and are often misdiagnosed by physicians, as reports have revealed that about 25% of patients are idiopathic cases. Even if an anti-Parkinson treatment is not helping the patient so much, a patient will still be faced with the side effects brought about by the medication. Computer-Aided Diagnosis (CAD) involves the use of Machine Learning (ML) methods and has been implemented regularly to help doctors. The approaches we took, in effect, aimed to break down the dataset into two groups, numeric or categorical properties. As the field of Deep Learning (DL) continued to develop, and events such as the Microsoft competition proved that the field had acquired the ability to classify medical images such as X-Ray images with 96% accuracy. A lot of researchers began to try to apply DL techniques for the medical image segmentation process.

2 Medical Examinations

For most neurodegenerative disorders, one of the most important problems is that there is no single test that determines and identifies the disease accurately. Having said that there are many screening measures and methods of brain imaging that can support the doctor make his or her determination [17].

In the past, the fast growth of non-invasive medical imaging techniques for decades, the study of the brain has opened up new possibilities. Multiple approaches of the brain-related disorder use different imaging techniques [18].

2.1 Film Labeling

No non-invasive imaging of the brain uses a magnetic field for MRI. Cells are modified to be able to soak up the energy of the radio waves. The iron target is placed under a magnetic field that polarizes his governing bodies. A radio wave pattern is delivered into the brain to change the orientation of the electric field (and thus the alignment of the protons). At first, the heart skips a beat (a signal of fatigue) and makes the nearest electrical circuit realign itself to the shape of a radio wave. In different tissues, the amount of time it takes to restore the electrical charge is different. There are in total three anatomical planes: frontal, parietal, and occipital. There are two hand and arm locations, in line with your left arm, and your back. Both the left and right sides (horizontal), for use in different orientations. A radio emission can be produced that is a variable in pulse rate (i.e., pulse length, from one pulse to the next, relaxation time). Each set consists of two pairs of slices, with one looking at brain structures while the other looks at the digestive tract. MRIs are used to image the atrophy of the cortex of schizophrenia [19]. Table 1 shows a comparison of different modalities.

Table 1 Comparison of imaging modalities

Medical imaging	Problem	Approach	Disease	Expense per visit	Accuracy (%)
MRI	Partial volume, Rf noise, intensity homogeneity, gradient, motion, wrap around, Gibbs ring, susceptibility	Structural, anatomical, Pictures	AD, Parkinson, Dementia	7000	80
FMRI	Severe cognitive impairment limited	Blood oxygen level-dependent (BOLD) MR signal	Schizophrenia, autism	16,000	88
PET	Fluid	Amyloid, binary diagnostic	AD, Parkinson's	15,000	89
SPECT	Spatial resolution of the contrast is reduced because the high lesion activity blurred into the neighborhood by the detector response	Conventional nuclear imaging technique and tomographic reconstruction methods	Parkinson, autism, AD	17,000	85

2.2 Computerized Tomography Imaging (CT)

An X-ray unit is used to generate a set of X-ray images of the patient toward the patient's lower abdomen. CT is a diagnostic method of ALZ. We are learning that as our childhood years grow shorter, our long-term memory grows short, as well. With Alzheimer's patients, the third and directional ventricles are larger in size than in control subjects [20].

2.3 CT for Single-Photon Emission (SPECT)

SPECT is another form of scanning in which gamma rays are used to scan the human body. Gamma rays could be used to treat the disease. The radioactive particles were of the injection type. MRI creates different images depending on how the radioactive isotope is administered into the body. Dopamine Transporters (DaT) scans can be used for Parkinson's Disease (PD). A computed tomography scan uses a contrast agent that is iodine-123. This test enables differentiation of a real Parkinson's disease from other cases of Parkinsonism, even in case where there is no history of dopamine deficiency or brain injury [21].

2.4 PET for Positron Emission (PET)

PET imaging is scanning the brains of patients to measure their metabolic activity. Gamma rays are radionuclides, and they are capable of being injected through the radiogenic. PET scans are used to diagnose and measure the duration of the disease. Using PET scans that look at glucose metabolism rates, it is the rate of glucose metabolism in the temporal cortex that is most predictive of the worsening of dementia (see Table 1).

3 Medical Scales

Numerous ranking scales, questionnaires, and measures to measure patient status.

3.1 Mental State Score Card Test (MMSESC)

The MMSESC is a test used to measure a person's cognitive abilities. Cognitive abilities are tested. Scoring vary based on whether the test is based on memory, attention, or some other cognitive domain. ALZ dementia and PD-associated cognitive disability are both measured with MMSESC.

3.2 Unified Classification Scale for PD (UPDRSI)

The most important measure used for Parkinson's disease, UPDRS is a good measure. The analysis has four sections.

3.3 Unified Dyskinesia Scale (UDysRSS)

The scale is for the assessment of the involuntary movements P.D. "ON" process has two parts corresponding to dyskinesia.

For the "OFF" phase: Toggle ON.

Patients with this type of dystonia display hyperkinesia.

Once you are under the control of a psychotropic drug, you must cope with it. In this group, there are 11 elements from 0 to 4.

(0 = healthy, 4 = symptomatic).

Step "OFF".

Dystonia is the involuntary twitching of muscles. The patient is not harmed by the medications. The dosage of 15 is similar.

3.4 Questionnaire on Freezing of Gait (FOGQ)

The purpose of this survey is to collect information on how many and the severity of the patients' PD symptoms. There are 16 questions in a timed test.

Tests for Parkinson's disease are very common and include tests of hand-limb actions (in essence, reaching for an object with the hand), and also tests of walking.

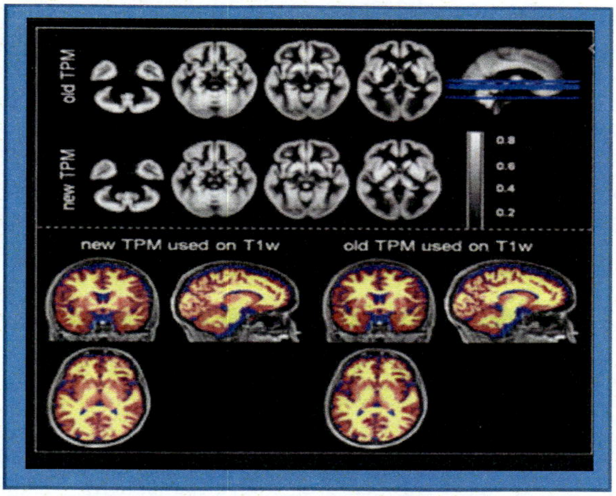

Fig. 1 Tissue segmentation [5]

4 Biomarkers

Biomarkers are a quantitative measure of the external reported physiological indicators. Biomarkers are neurochemicals used in Alzheimer's disease to determine risk, success, or effect. One of the most famous examples is the identification of amyloid beta peptides in the cerebrospinal fluid and plasma. Figure 1 shows the tissue classification using SPM.

5 Unsupervised Techniques

5.1 SOM

SOM mapping was introduced by Kohonen et al. [22] as an artificial neural network for the representation of high-dimensional data. In order to segment the MR brain image, [23] uses the SOM (self-organizing map) network. In this procedure, the data to be traced is first read or initialized [20]. The map of some shape to be shaped is then made. A neuron is initialized here and then the neighborhood is measured using the Euclidean distance. The data is then trained using a sequential or batch process [21] and mapped over the net afterward. The process is replicated until established clusters are formed shown in Fig. 1 [24, 25].

5.2 Wavelet-SOM

Using wavelet transform and statistic features that improved the classification were combined here with function vector.

Discrete Wavelet Transformation (DWT) is a type of image processing [20] that provides a multi-scale representation of the signal or image specified [21]. Normal DWT is unable to adjust the problem of variation and has horizontal and vertical dimensional selectivities.

5.3 FCM-SOM

Ortiz et al. [6] the author implemented self-organizing mapping with fuzzy logic using 3-D statistical features that enhanced the features. The weight factors are updated through Euclidean distance and then the neuron vectors are updated [21].

5.4 ASGHSOM

Zhang et al. [26] proposed an Adaptive Growing Hierarchical SOM (ASGHSOM) in which multi-scale segmentation is coupled with the algorithm of competitive learning clustering to address the question of overlapping grayscale intensities on boundary regions. To decrease the noise effect and the grouping uncertainty, an adaptive spatial gap is implemented with ASGHSOM. From low resolution to high resolution, ASGHSOM uses several SOMs, but the number of neurons in each layer is fixed. SOM algorithm with adaptive spatial distance enhances clustering technique [25].

5.5 GHSOM

By incorporating Growing Hierarchical Self-Organizing Map (GHSOM) and multi-objective-based feature selection strategies to maximize segmentation efficiency, [6] enhanced SOM performance. SOM's biggest downside is that it is important to pick the scale of the performance chart before classification. GHSOM is a variant of SOM that dynamically expands and helps the data to explore underlying hierarchies. GHSOM comprises several variable-size SOM layers. The number of SOM maps and map sizes are determined during the training process. In the segmentation process, the feature vectors selected from an image have a greater effect since the unusual characteristics can cause misclassification. The collection of discriminatory characteristics can enhance classification efficiency. Once GHSOM is qualified, classification is carried out using the method of probability labeling [25].

5.6 EGS-SOM

EGS-SOM (entropy gradient segmentation), which relies on statistical features such as first order, second order, etc. Under noisy and low strength conditions, EGS-SOM is stable and offers high-resolution images with good segmented outcomes. For segmentation, actual brain MR images from the Internet Brain Segmentation Repository (IBSR) are used.

5.7 SOM-FKM

In the method of segmenting MR brain image [25], the proposed hybrid SOM-FKM algorithm exceeds the FCM algorithm. In order to show the feasibility of the proposed algorithm, numerous clinical databases along with regular database images have been used. The similarities and drawbacks of earlier mapping methods applied are seen in Table 1.

5.8 AMSOM-FKM

The adaptive moving SOM is an extension of the SOM algorithm that allows a related competitive strategy used in testing to allow neurons to change locations. Suwalka et al. [27] proposed a hybrid technique in which neurons can be increased or decreased during the training phase in this strategy, allowing for a more versatile framework grid that can more effectively map the dataset with high accuracy results [28].

5.9 CNN

The images are first preprocessed using a filtering and enhancement process where noise is eliminated. The features are extracted to create a vector collection for large MRI data. The GLCM features are collected and fed as vector data. It transforms the image to a gray scale and then uses the skull removal algorithm to delete the skull and scalp. The region area of interest is extracted for which the features are performed using 3×3 voxel. The extracted features are first order and second order using the

256 × 256 Gy-level co-occurrence matrix (GLCM) from the MRI image. GLCM means the frequency of a gray-level pixel (intensity of grayscale) value I appears horizontally next to a pixel with the value j [29].

Each GLCM vector p(i, j) determines the number of times the i-value pixel has appeared horizontally adjacent to the j-value pixel. An extension to the 3-D CNN algorithm that requires neurons to change locations according to a similar competitive strategy employed in testing is the proposed solution. Using 3-D CNN, data reduction is conducted and the cluster is then fed to FKM. Neurons may be increased or decreased during the training phase in this technique, creating a more versatile structure grid and increasing the classification process. Using 24 different features, the dataset is generated. The data subjects used for the studies are actual OASIS collected datasets that include male and female subjects of various age groups [1, 22, 23, 30, 31].

6 Conclusion

Different comparative approaches differ in the process of precision and extraction of features. Machine learning has made notable progress in the process of data classification for a long time, largely thanks to the scientific community's efforts to improve. Better algorithms, whether binary methods or multi-class ones or machine learning, approach the gait classification of neurodegenerative disorders, hitting strong percentages in multiple controlled trials. With values ranging from 65 percent to 96.83 percent accuracy varies with the approach used. We demonstrated that the limit has been set with the conventional classification techniques using one machine learning algorithm, about percentages of proper classification. Certain unsupervised methods are monitored. The proposed multimodal CNN is for image analysis using mathematical modeling for ALZ diagnosis of non-local strength associations, the outcome of which is better in terms of modeling accuracy and predictive power than classical SOM-based multimodals.

References

1. Mandal Pravat K, Deepika S (2018) Brain metabolic, structural, and behavioral pattern learning for early predictive diagnosis of alzheimer's disease. 1 Jan 935–939
2. Anwar A, Iqbal A (2013) Image processing technique for brain abnormality detection. Int J Image Process 7:709
3. Fung Gleng S (2007) SVM feature selection for classification of SPECT images of alzheimer's disease using spatial information. Knowl Inf Syst 11:243–258
4. Gunna (2016) Implementation and comparision of different segmentation techniques for medical images. Int J Comput Appl 134:5–9
5. Isha S, Navneet A (2016) Assessment of segmentation techniques for neurodegenerative disease detection. In: International conference on circuit power computer technology (ICCPCT) IEEE explore digital diary, 1–5

6. Ortiz G, Ramirez L-E, Salas González D (2013) Two fully-unsupervised methods for MR brain image segmentation using SOM-based strategies. Appl Soft Comput Sci Direct 13:2668–2682
7. Ortiz P, Gorriz R, Salas G (2013) Segmentation of brain MRI using SOM-FCM based method and 3D statistical descriptors. Comput Math Methods Med Article ID 638563
8. Ortiz G, Ramirez SG (2013) Improving MRI segmentation with probabilistic GHSOM and multiobjective optimization. Neurocomput Elsevier 114:118–131
9. Dinesh P, Deore SG (2015) Medical image segmentation: a review. In: Proceedings of 17th international conference on system man & cybernatics held at New York during, 4–5, 22–27 June
10. Padilla P, Lopez M, Gorriz J, Ramirez J, Salas Gonzalez D, Alvarez I (2012) NMF-SVM based cad tool applied to functional brain images for the diagnosis of alzheimer's disease. IEEE Trans Med Imaging 31:207–216
11. Rafael CG, Richard EW (2007) Digital image processing. Beijing Publishing Housing of Electronics Industry, pp 689–785
12. Ramrez JM, Górriz F, Segovia R, Chaves D, Salas-Gonzalez M, López I, Álvarez PP (2010) Computer aided diagnosis system for the Alzheimer's disease based on partia. Neurosci Lett Elsevier 472:99–103
13. Rowayda S (2012) An improved MRI segmentation for atrophy assesment. Int J Comput Sci 6:4
14. Rowayda S (2013) Regional atrophy analysis of MRI for early detection of alzheimer's disease. Int J Signal Process Image Process Pattern Recogn 6:49–58
15. Rueda F, Gonzalez ER (2009) Extracting salient brain patterns for imaging-based classification of neurodegenerative diseases. IEEE Trans Med Imaging 31:1262–74
16. Dahshan S, Abdel-Badeeh MS, Tamer HY (2010) A hybrid technique for automatic MRI brain images classification. Digit Signal Process 20:433–441
17. Tagaris A, Kollias D, Stafylopatis A, Tagaris G, Kollias S (2018) Machine learning for neurodegenerative disorder diagnosis—survey of practices and launch of benchmark dataset. Int J Artif Intell Tools 27(3):1850011. https://doi.org/10.1142/S0218213018500112
18. Saini S, Arora K (2014) Study analysis on the different image segmentation techniques. Int J Inf Comput 4:1445–1452
19. Selvaraj DB, Kim T-H (2011) MRI brain image segmentation techniques—a review, Indian. J Comput Sci Eng 4:364-381
20. Segovia CB, Salmon E, Gorriz JM, Ramırez J, Phillips C (2014) Combining PET images and neuropsychological test data for automatic diagnosis of alzheimer's disease. 3PLOS ONE 9:e88687
21. Kalpana S, Neelesh G (2014) Medical image segmentation using modified K means clustering. Int J Comput Appl 103:16
22. Kohonen T (2001) Self-organizing maps. Third, Extended Edition. Springer series in information sciences, vol 30. Springer, Berlin, Germany. ISBN 978-3-540-67921-9
23. Li Y, Chi Z (2005) MR brain image segmentation based on self-organizing map network. Int J Inf Technol 11:45–53
24. Sonka, HB, Digital image processing and computer vision, pp 56–110
25. Suwalkal A (2017) An improved unsupervised mapping technique using AMSOM for neurodegenerative disease detection. Int J Comput Syst Eng
26. Zhang J, Sokal I, Peskind ER, Quinn JF, Jankovic J, Kenney C, Chung KA, Millard SP, Nutt JG, Montine TJ (2008) CSF multianalyte profile distinguishes alzheimer and parkinson diseases. Am J Clin Pathol 129(4):526–529. https://doi.org/10.1309/W01Y0B808EMEH12L
27. Suwalka I, Agrawal N (2019) A machine learning technique for tumour detection around edema region. In: IEEE Conference Proceeding, Second International Conference on Advanced Computational and Communication Paradigms (ICACCP-2019), 25–28 February 2019, SMIT, Sikkim

28. Govinda V (2016) Short notes on unsupervised learning method with clustering approach for tumor identification and tissue segmentation in magnetic resonance brain images. J Clin Exp Neuroimmunol 1:101
29. Gerhand W (2016) AMSOM: Adaptive moving self-organizing map for clustering and visualization. In: Proceedings of 8th international conference on agents and artificial intelligence
30. Raghavendra U, Bhat NS, Gudigar A, Acharya UR (2018) Automated system for the detection of thoracolumbar fractures using a CNN architecture. Futur Gener Comput Syst 85:184–189
31. Rajendra Acharya U, Fernandes SL (2019) Automated detection of alzheimer's disease using brain MRI images–a study with various feature extraction techniques. J Med Syst 43:302. https://doi.org/10.1007/s10916-019-1428-9
32. Lorio S, Fresard S, Adaszewski S, Kherif F, Chowdhury R, Frackowiak RS, Ashburner J, Helms G, Weiskopf N, Lutti A, Draganski B (2016) New tissue priors for improved automated classification of subcortical brain structures on MRI. NeuroImage 130:157–166. ISSN 1053-8119

Chapter 18
An Ensemble Approach for Classification and Prediction of Diabetes Mellitus Disease

Dhyan Chandra Yadav and Saurabh Pal

Abstract The machine learning methods are used to model several complex field datasets. Predictive analytics can help physicians make timely decisions about the health and care of patients. This research paper examined three different algorithms for machine learning and addressed finding the best classifier for this dataset for diabetes. For pattern recognition and data visualization, we have used box–whisker plots. We have used function-based classifier algorithms for data classification, such as multilayer perception and SGD. Two separate consequences are planned by these two classifiers. So the main objective of this paper by AdaBoostM1 ensemble techniques is to find better test results. Classification accuracy, precision, recall, and F1-score for the diabetes UCI dataset are calculated by these ensemble methods. Different learning strategies assist physicians and clinicians in the early prediction of diabetes in the prepared outcomes. After all the observation, we find the final effects of the AdaBoostM1 Ensemble process contained 98% accuracy.

Keywords Feature selection methods · Plot correlation matrix · Functional classifier algorithms · Multilayer perception · SGD and ensemble methods · AdaBoostM1

1 Introduction

For each human body, the right amount of glucose is most required. Diabetes mellitus is split into two sections:

- Type-1 and
- Type-2.

D. C. Yadav · S. Pal (✉)
Veer Bahadur Singh Purvanchal University, Jaunpur, UP, India

R. Mathur et al. (eds.), *Emerging Trends in Data Driven Computing and Communications*, Studies in Autonomic, Data-driven and Industrial Computing,
https://doi.org/10.1007/978-981-16-3915-9_18

1.1 Type-1 Diabetes Mellitus

Type-1 Diabetes Mellitus depends on the level of insulin and is known as diabetes. In the human body, insulin development relies on the pancreas. The pancreas does not manufacture type-1 Diabetes Mellitus insulin and the body's immune system is therefore unable to stay healthy, i.e., it is killed.

1.2 Type-2 Diabetes Mellitus

Type-2 Diabetes Mellitus does not regulate the amount of insulin such as type-1. The pancreas produces any amount of insulin, but this quantity is still not adequate to satisfy the needs of the body, so type-2 Diabetes Mellitus is considered to induce insulin resistance diabetes, also known as insulin-resistant cells. The below are some signs of Diabetes Mellitus:

- Excess of hunger, urine, fatigue, and incidence of being overweight.
- Excessive malnutrition is why cells are starving.
- Yeast illness is also increased.

1.3 Complications Associated with Diabetes Mellitus

The blood vessels collapse very rapidly due to excessive levels of sugar within the blood. The life of the tissue of the body is not improved due to a reduction in the blood supply of the body and the main body organs such as the heart, kidney, and nervous system are also not very healthy [1].

2 Related Work

In [2], thyroid disorder has been studied by machine learning algorithms. Blood pressure values reported by each patient were observed and accuracy is tested by linear regression based on the method of collection of features. The Random Forest tested 94.25% precision with the feature selection technique [2].

In [3], diabetes condition is investigated by using machine learning algorithms on the basis of glucose and blood pressure. SVM, KNN, DT, LR, and DA expected diabetes condition and measured 75% accuracy of SVM by SVM [3].

In [4], around type-2 diabetes by data mining algorithms is discussed. The decision tree and random forest algorithms analyzed the complications of collected type-2 Diabetes Mellitus disease dataset. They measured disease complications and found that in chosen algorithms, random forests work well [4].

In [5], the classifier decision tree and meta-classifiers addressed diabetes disorder. For stronger estimation, they used Naïve Bayes, Random Tree, ID3, and AdaBoost. Finally, the high precision measured by Random Forest is compared to ID33 [5].

In [6], diabetes by algorithms of neural network and regression has been observed. For the better students, they used ANN, GMM, SVM, Logistic regression, and ELM. Authors considered high precision determined by ANN relative to other algorithms [6].

In [7], fuzzy, CART, and cluster algorithms analyzed large diabetes datasets. To eliminate noise and improve performance, they used CART, PCA, and EM [7].

In [8], the tree classifiers, neural network, and filtered classifier algorithms addressed characteristics and instances of diabetes disorder. On various datasets, they measured 77.01% precision Naïve Bayes [8].

In [9], different diabetes datasets were evaluated by selection features and classifier algorithms. Hoeffding, J48, and tenfold cross validation were used by them. 77.0% precision of search algorithms was determined by Hoeffding [9].

In [10], various disease datasets using diabetes were discussed. HMV, NB AdaBoost, RR, SVM, KNN with LR, and the measured HMV algorithm are used to achieve a high precision of 78.08% compared to all other classifiers [10].

In [11], high-precision algorithms were analyzed by Machine Learning Random Forest on diabetes disease. For stronger estimation, they used the Random Forest and Decision Tree. Finally, the author observed the high precision measured by Random Forest relative to Decision Tree [11].

3 Methodology

The machine learning algorithms in a huge dataset are used to identify the pattern. In this research paper, to predict the diabetes disease dataset, we used different functional classifiers, Python language, and Weka tools. Study work is covered by

1. Data collection and attributes selection.
2. Functional classifiers.
3. Ensemble methods.
4. Results evaluation.

3.1 Data Description

Diabetes dataset is collected from UCI repository records. The class-level categories are as follows:

Class
0 500
1 268
dtype: int64

Total number of instances and their attributes with target variables are described below: (no. of instances: 768, no. of attributes: 9).

Used total attributes and their representation are described below:

'times_pregnant', 'plasma_glucose', 'diastolic_blood_pressure', 'tricep_skin_fold_thickness', 'serum_insulin', 'body_mass_index', 'diabetes_pedigree_function', 'age', 'class'

3.2 *Box and Whisker Plot*

The box and whisker plots display complex attributes of the dataset in graphical format and the distribution of measured attributes with whisker min, median, and max values. Standard values are provided as central values during motion with dynamic details and interquartile range attributes [12] (Fig. 1).

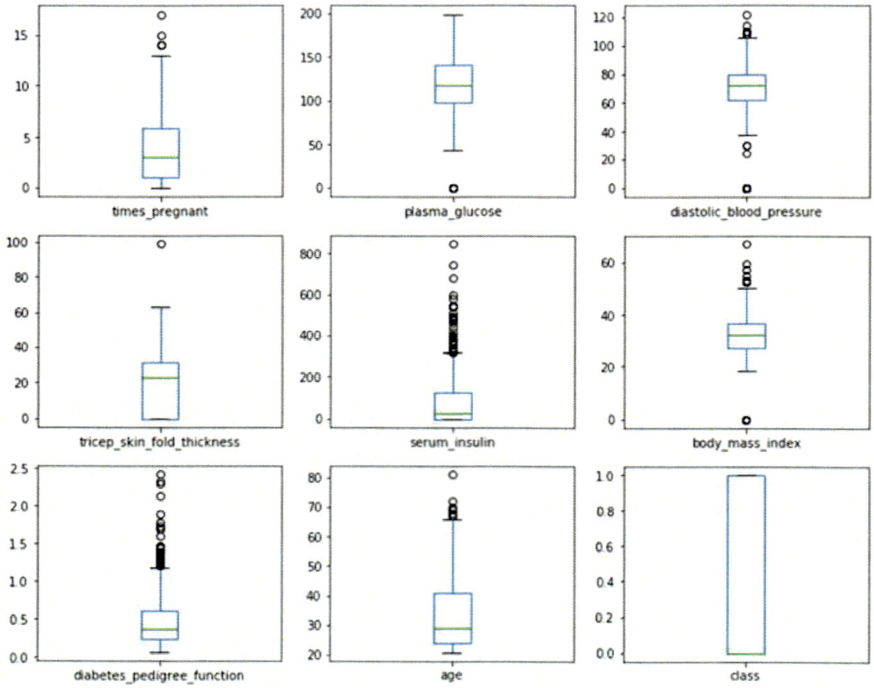

Fig. 1 Representation of box and whisker plot for attributes of diabetes patients

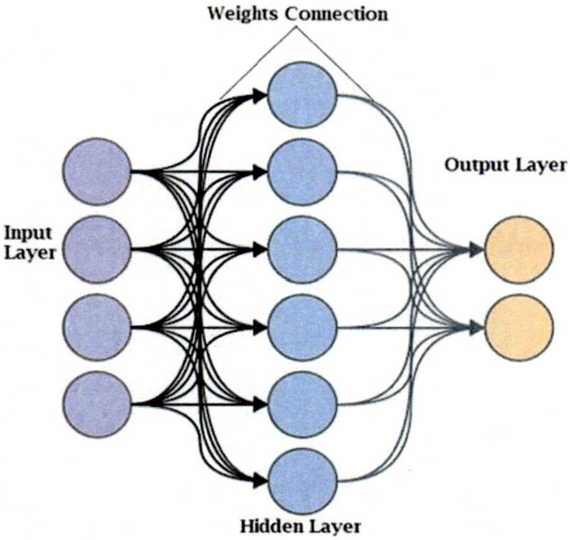

Fig. 2 Representation of multiplayer perception analysis

3.3 Algorithm Description

In this research work, we identify some important features with score values as follows.

3.3.1 Multilayer Perception

The multiplayer perception algorithm is developed by feedforward neural network classification techniques. The algorithm has several layers and a single perception layer. This solves a linear question but does not solve a nonlinear problem. The methods of the neural network include one or more hidden layers for pattern and other division recognition [13] (Fig. 2).

3.3.2 Stochastic Gradient Descent

On sparse data, gradient descent works. By descending down to the lowest point on the gradient, it creates a slope. The gradient descent involves approaches to batch, stochastic, and mini batch gradient descent. For the exact classification, each iteration produces the variance of the sample, preventing noise in the whole dataset [14] (Fig. 3).

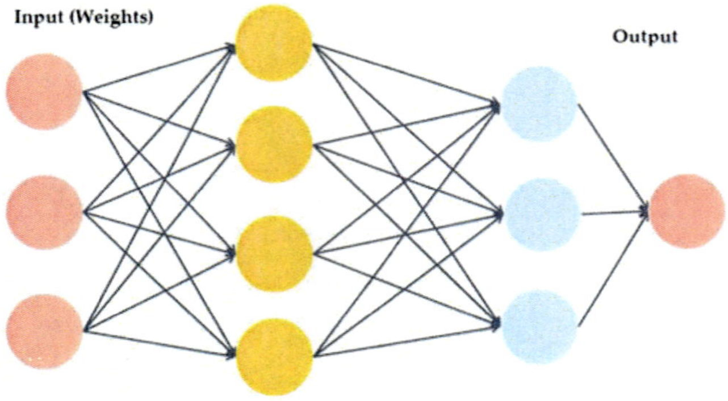

Fig. 3 Representation of stochastic gradient descent analysis

3.3.3 AdaBoostM1

AdaBoostM1 produces slow simple learners and a given dataset. Judge incorrect for greater weight after the more related data sample algorithms given. Each round is repeated as many rounds and then each prediction model has a high-precision bin [15] (Fig. 4).

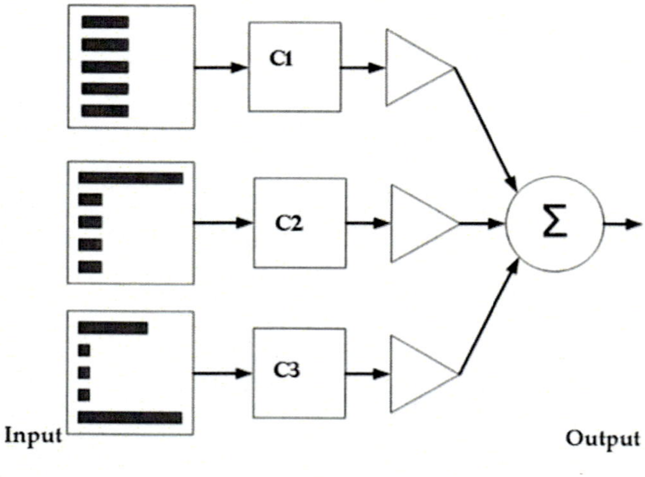

Fig. 4 Representation of AdaboostM1analysis

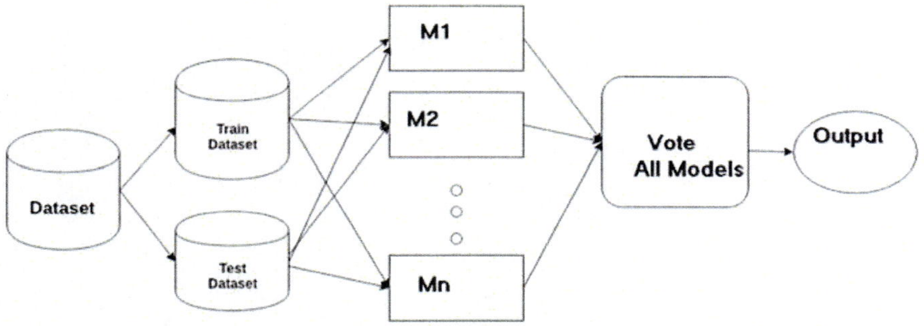

Fig. 5 Representation of proposed analysis model

3.4 Proposed Method

We have used the diabetes dataset from the UCI repository for this research model. By avoiding missing values, the data preparation was completed. The significant characteristics selected in the dataset with high score values for each attribute in sequence, box and whisker plot visualized the distribution of data frequency (Fig. 5).

For important features in the dataset, we use the heatmap. The heatmap generates links and priorities with appropriate correlations for better attributes. We have used functional algorithms: multilayer perception, SGD, and ensembles for better prediction (accuracy, precision, recall, and F1-score).

In statistical analysis, each model produces different values, so we use the voting algorithm to pick learners for better results.

4 Results and Discussion

In this research study, with AdaboostM1 ensemble methods, we have chosen best strategies for classifier algorithms. We picked datasets from the diabetes UCI registry with 768 instances and 9 attributes for the prediction. The result was evaluated by the various functional algorithms.

4.1 Correlation Heatmap

The heatmap shows the associations between data values in the color format. The first and second dimensions are represented by a row and a column. For all attributes and instances and calculated instances allocated by bin, the heatmap designs a mix of features associated with each other [16] (Fig. 6).

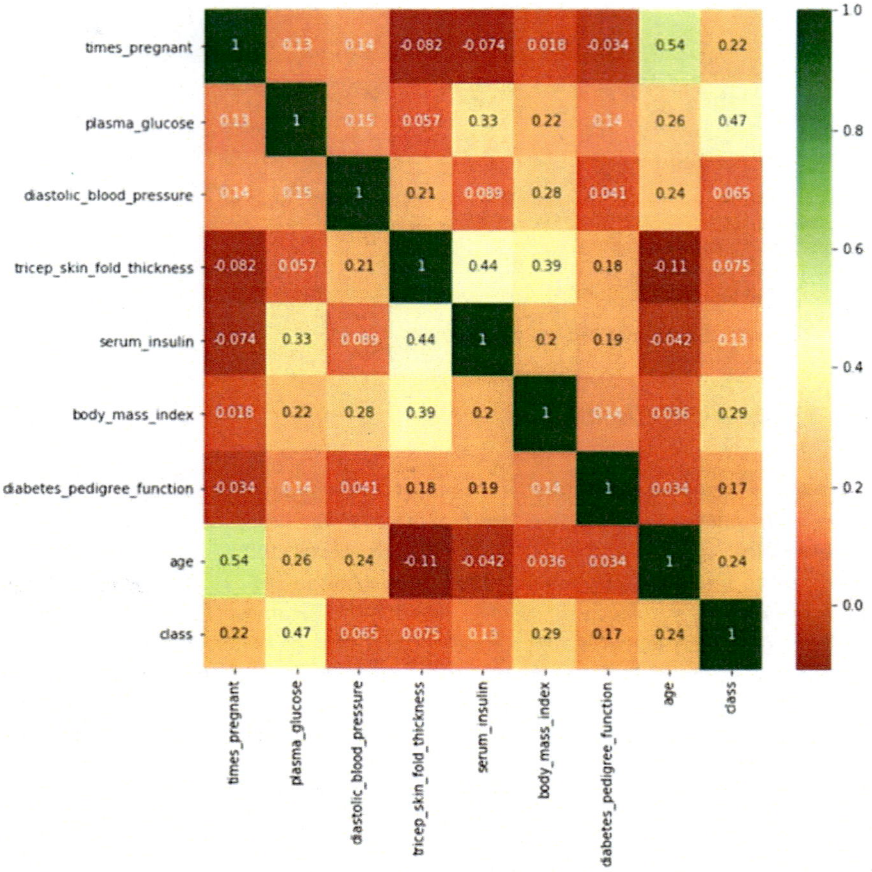

Fig. 6 Representation of heatmap for diabetes instances

After all the analysis, the accuracy determined by the classifier for AdaBoostM1 (98%) in Table 1 reflects the accuracy of classification for different model algorithms. Each model has high- and low-accuracy values, but compares to AdaBoostM1 they did not predict high accuracy. Compared to other chosen algorithms, Table.1 reflects

Table 1 Computational table for classification algorithms

Algorithms	Accuracy (%)	Precision (%)	Recall (%)	F-measure (%)
Multilayer perception	0.93	0.93	0.93	0.92
Stochastic gradient descent	0.92	0.92	0.92	0.91
AdaboostM1(ensemble)	0.98	0.98	0.98	0.97

Table 2 Representation of previous year papers

Authors	Algorithms	Finding accuracy (%)
Manoj et al. [17]	LR NB DT KNN SVM	82.35 76.62 75.97 66.23 64.28
Idemudia et al. [18]	PCA + k-Means K-Means	51.53 49.13
Aada et al. [19]	DT SVM AdaBoost L R	79.89 94.44 94.44 93.79
Saru and Subashree [20]	L R + SVM J48 KNN(K = 1) KNN(K = 3)	74.89 94.40 93.79 76.69
Abirami et al. [21]	M L P	97.61
Yang et al. [22]	H P M R F B N Logistic J48 MLP	93 89 88 85 85 85
Kaur and Kumari [23]	L K + SVM R B K SVM KNN ANN MDR	89 84 88 86 83
Sisodia et al. [24]	N B SVM D T	76.30 65.10 73.82

the importance of AdaBoostM1 and Judgment Stump for accuracy, recall, and F1-score are often higher.

From Table 1, algorithms such as multilayer perception and SGDs measured accuracy, precision, recall, and F-measure. With the results that it is obvious that AdaBoostM1 calculated high accuracy (98%) in the experiment compared to stochastic gradient descent and multilayer perception. The estimated value is not 100% but strongly examined relative to the list in the process (Table2).

With their forecast quality, we have listed (2018–19) previous paper in tabular form. All the papers mentioned have a mixture of precision separately and individually, but it did not cover 100% accuracy.

5 Conclusion

We also used diabetes condition datasets for UCI repositories with (768) instances and 9 attributes in this experiment. The correlation heatmap information approaches endorse a structured dataset format and dimension-related attributes. With the findings of statistical analysis, we used two functional-based classifiers multilayer perception, SGD, and ensemble method AdaBoostM1. We considered high precision compared to other algorithms determined by AdaBoostM1. For future work, with stacking ensemble techniques, we can analyze and evaluate Random Forest's complex dataset.

Acknowledgements The author is grateful to Veer Bahadur Singh, Purvanchal University Jaunpur, Uttar Pradesh, for providing financial support to work as Post Doctoral Research Fellow.

References

1. Henkel R (2020) Infection in infertility in male infertility. Springer, Cham, pp 409–424
2. Maniruzzaman M, Rahman MJ, Ahammed B, Abedin MM (2020) Classification and prediction of diabetes disease using machine learning paradigm. Health Inf Sci Syst 8(1):7. https://doi.org/10.1007/s13755-019-0095-z
3. Prasad CS, Hossain G, Goyal A, Bhan A, Bhattacharya S, Gaurav D, Tiwari SM (2020) Smart home health monitoring system for predicting type 2 diabetes and hypertension. J King Saud Univ-Comput Inf Sci 1–9. https://doi.org/10.1016/j.jksuci.2020.01.010
4. Muhammad Y, Munna MTA, Alam MM, Allayear SM, Ara SJF (2020) Prediction model for prevalence of type-2 diabetes mellitus complications using machine learning approach. In: Data management and analysis. Springer, Cham, pp 103–116
5. Weifeng X, Zhang J, Zhang Q, Wei X (2017) Risk prediction of type ii diabetes based on random forest model. In: 2017 third international conference on advances in electrical, electronics, information, communication and bio-informatics (AEEICB). IEEE, pp 382–386
6. Messan K, Li J, Zhai Y, Zhang X (2017) Application of data mining methods in diabetes prediction. In: 2017 2nd International conference on image, vision and computing (ICIVC). IEEE, pp 1006–1010
7. Mehrbakhsh N, Ibrahim OB, Ahmadi H, Shahmoradi L (2017) An analytical method for diseases prediction using machine learning techniques. Comput Chem Eng 106: 212–223
8. Rani AS, Jyothi S (2016) Performance analysis of classification algorithms under different datasets. In: 2016 3rd international conference on computing for sustainable global development (INDIACom). IEEE, pp 1584–1589
9. Mercaldo F, Nardone V, Santone A (2017) Diabetes mellitus affected patients classification and diagnosis through machine learning techniques. Procedia Comput Sci 2519–2528:112
10. Saba B, Qamar U, Khan FH, Naseem L (2016) HMV: a medical decision support framework using multi-layer classifiers for disease prediction. J Comput Sci 13:10–25
11. Prajwala TR (2015) A comparative study on decision tree and random forest using R tool. Int J Adv Res Comput Commun Eng 4(1):196–199
12. Yadav DC, Pal S (2020) Prediction of thyroid disease using decision tree ensemble method. Hum-Intell Syst Integr 2(1):89–95
13. Yadav DC, Pal S (2020) Calculating diagnose odd ratio for thyroid patients using different data mining classifiers and ensemble techniques. Int J 9(4):5463–5470

14. Mohammadi AM, Gündüz D (2020) Machine learning at the wireless edge: distributed stochastic gradient descent over-the-air. IEEE Trans Signal Process 68:2155–2169
15. Yadav DC, Pal S (2019) To generate an ensemble model for women thyroid prediction using data mining techniques. Asian Pac J Cancer Prev APJCP 20(4):1275–1281
16. Yadav DC, Pal SA (2020) Prediction of heart disease using feature selection and random forest ensemble method. Int J Pharm Res 12(4):56–66
17. Varma, KM, Panda BS (2019) Comparative analysis of predicting diabetes using machine learning techniques 6(6):522–530
18. Changsheng Z, Idemudia CU, Feng W (2019) Improved logistic regression model for diabetes prediction by integrating PCA and K-means techniques. Inform Med Unlocked 100179(17):1–7
19. Aada MTSA, Tiwari S (2019) Predicting diabetes in medical datasets using machine learning techniques. Int J Sci Res Eng Trends 5(2):257–267
20. Saru S, Subashree S (2019) Analysis and prediction of diabetes using machine learning. Int J Emerg Technol Innov Eng 5(4):167–175
21. Sengamuthu MR, Abirami MR, Karthik MD (2018) Various data mining techniques analysis to predict diabetes Mellitus. Int Res J Eng Technol (IRJET) 5(5):676–679
22. Wu H, Yang S, Huang Z, He J, Wang X (2018) Type 2 diabetes mellitus prediction model based on data mining. Inform Med Unlocked 10:100–107
23. Harleen K, Kumari V (2018) Predictive modelling and analytics for diabetes using a machine learning approach. Appl Comput Inform Learn 5(23):1–6
24. Deepti S, Sisodia DS (2018) Prediction of diabetes using classification algorithms. Procedia Comput Sci 132:1578–1585

Chapter 19
A Novel Approach for Identification of Brain Tumor by Combination of Intelligent Water Drop Algorithm and Convolutional Neural Network

Ashish Kumar Dehariya and Pragya Shukla

Abstract Brain tumor consists of malignant cells that harm brain, and hence its diagnosis and treatment are required for human wellness. In computer vision, tumor detection algorithms have been developed for identifying the tumor cells from normal brain cells. This work proposed a magnetic resonance image segmentation-based classification model for brain tumor detection. Training data for Convolutional Neural Network (CNN) was obtained from Intelligent Water Drop (IWD) genetic algorithm. IWD provides segmented cluster center pixel values of MRI image. Image was segmented as per cluster center pixels and blocked for training by CNN. Classification process was performed by CNN. Experiment was done on real brain tumor dataset. This model was compared with FCM clustering algorithm with rough set where obtained accuracy was 92.57%. Obtained accuracy of proposed model is 95.24% and hence result shows that proposed IWDCNN model has increased the pixel segmentation-based classification accuracy. Evaluated average values of some other metrics are precision = 0.971486, recall = 0.979, and F-measure = 0.975.

Keywords Digital image processing · Digital image segmentation · Intelligent water drop algorithm · Digital image classification · Convolutional neural network

1 Introduction

Biomedical image analysis helped doctors a lot in achieving perfect diagnosis [1] and it also helps several biologists in their discoveries [2, 3]. A brain tumor is a condition when there is the formation of unusual cells in the brain. These cells affect the functioning of the brain and create problems in the overall health of a person [4]. The main focus of several researchers nowadays is to use the technique of medical imaging to study the brain tumor patient so that it can be cured at an early stage as late detection of such cells has killed many patients in the entire world [5]. Till now detection of tumor cells in the brain of the patients is done

A. K. Dehariya (✉) · P. Shukla
Institute of Engineering and Technology, Devi Ahilya Vishwavidyalaya, Indore, India

R. Mathur et al. (eds.), *Emerging Trends in Data Driven Computing and Communications*, Studies in Autonomic, Data-driven and Industrial Computing,
https://doi.org/10.1007/978-981-16-3915-9_19

manually by experts who take a lot amount of time. One more problem with this is the differences in the MR image obtained by different experts. Because of these reasons, it was important to develop an automatic detection process to detect the brain tumor accurately and timely. Many approaches were proposed to achieve success and accuracy in the image segmentation, classification, and analysis process [9]. Proposed model works for brain MRI image segmentation using intelligent water drop algorithm then classification of tumor and non-tumor region is done by using convolutional neural network. IWD is a genetic algorithm which finds cluster center pixel values of the supplied brain MRI image. Based on these values MRI image get partitioned into clusters. Image clusters then get processed by CNN for classifying tumorous and non-tumorous brain.

2 Literature Survey

Nilesh Bhaskarrao Bahadure et al. [5] made an attempt to solve the complexities-related problem that comes during the image segmentation process. He worked on Berkeley wavelet transformation that helps to solve problems during the image segmentation process. He also proposed SVM (support vector machine) to improve the accuracy further to obtain better features from the segmented tissue. Mohammed Sahib Mahdi Altaei et al. [6] introduced a SIFT descriptor here to obtain MRI features of the brain to identify the tumor. An excellent brightened image was obtained by using this technique when the MRI image was rotated. He adopted two models to classify levels, i.e., use of Naïve Bays classifier to detect abnormal and normal tumor and use of J48 classifier to diagnose the abnormal ones that was found on the above level (benign or malignant cancer). Ashish et al. [7] used bio-geography-based optimization (ISBGO) to obtain image segmentation of brain MRI image. In this algorithm, every time accuracy of the random segment was reduced as compared with the same image set. Chithambaram et al. [8] proposed artificial neural network which works in vector quantization manipulation process to conduct the automated MRI scans in finding the tumors. Computational time, training performance, and classification accuracies were some of the parameters that were studied in this modified ANN. Two techniques were introduced by him for tumor detection, i.e., histogram thresholding and artificial neural network. Chenyu et al. [9] discussed various techniques that were proposed earlier related to the reduction of noise during ionizing radiation and then he proposed his own algorithm which can give high-quality images along with low dose and with improved noise reduction. He provided six different methods related to CT(LDCT). SMGAN-3d was also proposed by him which gives high-quality images than previous process. Taghizadeh et al. [10] proposed that the direct use of stereo radio surgery on the patient causes a harmful and noisy impact on the body in form of ionizing radiation. His work revolves within comparison of MRI sequence with SRS and CT technique. Hong huang et al. in 2019 used FCM clustering algorithm with rough set for the image segmentation. They constructed the attribute table by the values obtained from the FCM segmentation result and image is divided into the small areas on attribute basis. Weighted values were obtained by value reduction and used for the calculation of difference between the region and similarity of region. Later, model was realized through equivalence difference degree. Final values of equivalence degree were used to evaluate segmentation of images and merge regions.

3 Proposed Model

In proposed model, segmented image (Fig. 3) shows tumor segment, i.e., tumor and non-tumor segments were obtained by intelligent water drop algorithm that was blocked into mxm, two-dimension matrix. Each matrix was used for the training of convolutional neural network.

3.1 Image Pre-processing

Small dots noise of image can be removed by Wiener filter [11] so we used this filter for noise pixel value adjustment. Another image pre-processing step was skull stripping. The aim of this process is to separate the brain tissues from the non-brain intracranial tissues and skull. Figure 1 shows the sample of input MRI image and Fig. 2 is showing the pre-processed image.

Fig. 1 Input MRI image

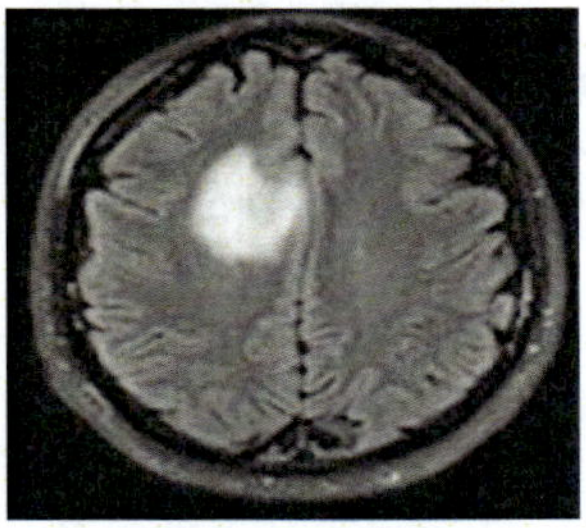

Fig. 2 MRI image after pre-processing

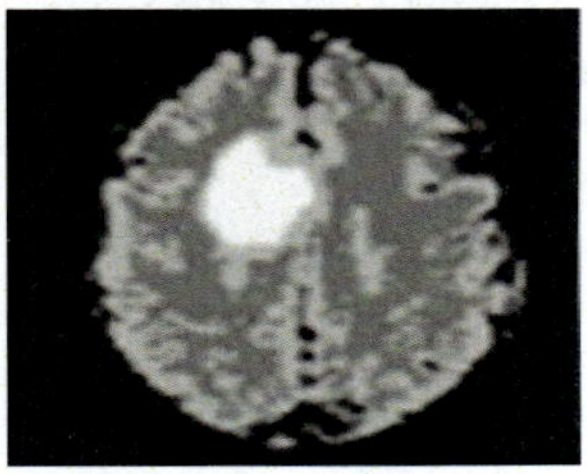

Fig. 3 Tumor region in MRI image

3.2 Intelligent Water Drop (IWD) Algorithm

In IWD, one drop moves toward other drop and forms a group as per low soil between the two merging drops. In this paper, drop is pixel and soil is pixel value difference. So soil values between each drop were summarized in a graph. As graph has nodes so unique pixel counts are presented as graph nodes. While distance of each node from other node acts as weight or soil in graph (calculated using Eq. 1)

$$Soil(x, y) = Euclidian(x, y) \tag{1}$$

where x, y are node values in graph Soil and Soil (x, y) is distance between pixels.

Static and Dynamic Parameter: In this step, some of constants were initialized before the start of algorithm such as soil updating parameters S1 = 1, S2 = 0.01, and S3 = 1 and velocity updating parameters V1 = 1, V2 = 0.01, and V3 = 1. Finally, global and local soil constants β_L and β_G are initialized by 0.9. The values of constants may vary as per algorithm requirement.

Population Generation: Random set of pixels were collected as cluster center candidates. Collection of these candidates were termed as population. Each candidate has two pixel value sets known as chromosome [12]. If population PW has n number of candidates, then Eq. 2 gives population set for processed image PI.

$$PW \leftarrow Population\_Generation(PI, n) \tag{2}$$

Drop Movement Probability: Association of a drop toward another drop depends on movement probability. So as per soil, weight value obtained from $WS(i, j) = Soil(i, j)$ if minimum (Soil (i, all element) > 0 Otherwise $WS(i, j) = Soil(i, j)-$ minimum (Soil(i, all element)). Movement probability DMP was evaluated by Eqs. 4 and 3 and feasible value solution (FS) is calculated as per soil.

$$FS(i, j) = \frac{1}{\delta + WS(i, j)} \tag{3}$$

$$DMP(i, j) = \frac{FS(i, j)}{\sum_{K=1}^{N} FS(i, j)} \tag{4}$$

Update Drop and Soil Values: Drop movement changes velocity of drops as per soil and n merging drop velocity. Equation 5 gives velocity update value for t^{th} iteration. Similarly, soil values were updated by Eq. 6. HD is heuristic durability, a constant value ranged between 0 and 1.

$$DV(t+1) = DV(t) + \frac{v_1}{v_2 + v_3 + Soil(i, j)^2} \tag{5}$$

$$\Delta S(i, j) = \frac{S_1}{S_2 + S_3 + T(t+1)^2} \tag{6}$$

$$T(t+1) = \frac{HD}{DV(t+1)}$$

$$Soil(i, j) = (1 - \beta_L) * Soil\ (i, j) - \beta_L * \Delta S(i, j)$$

Fitness Function: Each chromosome fitness was evaluated by Eq. 7. As per cluster center, other pixel values were grouped into minimum value cluster center. Summation of each pixel difference was done to get single fitness value, where F is fitness value and Pc,s is cluster center pixel in chromosome for s segments.

$$F_c = \sum_{x=1}^{Column} \sum_{y=1}^{Row} Min(P_{c,s} - I(x, y))^s \tag{7}$$

IWD Crossover: Some groups of chromosomes were prepared and good solution in a group was considered as the local parent which crossover with other chromosome in the group [13]. In crossover, one of the random position pixel values was copied from the local parent chromosome and replaces the same position of other chromosome pixel value of same group. Obtained new chromosome further evaluated to get its fitness value, if new chromosome fitness value was better than group chromosome then it replaces one of poor parent in population otherwise parent get continued.

3.3 Convolutional Neural Network (CNN)

Segmented image by IWD is further passed to CNN. CNN's two main operations are convolution and max-pooling. The main idea of these two operations is to leverage the geometric transformation invariant property of the image data [14]. Block b of segmented image was considered here as input.

$$B \leftarrow Block(SI, b) \tag{8}$$

Convolution: Image features extracted by the network should be position invariant. Thus, it is mostly unnecessary to use different weights for different parts of the image in the neural network. In CNN's convolution operation, we applied a small convolution mask on the 2D input via convolution. There are two main advantages of the convolution operation. (1) The 2D structure of image data is preserved. Thus, the convolution-based feature extraction is more effective than the fully connected operation. (2) Because different image areas share the same weight parameters, it greatly reduces the number of free parameters in the network. Hence, CNN is easier to train and less vulnerable to overfitting.

$$C \leftarrow Convolution(B,\ s,\ p,\ F_c) \tag{9}$$

Stride is movement speed control variable having integer values. Padding is null row or column added in the block if required. F is filter applied to the B.

Max-pooling: Objects in the images have various sizes, the receptive field of the extracted image features needs to be large enough to handle objects with different sizes. In biomedical application, we usually require a receptive field size around 200 $\times$ 200 pixels. A simple way to achieve large receptive fields is to stack the convolution operations. Within 3 $\times$ 3 convolution, we can achieve a receptive field of size (2 *n + 1) x (2 * n + 1) pixels. But, one can see that, to reach 200 $\times$ 200, we need to stack around 100 convolution operations, which is quite inefficient. To address this issue, the pooling operation is proposed. Its basic idea is to enlarge the receptive field by downsampling the image feature maps. In CNN, this downsampling is commonly achieved by maximum pooling or average pooling. After pooling, the image feature maps by a factor a, and the convolution operation is s times more effective in enlarging the receptive field. To gradually encode high-level image features, the pooling operations usually use s = 2 and the convolution operations and the pooling operations often work together in groups.

$$C \leftarrow Maxpooling(C,\ s,\ p,\ F_m) \tag{10}$$

Fully Connected Neural Network: Convolution operations and the pooling operations allow CNN to power the spatial information from the input space, and the per-pixel classification formulation prevents the network to utilize the spatial correlation in the output space. To address this problem, fully connected neural network (FCNN) [15, 16] with sigmoid activation function was proposed. Instead of computing the classes of the pixels individually, FCNN directly outputs the segmentation of the entire input image. Notice that the convolution operations and the pooling operations preserve the image structure in every hidden layer prior to the fully connected operations and these hidden feature maps have complete information for every pixel in image. Input block B obtained after convolutional operation was passed as training vector in FCNN and desired output was tumor or non-tumor class. Trained CNN directly accepts processed blocked image as input and predicts tumor or non-tumor class.

4 Experiment and Results

For experiments, brain tumor MRI dataset was taken from the database maintained on URL https://ijsret.com/2017/12/14/computer-science/. Here actual brain MRI images and related ground truth images are stored. Combining both type of images was considered as 1 set. The model was developed using MATLAB. FCMRS model in [17] was considered for comparative analysis. As results are concerned, accuracy, precision, recall, and F-measure value-based comparison are shown in Tables 1, 2, 3, and 4, respectively. It is found that proposed IWDCNN result was improved as compared to FCMRS model. Reduction of noise by Wiener filter increased the work performance. Use of convolution model for image segment classification improved the accuracy, precision, recall, and F-measure value-based performance.

Table 1 Accuracy value-based comparison

Images (Set)	FCMRS	IWDCNN
1	90.0467	90.825
2	97.1351	98.2105
3	91.1805	93.4383
4	91.6686	97.8317
5	92.805	95.9096

Table 2 Precision value-based comparison

Images (Set)	FCMRS	IWDCNN
1	0.9163	0.91908
2	0.98271	0.985649
3	0.97098	0.973736
4	0.99285	0.993247
5	0.92263	0.985716

Table 3 Recall value based comparison

Images (Set)	FCMRS	IWDCNN
1	0.98203	0.9863
2	0.988461	0.9963
3	0.936821	0.958
4	0.922492	0.9848
5	0.993147	0.9722

Table 4 F-measure value-based comparison

Images (Set)	FCMRS	IWDCNN
1	0.947231	0.9515
2	0.985446	0.9909
3	0.953573	0.9658
4	0.956377	0.989
5	0.956583	0.9789

5 Conclusion

Intelligent water drop algorithm was utilized for identifying representative pixel sets for segmenting tumor/non-tumor region of the brain MRI images. Using representative pixel set, segmented image was obtained that further gets blocked for preparing the training vector of convolutional neural network. In CNN, convolution and max-pooling operator increased the efficiency for learning. Experimented result revealed that proposed model has increased the accuracy of tumor detection by 2.68% as compared to FCMRS algorithm. In future, scholars can adopt other genetic algorithm for increasing the segmentation accuracy of work. CNN capability of learning can be increased by increasing number of layers in its deep architecture.

References

1. Filho PPR, da Silva Barros AC, Almeida JS, Rodrigues JPC, de Albuquerque VHC (2019) A new effective and powerful medical image segmentation algorithm based on optimum path snakes. Appl Soft Comput J 76:649–670
2. Brain tumor: introduction cancer.net. https://www.cancer.net/cancer-types/brain-tumor/introduction. Accessed 25 Feb 2019
3. Brain tumors-classifications, symptoms, diagnosis and treatments. https://www.aans.org/Patients/Neurosurgical-Conditions-and-Treatments/Brain-Tumors
4. Li M, Kuang L et al (2019) Brain tumor detection based on multimodal information fusion and convolutional neural network. IEEE Access 7:180134–180146
5. Bahadure NB, Ray AK, Thethi HP (2017) Image analysis for MRI based brain tumor detection and feature extraction using biologically inspired BWT and SVM. Int J Biomed Imaging Hindwai
6. Altaei MSM, Kamil SY (2020) Brain tumor detection and classification using SIFT in MRI images. AIP conference proceedings, p 2292
7. Dehariya A, Shukla P (2020) Brain cancer prediction through segmentation of images using bio-geography based optimization. Int J Adv Res Eng Technol 11(11)
8. Chithambaram T, Perumal K (2016) Edge detection algorithms using brain tumor detection and segmentation using artificial neural network techniques. Int Res J Adv Eng Sci 1(3):135–140
9. Chenyu Y (2018) Structurally-sensitive multi-scale deep neural network for low-dose CT denoising. IEEE access 6:41839–41855
10. Taghizadeh S et al (2019) Optimizing MRI sequences and images for MRI-based stereotactic radiosurgery treatment planning. Rep Pract Oncol Radiother 24(1):12–19

11. Benesty J et al (2010) Study of the widely linear Wiener filter for noise reduction. In: International conference on acoustics, speech and signal processing. Dallas, TX, 20104
12. Dadaneh BZ, Markid HY, Zakerolhosseini A (2015) Graph coloring using intelligent water drops algorithm. In: 23rd Iranian conference on electrical engineering. Tehran
13. Alijla Basem O, Lim CP et al (2018) An ensemble of intelligent water drop algorithm for feature selection optimization problem. Appl Soft Comput 65
14. Chen H et al (2016) DCAN: deep contour-aware networks for accurate gland segmentation. In: IEEE conference on computer vision and pattern recognition, pp 2487–2496
15. Baur C, Albarqouni S, Navab N (2017) Semi-supervised deep learning for fully convolutional networks. In: International conference on medical image computing and computer-assisted intervention, pp 31–319
16. Chen J, Yang L, Zhang Y, Alber M, Chen DZ (2016) Combining fully convolutional and recurrent neural networks for 3D biomedical image segmentation. In: Advances in neural information processing systems, pp 3036–3044
17. Huang H, Meng F, Zhou S, Jiang F, Manogaran G (2019) Brain image segmentation based on FCM clustering algorithm and rough set. IEEE access special section on new trends in brain signal processing and analysis, Feb 6

Chapter 20
Time Restricted Frame Aggregation for Better QoS in IEEE 802.11ac

Mayank Patel, Saurabh Srivastava, Harshita Jain, and Milind D. Jain

Abstract The audio transmission over WLAN requires the better Quality of Service (QoS). In this paper, we are proposing and implementing the novelty adaptive mapping of cross-layers scheme for height the Quality of Services for transmitting the scalable video streams in IEEE 802.11ac wired less environment tends to approximating of the AC's queue length and delaying of the timely constraining of the streams for which it is to be transmitted. The results clearly states that performing of propose dynamic cross-layers mapping is better with respect to the mapping of static cross-layers scheme.

Keywords IEEE 802.11ac · EDCA · H.264/SVC · Mapping of cross-layers · Quality of services

1 Introduction

The Aggregation mechanism is used to accumulate many of the frames before transmitting into single large frames which reduces overhead and increase the efficiency and throughput. IEEE 802.11n WLAN has four Access Categories (AC) at its MAC layer [1]. Each AC has individual buffers forming the queue. Each AC is reserved for different traffic and works as an independent queue [2]. The queues of different traffic class have different priorities [3]. The queue with higher priority gets the transmission medium access first. The aggregation headers cause supplementary delaying particularly when aggregate in the frequent packets with the small sizes.

M. Patel · S. Srivastava · H. Jain (✉) · M. D. Jain
Geetanjali Institute of Technical Studies, Udaipur, RJ, India

Rajasthan Technical University, Kota, RJ, India

R. Mathur et al. (eds.), *Emerging Trends in Data Driven Computing and Communications*, Studies in Autonomic, Data-driven and Industrial Computing,
https://doi.org/10.1007/978-981-16-3915-9_20

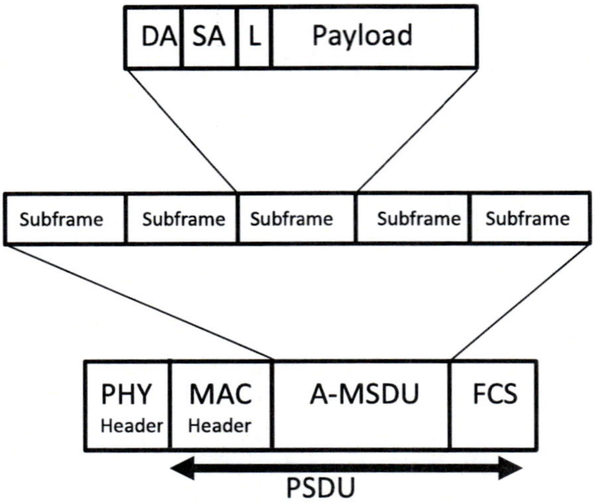

Fig. 1 MSDU aggregation

1.1 Frame Aggregation

The Frame aggregation increases the throughput as it helps in sending multiple data frames in a single transmission [4]. Techniques used by single-block of aggregation algorithm performed by using single-block of acknowledgement (ACK) frame to exchange multiple MPDUs [5].

1.2 MSDU Aggregation

Principle allows sending of multiple MSDU to same receiver such that it can be concatenated in a single MPDU. The maximum size of conventional and high throughput (HT) are either 3839 bytes or 7935 bytes, respectively, while for VHT there is no con-strain to them a size of AMSDU [6] (Fig. 1).

2 Proposed Methodology

The aggregation of frames effectively works in terms of saturating traffics, else frames wait for the next or upcoming packet increasing delivery delays [7]. We propose EFAS Method for best and efficient delivery of audio packet with minimal delay [8]. This method describes flow of methodology in which new frame aggregation limit is fixed according the traffic load. At first, A-MPDU is constructed and NFA (ar) fixed after MAC traffic being analyzed [9]. On comparing frame sequence with frame aggregation count, final A-MPDU transmitted once when the condition becomes true [10] (Fig. 2).

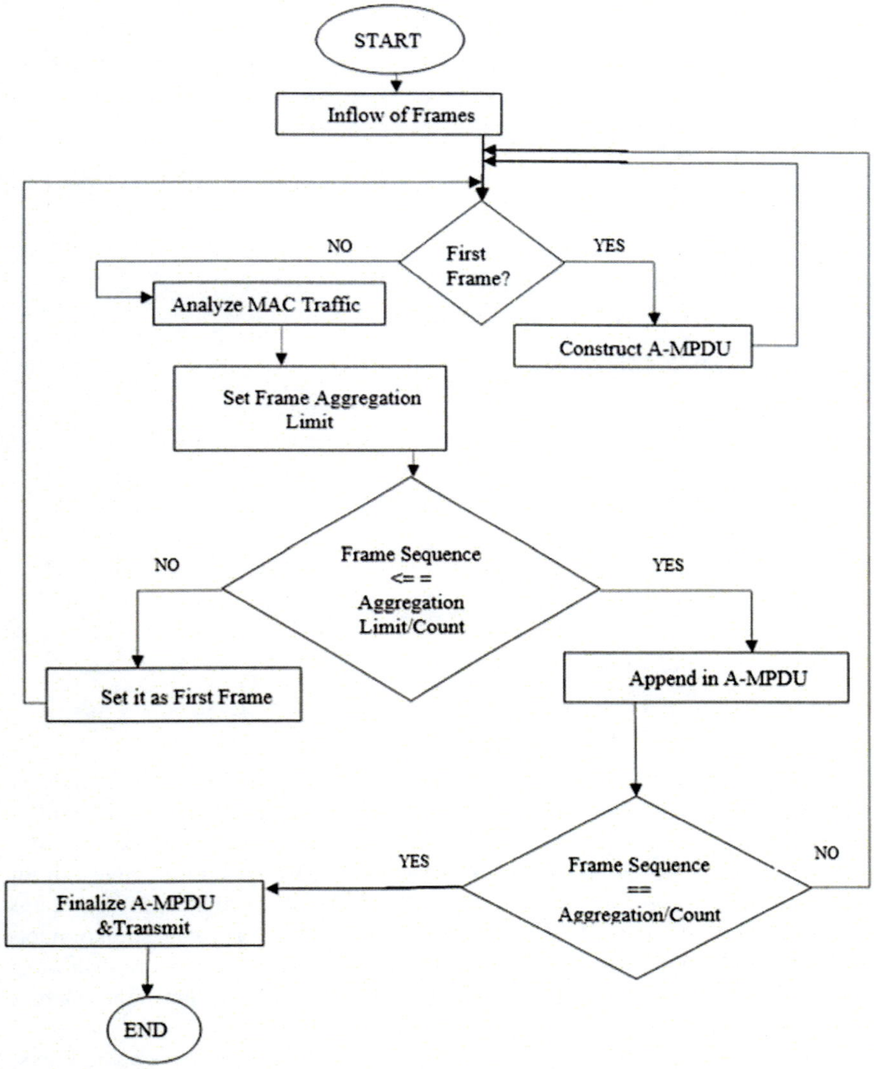

Fig. 2 Flowchart of EFAS Method

3 Algorithm

Step 1: Start

Step 2: Store Inflow Frame

Step 3: Check A-MPDU

Step 4: If A-MPDU is empty
Insert frame in A-MPDU;
Go to Step 2;

Sep5: Else
Analyze MAC Traffic;
Set New Frame Aggregation Limit (Ar)
NFA (ar) = Max. A-MPDU limit – Buffer size

Step 6.A: IF A-MPDU Size < Ar
Insert Frame in A-MPDU;
A-MPDU Size + +;
Else if
Frame Sequence = = Ar
Finalize A-MPDU and Transmit;
End

Step 7: Goto Step 2;

4 Experimental Results

We classify the network traffic in generic terms as below: Light Load, Medium Load, Heavy Load & Highly Heavy Load [11]. The above suggested network congestions scenarios are mainly classified by measuring the traffic [12]. The network handling only one audio stream is considered as having no traffic congestion or having minimum network traffic load [13]. Transmitting of two or three audio streams concurrently over the network is classified as moderate network congestion or intermediate traffic load that is in permissible limit [14]. Transmitting more than three audio streams concurrently are classified as heavy network congestion or the network load that is not suitable for the better audio QoS [15].

In our proposed methodology, the dropping of audio frame from the audio access category due to unavailability of audio buffers space has been rectified. The more audio buffers in turn leads to less re-transmission of arrived audio frames at MAC [16]. In very high load, the increased buffer space along with higher priority frame transmission policy leads to fluent transmission from audio queue. The above features of proposed methodology shows better Jitter result in comparison with the standard protocol. In highly heavy load, the methodology reduces Jitter compared to standard protocol. In Heavy load and medium load, the proposed methodology performs better, where it reduces Jitter respectively [17]. Figure 3 is showing the comparing of PDR termed as packet delivery ratio in standard protocol and methodology at highly heavy, Heavy, Medium, Light load.

In Highly Heavy load and Heavy load, the methodology performs better than standard IEEE 802.11ac protocol due to swift transmission of data packets which in results reduced End-to-End delay respectively by EFAS methodology compared to Standard protocol [18]. In medium load and light load, the proposed methodology reduces the End-to-End delay respectively with compared to standard protocol. Therefore, End-to-End delay for proposed EFAS methodology is closer to standard protocol. Figure 4 clearly shows the comparing of Jitter in standard protocol and EFAS methodology at highly heavy, heavy, medium and light load.

Due to unavailability of audio buffers the dropping of audio frame from the audio access category has been rectified. The more audio buffers in turn leads to less retransmission of arrived audio frames at MAC. In highly heavy load, the increased buffer space along with higher priority frame transmission policy leads to fluent transmission from audio queue. The above features of proposed methodology shows better Jitter result in comparison with the standard protocol. In highly heavy load, the methodology reduces Jitter compared to standard protocol. In Heavy load and medium load, the proposed methodology performs better, where it reduces Jitter. Figure 5 shows the comparison of PDR (packet delivery ratio) in standard protocol and methodology at highly heavy, Heavy, Medium and Light load.

By predicting the current traffic flow and future audio injection rate, the decision is taken to borrow buffers from other queues. In Highly Heavy and heavy traffic load, the methodology increases packet delivery ratio on average with standard protocol.

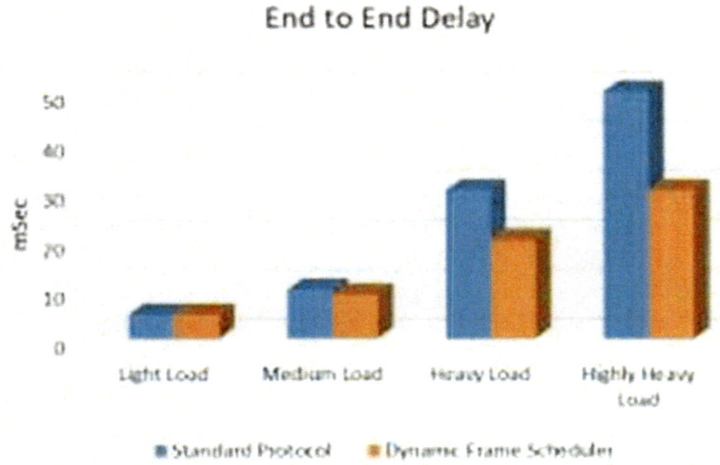

Fig. 3 Comparing the end-to-end delay of standard protocol and DFS at highly heavy, heavy, medium, light load

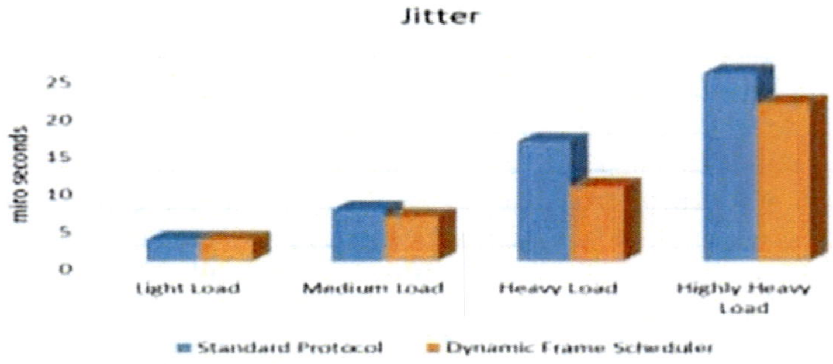

Fig. 4 Comparing of jitter of standard protocol and DFS at highly heavy, heavy, medium and light load

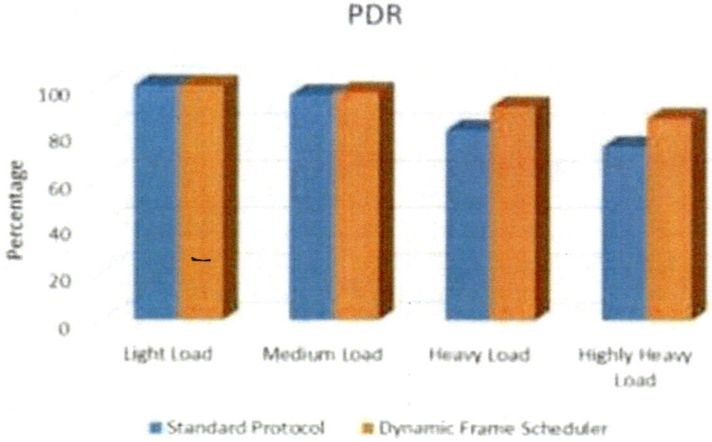

Fig. 5 Comparison of packet delivery ratio (PDR) of standard protocol and methodology at highly heavy, heavy, medium, light load

Thus by saving the audio frames from being dropped out from audio queue, the packet delivery ratio of the proposed methodology shows better results than standard protocol of IEEE802.11ac. In medium and light load on traffic, methodology increases packet delivery ratio on average compared to standard protocol.

As the traffic congestion decreases, audio buffer length becomes same to the standard audio buffer length (Table 1).

Table 1 Table contents of the light, medium, heavy and highly heavy load

Methodology		Light load	Medium load	Heavy load	Highly heavy load
Standard protocol	End-to-End Delay (mili Seconds)	5	10	30	50
	Jitter (micro seconds)	3	7	16	25
	PDR (Percentage)	100	96	81	74
DFS	End-to-End Delay (mili Seconds)	5	9	20	30
	Jitter (micro seconds)	3	6	10	21
	PDR (Percentage)	100	97	91	86

5 Conclusion

We made an approach to make the dynamic scheduler for high throughput WLAN. We will deeply study the advantages and limitations of aggregation mechanism. Our main focus is reducing waiting time delays and optimizing aggregated frame size.

References

1. Charfi E, Chaari L, Kamoun L (2013) PHY/MAC Enhancements and QoS mechanisms for very high throughput WLANs: a survey. IEEE Commun Surv Tutor 15(4). The fourth quarter of 2013
2. Nojima D, Lanante L, Jr Nagao Y, Kurosaki M, Ochi H (2012) Performance evaluation for multi-user MIMO IEEE 802.11 ac wireless LAN system. In: Proceedings of the IEEEICACT, pp 804–808
3. Choudhary N, Patel M (2014) QoS enhancements for video transmission over high throughput WLAN: a survey. Int J Res Sci Innov 1(7):43–50
4. Perahia E, Gong MX (2011) Gigabit eireless LANs: an overview of IEEE 802.11ac and 802.11ad. ACMSIGMOBILE Mob Compu Commun Rev 15(3):23–33
5. Gong MX, Perahia E, Stacey R, Want R, Mao S (2010) A CSMA/CA MAC protocol for multi-user MIMO wireless LANs. IEEE Globe com, pp 1–6
6. Patel M, Choudhary N (2017) Adroit buffer augmentation for reliable video trans-mission over WLAN. Indian J Sci Technol 10(22):1–13. 10.17485/ ijst /2017/v10i22/113962
7. Gong MX, Hart B, Xia LF, Want R (2011) Channel bounding and MAC protection mechanisms for 802.11ac. IEEE Globecom2011, pp 1–5
8. Ong EH, Kneckt J, Alanen O, Chang Z, Huovinen T, Nihil T (2011) IEEE802.11ac:enhancements for very high throughput WLANs. IEEE PIMRC, pp 849–853
9. Park M (2011) IEEE802.11ac: dynamic bandwidth channel access

10. Chang Z, Alanen O, Huovinen T, Nihtila T, Ristaniemi EHOT (2021) Performance analysis of IEEE802.11ac DCF with hidden nodes
11. Patel M, Choudhary N (2017) Time restraint tractable aggregation scheme for beneficial video transmission over WLAN. J Eng Appl Sci 12(23), 7170–7179. https://doi.org/10.36478/jeasci.2017.7170.7179
12. Madan R, Sampath A, Khude N (2012) Enhancing 802.11 carrier sense for high throughput and Qo Sin Dense user settings
13. Charfi E, Chaari L, Kamoun L (2012) Upcoming WLANs MAC access mechanisms: an overview
14. Patel M, Choudhary N, Performance evaluation of MPEG-4 and H.264/SVC video codec over WLAN. Int J Res Appl Sci Eng Technol (IJRASET) 5(VI):899–905. ISSN :2321–9653
15. Zhu CH, Ngo C, Bhatt A, Kim YS (2013) Enhancing WLAN back off procedures for down link MU-MIMO support
16. Patel M, Choudhary N (2017) A dynamic cross-layer mapping to support real time multimedia applications over IEEE802.11n. In: 2017 international conference on energy, communication, data analytics and soft computing (ICECDS), Chennai, 2017, pp 3474–3478. https://doi.org/10.1109/ICECDS.2017.8390107.
17. Milos J, Polak L, Slanina M (2017) Performance analysis of IEEE 802.11ac/ax WLAN technologies under the presence of CFO,©2017 IEEE
18. Sharon O, Alpert Y (2017) Single user MAC level throughput comparison: IEEE 802.11axvs.IEEE802.11ac

Chapter 21
Self-Powered Intelligent Wastewater Treatment System with IOT-Based Monitoring for Smart Cities

Mohammad Ateek Samma, Shubham Vagrecha, Rajeev Mathur, Mr. Latif Khan, Lokesh Sharma, and Sheetal Sharma

Abstract Now-a-days, the availability of potable or clean water is a major concern all over the world. All the water resources like rivers, lakes, ponds etc. are polluted by the industrial wastes and sewerage. Few wastewater treatment techniques are used to clean the water using physicochemical treatment and biological treatment processes. Coagulation and flocculation processes are used to remove the pollutants. We propose this project to have a water treatment plant in every house, so that reuse of wastewater could become a reality and water scarcity problem could be solved.

Keywords Sewage · Wastewater treatment · IOT · Wi-Fi · MLD

1 Introduction

1.1 Background

The World Water Development Report stated in the 2018 edition that nearly 6 billion people will suffer from scarcity of clean water by 2050. It will be because of the ever-increasing demand for water, depletion of water resources, and increasing pollution of water, which is escalated by increase in world population and infrastructural development of many countries. The report promotes the adoption of nature-based solutions, but there is an earnest need to manage demography and economy while upholding clear principles to restrict contamination, protect springs, and save water wherever on the planet [1].

The greater part of this populace development is normal in Asian and African agricultural nations. A review of the world water problem is made based on the literature available for the year 2018. Study related to the scarcity of water and water pollution is made, which was a motivation factor to design and develop a wastewater treatment system [2].

M. A. Samma (✉) · S. Vagrecha · R. Mathur · Mr. L. Khan · L. Sharma · S. Sharma
Geetanjali Institute of Technical Studies, Udaipur, Rajasthan, India

Geetanjali Institute of Technical Studies, Udaipur, RTU, Kota, Rajasthan, India

R. Mathur et al. (eds.), *Emerging Trends in Data Driven Computing and Communications*, Studies in Autonomic, Data-driven and Industrial Computing,
https://doi.org/10.1007/978-981-16-3915-9_21

From: Reassessing the projections of the World Water Development Report

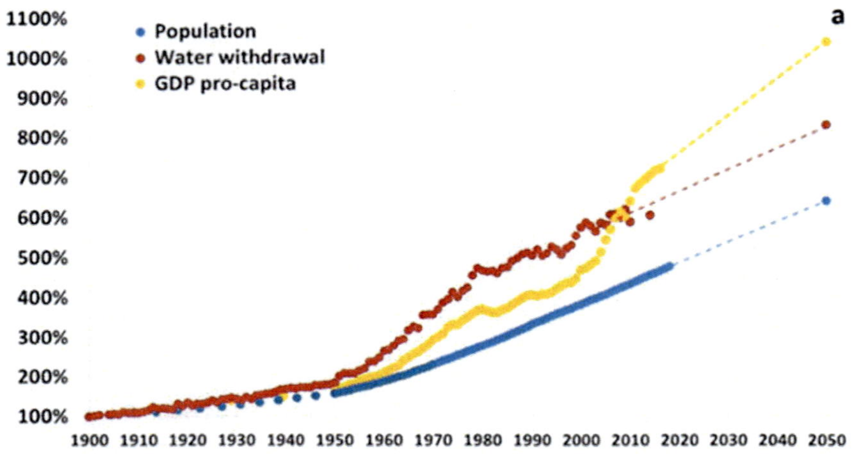

Fig. 1 Graph of water withdrawal, GDP pro-capita, and world population [1]

1.2 Water Demand

Numerous nations are as of now encountering water shortage conditions. A lot more count attempts will confront a decreased accessibility of surface water assets by 2050. About 73% of individuals influenced by water shortage as of now live in Asia [3].

A projection of population growth and water withdrawal in the world is as given in Fig. 1, which clearly depicts that the gap between the two is going to be very large and very dangerous to mankind [4].

Non-linear lines indicate that water crisis may also be there even before 2050 [1].

1.3 Water Pollution

Water pollution by mankind, at the same time, is increasing the problem. There is, thus, an urgent need to address the problem and find out ways and means to have an effective and cheap water treatment process. It is suggested here to have water treatment plant in every house, so that reuse of wastewater could become a reality [5].

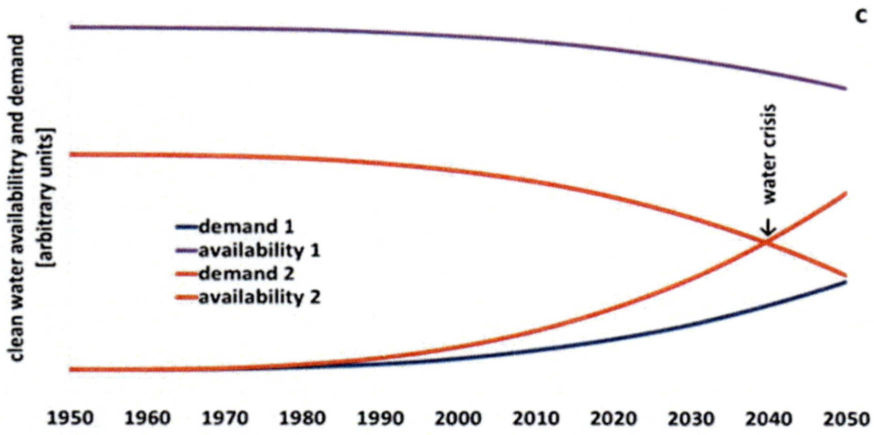

Fig. 2 Graphical concept of water scarcity and clean water availability [1]

1.4 Wastewater Treatment

In this paper, we have discussed the use of smart technology to develop a domestic wastewater treatment system with smart quality monitoring features [1] (Fig. 2).

2 Proposed Wastewater Treatment System

2.1 Self-Powered Intelligent Wastewater Quality Monitoring System (Real Time)

We can compute the physical and chemical parameters of the water such as temperature, pH, turbidity, conductivity, TDS etc. using the latest technology and various sensors, and real-time monitoring can be done based on wireless sensor network and Internet of Things (IoT). Furthermore, this data can be shared with the concerned local municipal authority to improve the water quality.

2.2 Methodology

The objective of the proposed system "Self-Powered Intelligent Wastewater Treatment System" is to purify domestic/municipal wastewater and reuse it. This system will be a fundamental component of a future smart city idea, which is a promising answer to resolve the issue of wastewater treatment [6]. Our goal is to address the wastewater treatment issues utilizing cutting-edge technology like Internet of Things (IoT).

We have proposed the project that has been selected as one of the smart cities under the ministry of Urban Development (MoUD), Government's Smart City Mission. The entire monitoring of all the relevant parameters of water and its online display is based on Internet of Things (IOT). In the proposed project, we will collect wastewater from colony or house in a container. This water will go through primary and secondary stages in the process of its purification. Different types of chemicals are also used such as:

1. Alum.
2. Calcium Hydroxide.
3. Bleaching Powder.

When we mix bleaching powder and water, it gives off chlorine and chlorine is toxic for the microorganisms present in water and kills them [7, 8]. This water purification process can be implemented at the place where sewage wastewater gathers. Very first the dirty water will pass through stainless steel mesh. It will filter large suspended bodies and the dust that is coming with water.

Simultaneously, by using the sensors, we will measure Turbidity, pH, Density, TDS, Temperature etc. in the LCD display as well as in Mobile. Then water will go to Primary Stage where Chemical Reaction takes place. Then, secondary stages will start and the water will pass through PP cotton, a type of cotton role that will stop the dust and powder particles, Pre Carbon Filter, Post Carbon Filter, Sediment Filter, and UV Stage. After passing through these chambers, we get the Pure Water, and the Pure Water will gather in another container. The pH, TDS, Turbidity, Temperature, Water Flow (in LPM) of purified water will be shown in LCD display as well as in Mobile. Energy required to run the system will be delivered through solar plate. So, it is a Eco-friendly System.

The complete process will be monitored through messages received on the mobile with the help of IoT applications. Following sensors along with the Ardiuno board will be used for online monitoring (Fig. 3).

3 Results of Prototype Model

First, we have taken two water collecting drums. In one drum, a stainless steel wire mesh screening system is employed as shown in Figs. 4 and 5. Screening removes coarse solid-suspended materials such as rags, paper, plastics, and metals to prevent damage and clogging of downstream equipment, sensors etc.

In the first tank, various sensors are installed, so that quality parameters of received wastewater are properly monitored and also characteristic parameters of water like pH; turbidity; TDS; temperature by using sensors like pH sensor, TDS sensor, turbidity sensor, temperature sensor are monitored. Measured parameters are

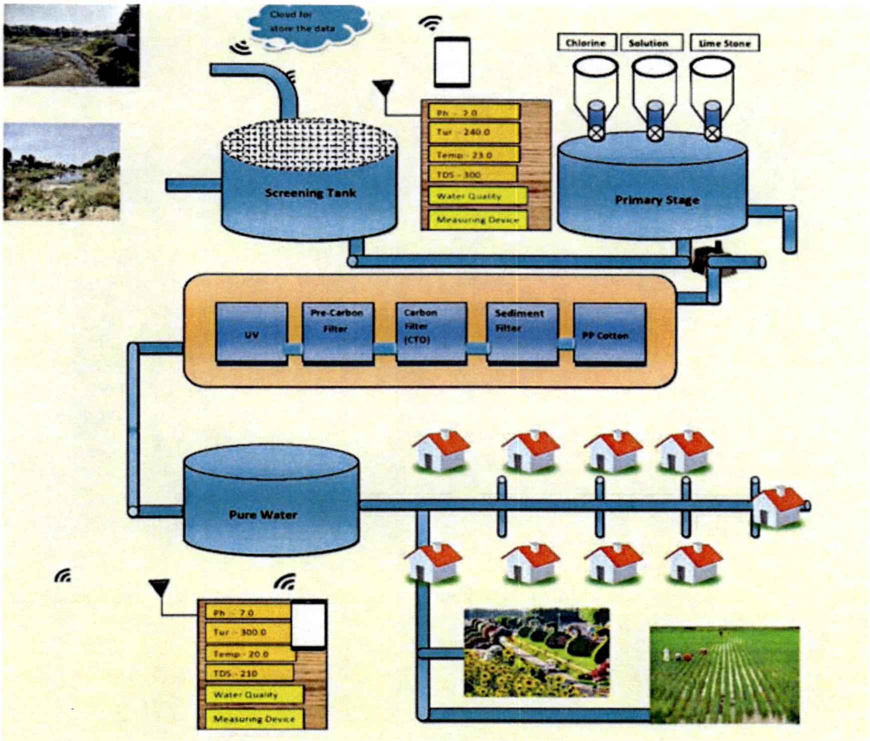

Fig. 3 Schematic of the self-powered intelligent wastewater treatment system

displayed on the four-line LCD panel, as shown in Fig. 6 and readings are noted as in Table 1.

The received wastewater is then pumped by a motor in the second tank where chemical treatment has taken place.

The chemical treatment of wastewater is then done in the second tank. Three Chemicals are used in the system. The structure to mix these chemicals uses electronics solenoid valves which automatically passes the required quantity of chemical through it. A fixed measured quantity is mixed in the screened and pumped water. The complete structure of the second tank is as shown in Fig. 7.

When water is pumped to the UV filter, it gets purified and this water is pumped to another tank where we have all the sensors to monitor the water parameters. We measure pH, TDS, Turbidity, and Temperature of the purified water as shown in Fig. 9. Smell, color and odor are checked manually. All the parameters we have measured after and before the process are as mentioned in Table 2 (Figs. 8 and 10).

All these parameters could be monitored at remote location by using GSM module and a SIM card, thereby making this system intelligent. Further quantity of chemicals

Fig. 4 Two tanks (drums) are used for prototype model of self-powered intelligent wastewater treatment system

Fig. 5 Screening system of steel wire mesh is used for primary stage in prototype model

Fig. 6 Display of parameters in LCD display for primary stage in prototype model

Table 1 Wastewater parameter measured on the designed LCD panel

Sensors parameters	Wastewater values
pH level	3.85
TDS	335 ppm
Turbidity	441.00 ppm
Temperature	27 °C

will be automatically controlled by the use of solenoid valve that is placed below the funnels used, as shown in Fig. 7.

4 Conclusions

In this paper, we have proposed a Self-Powered intelligent wastewater treatment system, which is developed for domestics and small industrial wastewater treatment. This system has been tested and the results are satisfactory. In this paper, first we have discussed the water problem around the world. To address the problem of water scarcity, we have been motivated to develop an intelligent wastewater treatment system that every household could own and install the system at home. This will reduce the water scarcity problem in the future. We have conceptualized an intelligent water treatment system that not only treat the wastewater but also monitors its quality

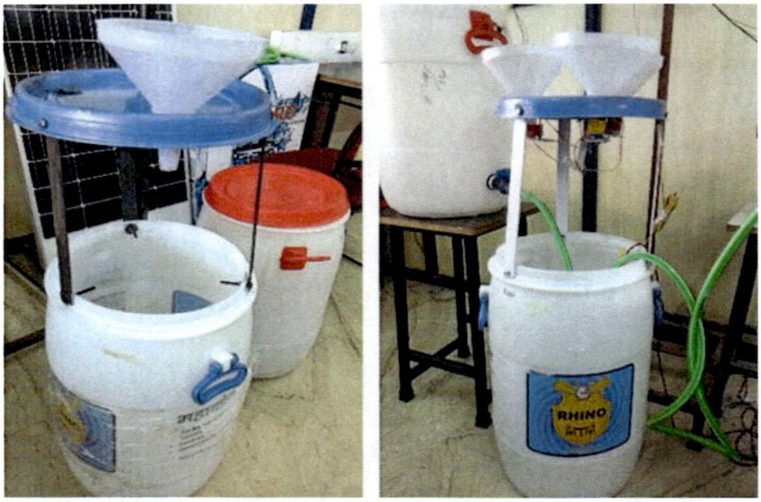

Fig. 7 Setup for chemical treatment in secondary stage for prototype model

Table 2 Comparison of parameters before and after the treatment process

Sensors parameters	Wastewater values	Treated water values
pH	3.85	6.4
TDS	335 ppm	95 ppm
Turbidity	441.00 ppm	256.21 ppm
Temperature	27 °C	27 °C

parameters in real time. The prototype model was successfully prepared and good results are obtained.

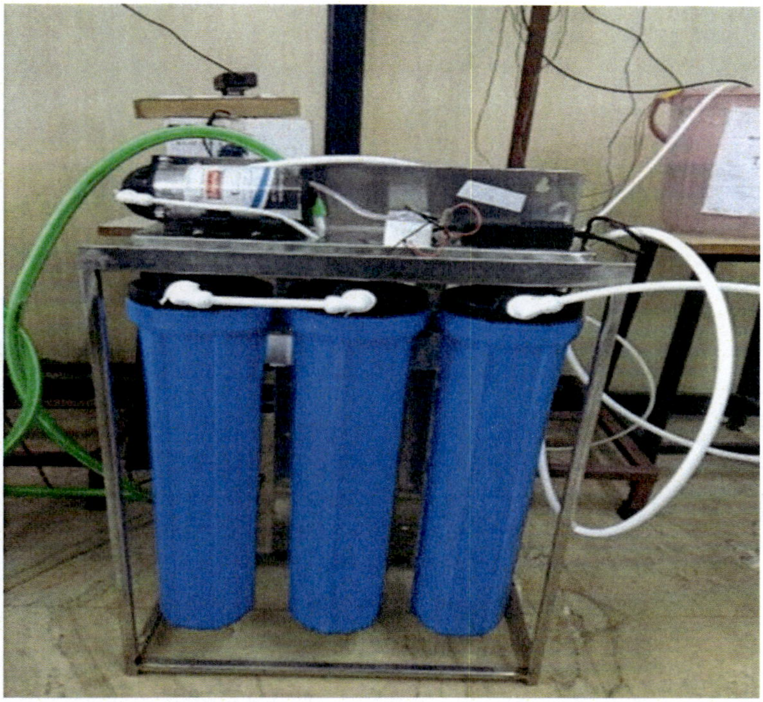

Fig. 8 Materials used in the filtration process

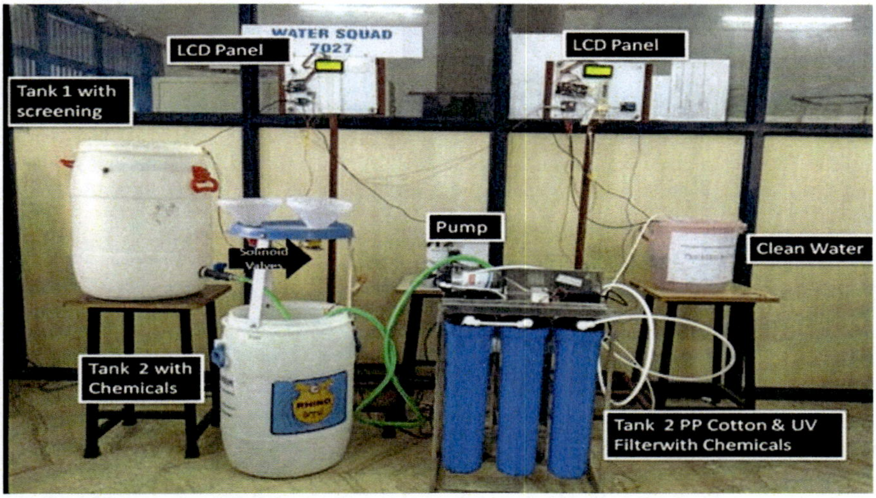

Fig. 9 Complete prototype model of self-powered intelligent wastewater treatment system

Fig. 10 Purified/treated water collected

Furthermore, this proposed system could also be used in municipal system in smart cities for real-time monitoring of the various parameters of quality of water through sensors and IoT.

References

1. A World Water Assessment Programme, The United Nations World Water Development Report 2018 (United Nations Educational, Scientific and Cultural Organization, New York, United States) (2018). www.unwater.org/publications/world-water-development-report-2018/
2. Burek P et al (2016) Water futures and solution: fast track initiative (Final Report). In: IIASA working paper (International Institute for Applied Systems Analysis (IIASA)), Laxenburg, Austria
3. Kopnina H, Washington H (2016) Discussing why population growth is still ignored or denied. Chin J Popul Resour Environ 14:133–143
4. Wada Y et al (2016) Modelling global water use for the 21st century: the water futures and solutions (WFaS) initiative and its approaches. Geosci Model Dev 9:175–222
5. European Environment Agency, Increasing Environmental Pollution (GMT 10) (2015). www.eea.europa.eu/soer-2015/global/pollution
6. Rosegrant MW, Cai X, Cline SA (2002) World water and food to 2025: dealing with scarcity. International Food Policy Research Institute (IFPRI), Washington, DC
7. De A, Bose R, Kumar A, Mozumdar S (2014) Targeted delivery of pesticides using biodegradable polymeric nanoparticles. Springer Briefs in Molecular Science, India
8. Food and Agriculture Organization of the United Nations (FAO). Global Forest Resources Assessment (2015) How are the world's forests changing? Second edition (FAO, Rome, p 2016

9. Alexandratos N, Bruinsma J (2012) World agriculture towards 2030/2050: the 2012 revision. In: ESA working paper no. 12–03. Rome, Food and Agriculture Organization of the United Nations (FAO)
10. United Nations Department of Economic and Social Affairs (UNDESA), World Population Prospects: Key Findings and Advance Tables—The 2017 Revision (2017). In: Working paper no. ESA/P/WP/248 (UNDESA, Population Division, New York)

Chapter 22
IoT-based Smart Ultraviolet Disinfecting and Sterilizing Robot for Public Places

Ashmit Dabi, Darshan Seth, Divya Soni, Siddhi Chaplot, Khushi Sharma, Latif Khan, and Rajeev Mathur

Abstract Diseases spread by pathogens are a major cause of both illness and death of human beings. The danger of getting infectious diseases in every hospital throughout the world is always there and is a universal fact. This paper proposes an Ultra Violet Disinfection and Sterilizing (UVDS) robot, which helps to sterilize an area in hospitals and resolve the problem of acquiring infectious diseases through hospitals and other potential places. The aim to innovate this equipment is to estimate the disinfection capacity of a UVDS robot. This robot can be used in hospitals, public transport, airlines and any enclosed areas. This technique will help to increase the sanitation process of contaminated areas and dramatically reduce the need for workforce. The UVDS robot is designed with ultraviolet lights and sanitizer spray along with the controllers to program I according to the need of disinfecting areas, which is contaminated by bacteria, viruses and other harmful microorganisms. To disinfect a wide range of surfaces, the sanitizer sprays are used, which rotates at an angle of 180°. Our UVDS robot has six 20 watts of UV lights. The concept of Internet of Things (IoT) is utilized to control the robot as well as measure the parameters. The operator will operate the system through the mobile application from outside the room. Ultraviolet Disinfection and Sterilizing Robot will remotely disinfect hospital wards and the patient's private room. The objective of this UVDS robot is to reduce Healthcare-Associated Infections (HAI) in hospitals. This robot moderates the number of viruses, infectious diseases and deaths caused by them in healthcare services provided during hospitalization. The proposed UVDS robot is used for hospitals, institutions, industries, airports and other infected areas.

Keywords UV light · Sterilization · Robot · HAI · Healthcare · IoT

A. Dabi (✉) · D. Seth · D. Soni · S. Chaplot · K. Sharma · L. Khan · R. Mathur
Geetanjali Institute of Technical Studies, Udaipur, Rajasthan, India

Prof. Geetanjali Institute of Technical Studies, Udaipur, Rajasthan, India

R. Mathur et al. (eds.), *Emerging Trends in Data Driven Computing and Communications*, Studies in Autonomic, Data-driven and Industrial Computing,
https://doi.org/10.1007/978-981-16-3915-9_22

1 Introduction

1.1 Objective of the Work

During taking treatment in any hospital, Healthcare-Associated Infections (HAI) are caused due to healthcare workers like nursing staff and non-medico staff, which may cause more time in the recovery of the patient. Also, the patient may have to bear additional cost due to extended hospital stay. Moreover, families of the patient may also be at high risk of getting infection. HAIs belong to different groups, such as Gram-negative and Gram-positive bacteria and yeasts. These pathogens can be spread by the hands of healthcare personnel and by patient-to-patient contact. Some pathogens can prevail in the hospital environment, surfaces, i.e., doorknobs, electronic devices, hospital supplies, television remote controls or inactive objects belonging to patients, such as mobile phones. All these surfaces have never been thought of as potential threats to getting infections.

1.2 Solution to the Problem

Therefore, in addition to preventing infections, UV Disinfection and Sterilizing (DVDS) Robot is one of the innovative technologies, which investigate the procedure of disinfection, killing bacteria, viruses and other microorganisms [1]. It has been reported that [2–4] at a particular wavelength of 254 nm, UV rays are able to break the DNA or RNA structure molecular bond of the environmental viruses, bacteria and microorganisms or pathogens as shown in Fig. 1. A good number of UV-C photons which when directed toward bacteria, viruses and other pathogens

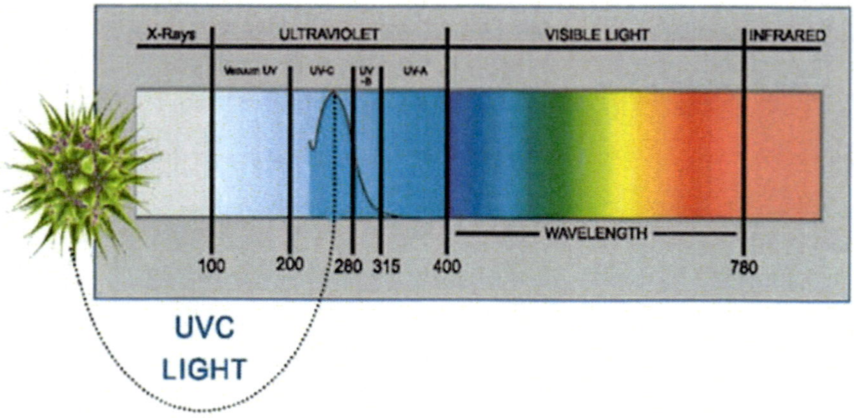

Fig. 1 Ultraviolet electromagnetic wave spectrum, 100–400 nm

render them harmless within seconds of exposure. The aim of the robot is to ensure definitive, precise measures and to remove the total number of bacteria and viruses and decrease the patient's discomfort.

1.3 Motivation

The present threat of pandemic due to the number of CORONA (COVID-19) cases rising in the world has really motivated us to propose a new technology for the sterilization of the room, hospitals and other public places. All over the world, healthcare system is struggling which the pandemic and infection are increasing day by day. Awareness of sterilizing even public places has really increased and institutions like education, malls, shops, corporate etc. have started using available manual disinfecting and sterilizing techniques. These infections increase by the hands with several germs, bacteria and yeast [5]. Some of the microbes may be effective in the environment, surfaces, doorways, electronic equipment, hospital, lifts or mobile phone accessories [5].

In this epidemic situation, the biggest problem is disinfecting the infected areas where any COVID-Infected patient has visited, so that chances of spreading COVID 19 infection are reduced, thereby preventing life from viruses. The sanitizing process done manually is risky and it could infect a person doing sanitization, in this situation, no one is ready to sanitize the places manually.

2 Methodology

2.1 Modeling of the UVDS Robot

We have first modeled and then developed a Robot according to the requirement to carry UV lights and sprayer with sanitizer as shown in Fig. 2. The Robot works on 12 V DC, which is provided by the rechargeable battery. Since UV light tubes work on AC, we have designed an inverter to convert DC into AC. Enough space is provided in these four-wheeled robots to carry a battery, spraying motor, a sanitizer and other circuit boards. The UVDS robot emits a UV-C light to kill the germs and microbes left after the process of manual cleaning on the surface. A good number of UV-C photons that when directed toward bacteria, viruses and other pathogens render them harmless within seconds of exposure. When UV rays hit the surface of a virus or bacteria molecule, they destroy the molecular bonds that hold the DNA. A video camera is provided on the robot, which can display the working of UVDS to the operator on the mobile screen through a mobile application developed. Camera placed on the top of the robot also checks the presence of any living being inside the room with the help of PIR Sensor. To measure distance and avoid obstacles, UVDS

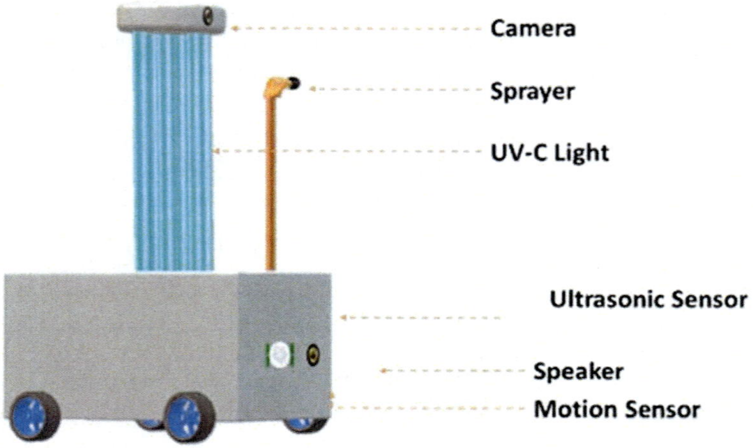

Fig. 2 Modeling of ultraviolet disinfecting and sterilizing (UVDS) robot

robot is equipped by ultrasound sensor. The robot is equipped with a sanitizer sprayer tool, which produces a dry spray mist that reaches all kinds of surfaces, even the inner surface of the drawers. Since it is a dry process, there is no residue left behind and there is no danger of corrosion or damage to the surfaces of medical equipment. This UVDS robot is secure, trustworthy and excludes human's interference and is made to be operated by health workers each and every day.

2.2 Controller Circuit Board of the UVDS Robot

Arduino Mega microcontroller board is used for the controlling of each and every function of the robot [6]. Through relays, switching of the higher voltage and current-controlled appliance is done. L298N is a Motor driver or Motor Driver IC, which allows to drive DC and steeper motor. It can control two or four DC motors with direction and speed. PIR sensor or motion sensor is used to detect the presence or absence of a living being by measuring the infrared light radiating from the object. ESP32 Camera with integrated Bluetooth is used to watch the live stream inside the room. A sound module and servo motor controller are also used to control the motion of the robot. To spray the sanitizer, a diaphragm pump is used.

Ultraviolet rays are used in industrial and healthcare departments to remove bacteria and other pathogens by decomposing their DNA structure and chemical bonds. It is a form of electromagnetic wave. The range the UV is up to 100–400 nm. This range explains the level of activity and energy of the photons and the size of wavelengths in each category [2].

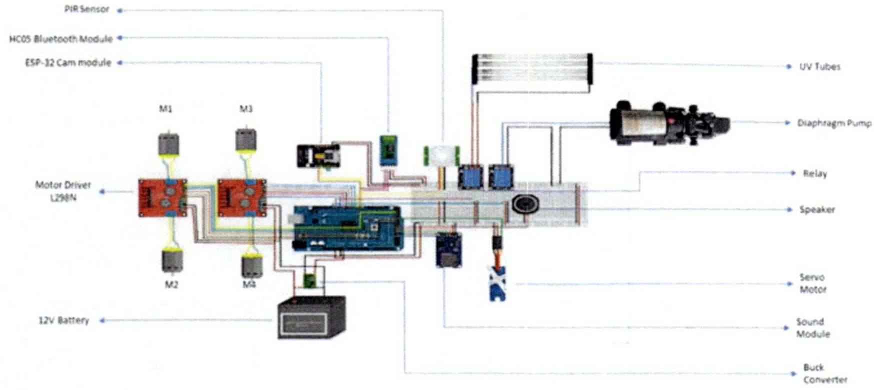

Fig. 3 Controller circuit board diagram of the proposed UVDS robot

The UVDS is a multifunctional robot, it includes both UV-C light and sanitizer spray. Because there is a specified wavelength of UV light, it cannot reach each and every place of the room. It is programmed to detect human beings /animals with the help of PIR sensors or IR motion sensors. If a person is detected while the disinfection process is going on, the UV light gets immediately turned OFF and warns the person "The room is under disinfection process please make a proper distance or leave the room". Once the person leaves the room or operating area, the UV lights which were OFF get turned ON automatically and the disinfection process goes on until the process is completed. The microcontroller controls the wheels of the robot by motor drivers in the path. Figure 3 shows the components used for developing the UVDS robot.

The UVDS is a multifunctional robot consisting of a smart sanitizing spray for disinfecting various or wide ranges of surfaces. The sanitizer spray will rotate at an angle of 180° because of the fact that UV has a specific wavelength, so that it can't reach each corner or part of the room. A complete flowchart of the designed and developed UVDS robot's work is given in Fig. 4. The UVDS robot is a smart IOT-based system; it can be operated through the mobile application. The live stream camera is mounted on the top of the robot to navigate the room as shown in Fig. 5.

3 Results and Discussions

As a matter of fact, all virus species of the comparable kind have the same kind of structure DNA and RNA strand length. UV light experiments conducted in the past were used to find the UV radiation dose required for 90% virus and bacteria depletion (i.e., the log-reduction dose). Dosages for a 90% destruction of most bacteria and viruses range from 2000 to 12,000 $\mu W/cm^2$. The robot utilizes four UV lamps in a circular pattern to cover 360° and maximize efficiency. Each lamp has 11-W output

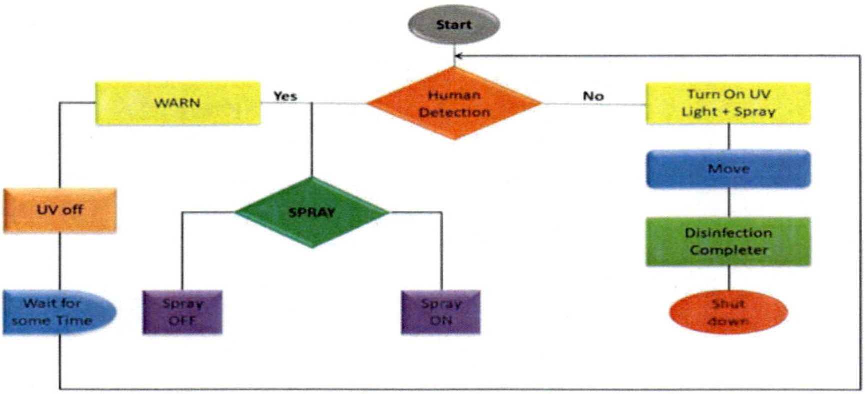

Fig. 4 Flowchart for the working of IoT-based ultraviolet disinfecting and sterilizing (UVDS) robot

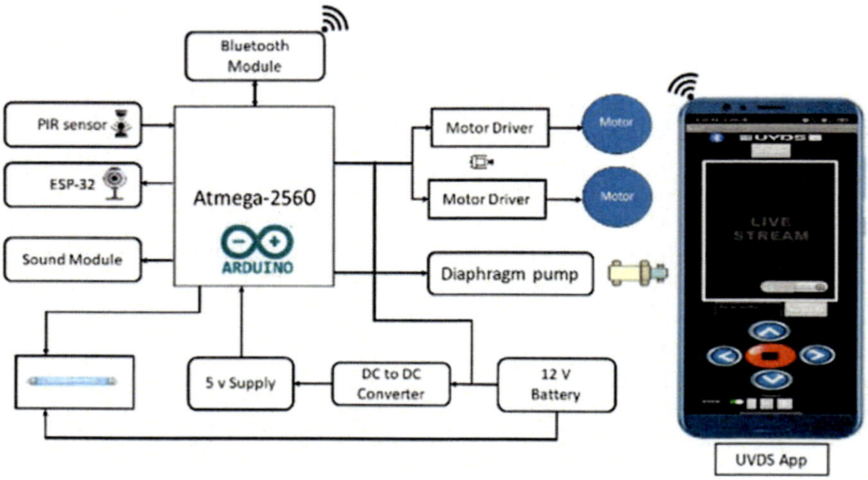

Fig. 5 **Mobile** Application (App) to control IoT-based UV disinfecting and sterilizing (UVDS) robot

power (as listed in the UV lamp datasheet). The amount of brightness for a distance of 2.5 feet away can be calculated as:

$$\text{Brightness} = \frac{11 * 4}{4 * \pi * (158.49)^2}$$

$$\text{Brightness} = 0.0003803539 \text{W/cm}^2$$

$$\text{Brightness} = 380.35 \mu\text{W/cm}^2.$$

A UV dose of 10,600 μW.sec/cm^2 is required for a 90% reduction in infection, which can be considered as the upper limit of log median log reduction (in low absorption media). Therefore, the time required to disinfect is expressed as:

$$\text{Time} = \frac{10,600(\mu\text{W}.\frac{\text{sec}}{\text{cm}^2})}{380.35(\frac{\mu\text{W}}{\text{cm}^2})} = 27.86\,\text{sec}.$$

The UVDS robot is designed and developed for both the air and the surface disinfection and sterilization of an area in the hospital or any place. The height of the robot is 2.5 feet. This UVDS robot is controlled through the mobile app developed by us. The UVDS robot can disinfect a $12' \times 12'$ rooms in 10 min without the presence of human being. The time taken for disinfection of a $15' \times 15'$ room is 16 min 49 s. In the case of humans present in the rooms, the robot stops working automatically, so that UV radiation does not harm. The disinfecting time depends on the relative distance between the robot and the human being and the speed with which the human being is moving. A good number of UV-C photons that when directed toward bacteria, viruses and other pathogens render them harmless within seconds of exposure. When UV rays hit the surface of a virus or bacteria molecule, they destroy the molecular bonds that hold the DNA. When the UV tubes are unable to disinfect objects out of their reach in the room, they can be disinfected using sanitizer there. The goal of the UVDS robot is to help prevent these infections in the first place. The main aim of UVDS robots is to make INDIA corona free.

4 Conclusion

In this paper, we have reported a design and development of an Ultraviolet Disinfecting and Sterilizing (UVDS) Robot. We have discussed the operating procedure of the UVDS robot, which we have conceptualized and developed due to the COVID situation. The robot will sterilize the areas such as hospitals, railways, airports, colleges, schools etc. This technique will help speed up the cleaning process of contaminated areas and dramatically reduce the need for workers.

The UVDS robot is designed and developed for both the air and the surface disinfection and sterilization of an area. The height of the robot is 2.5 feet. This UVDS robot is controlled through the mobile app developed by us. The UVDS robot can disinfect a $12' \times 12'$ room in 10 min without the presence of human being. The time taken for disinfection of a $15' \times 15'$ room is 16 min 49 s. In the case of human presence in the room, the robot stops working automatically, so that UV radiation does not harm. The areas where this tube and ultraviolet light could not penetrate can be sterilized using a sanitizer spray module.

References

1. Begic A (2018) Application of service robots for disinfection in medical institutions. Adv Technol, Syst, Appl II(28):1056–1065
2. https://www.sciencedirect.com/science/article/pii/S1684118217302001
3. https://www.mdpi.com/1660-4601/16/23/4747/htm
4. https://www.gogas.com/sites/default/files/downloads/gogas_uv_facts_uvdr-whitepaper_004.pdf
5. https://www.ncbi.nlm.nih.gov/pmc/articles/PMC7273323/
6. https://en.wikipedia.org/wiki/Wikipedia

Chapter 23
IoT-based Intelligent Intravenous (IV) Stand for Healthcare Sector

Mohammad Ateek Samma, Shubham Vagrecha, Mr. Latif Khan, Rajeev Mathur, Lokesh Sharma, and Sheetal Sharma

Abstract Intravenous stand are those medical care subordinate executes that associate in capturing and appropriating meds to patients via intravenous infusions. Intravenous imbuement depends on the heaviness of the Intravenous tubing salt. Interminable development of the intravenous stand is needed to supplant the liquid when abandoned. Additionally, activation of the intravenous shaft is needed from patients and attendants. This prompts distress and burden for the two patients and clinic staff. Regardless of the staying alive intravenous posts have tackled their challenge, notwithstanding they couldn't hold a wealth of weight clinched. Also, they come up short on any caution framework which betokens a vacuous sack (Jost G, Tseng CC (1990) Baxter International Inc., assignee. Home care intravenous stand. United States Patent US 4,905,944.; Larson GR (1980) Larson Godfrey R, assignee. Handle for mobile intravenous stand. United States Patent US 4,225,104;). To enhance current intravenous posts, we intend to build up a Robotic Keenly Intellective Smart IV Stand Predicated on IOT. Robotic Smart Intravenous Stand mounted with sensors and an alert framework (Qureshi and Syed in Saf Health Work 5:198–202, 2014). Trial results devoted to the plan showed that the novel planned post activation is improved by the utilization of Load Cell, Node MCU, Arduino UNO, and so on. Moreover, the caution framework has incorporated alarm to patients and attendants when the intravenous packs were emptied.

Keywords Intravenous poles · Alarm system · Internet of Things (IoT)

1 Introduction

Smart Intravenous stand treatment improves in general wellbeing by giving 100% of the nutrients and supplements needed by one's body to improve well-being, help energy levels, and improve indications of wretchedness. This is the explanation that

M. A. Samma (✉) · S. Vagrecha · Mr. L. Khan · R. Mathur · L. Sharma · S. Sharma
Geetanjali Institute of Technical Studies, Udaipur, Rajasthan, India

Rajasthan Technical Studies Kota, Kota, Rajasthan, India

R. Mathur et al. (eds.), *Emerging Trends in Data Driven Computing and Communications*,
Studies in Autonomic, Data-driven and Industrial Computing,
https://doi.org/10.1007/978-981-16-3915-9_23

these days, Intravenous (IV) posts are as a rule generally used in the medical care area [2]. IV shafts are suspended with packs to convey the meds required by the patients [4]. The IV medicines are acquainted with the patient through an IV infusion [6]. The current IV stands doesn't contain a characteristic caution which alert both the patient and attendant when the entire liquid taken care of is discharged [3]. We propose to prepare the Robotic IV Stand with a natural/implanted caution framework as well.

1.1 Existing Intravenous Stand

Herein, we have provided brief overview of the IV stand that exists today in every country. Their shape is subtle and has limited features. With time, most of them cannot even bear the weight of the bags [8]. Moreover, they don't have any alert system which could tell the nursing staff about the emptied IV drips solution which often leads to blood backflow (Fig. 1).

Fig. 1 Existing IV stand source GMCH

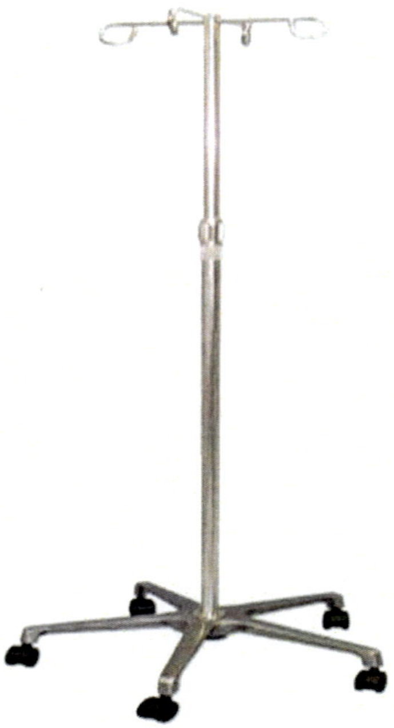

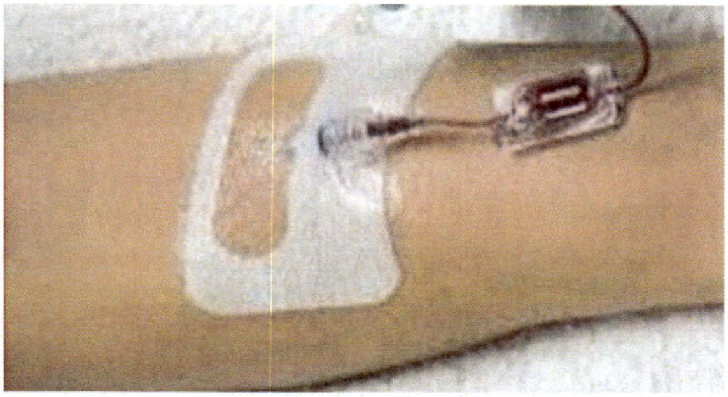

Fig. 2 Back flow of blood [5]

1.2 Problem Faced by Human

In hospitals, We saw that when the fluid inside the drip becomes completely empty and the doctors/nurses could not reach the patient timely to remove the needle from the veins, the pressure inside the veins overcomes the pressure inside the IV, allowing the blood to flow back into the line which is not desired, which is shown in Fig 2 [5].

2 Methodology

We have first developed an IV Stand according to our requirement. Then, we have designed a system Robotic Smart IV Stand based on IOT and Sensors [8]. The system started, when the IV fluid bottle is hanged to the hook which is connected with the sensor then it gives glow to the green LED's, ensuring that the veins are being infused with the fluid. When the drip solution is about to end, it gives an alert message to the doctor/nurse on the mobile phone. We can also check the system update on Android mobile application. Further, the red LED's indicate that drip solution is critically low so that doctor/nurse can reach the patient well on time to avoid the backflow of the blood in the patient's body.

This stand is portable because it is light in weight. There is also a feature of mobile charging in the Smart IV Stand. The mosquito mat system is also given to the Smart IV Stand. So, the patient is not affected by the mosquito. Block diagram is shown in Fig. 3.

2.1 Technology Used in Robotic Smart IV Stand

In this device, we are using Node MCU as a control unit. Two load cells of 20 kg are mounted on left and right arms of stand. Red and green led indicators are also mounted on both arms of the stand. We have also used buzzer for the sound alarm.

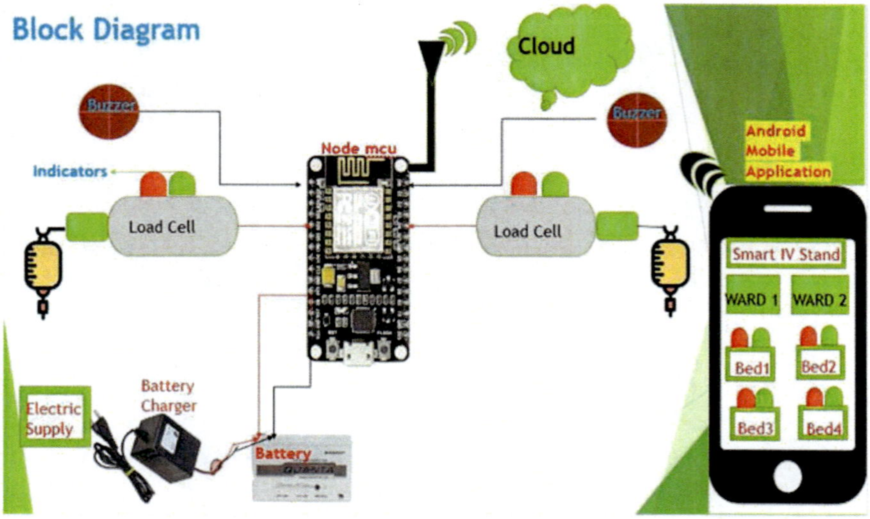

Fig. 3 Block diagram of robotic IV stand

Complete system is powered by 12 V and 7.5 Ah battery. Battery charger is also associated with the main unit for battery charging.

Load cell is continuously observing the level of fluid bag. Once it reached to critical condition stand will send alert message to ward attendee simultaneously, it will glow the red LED and also updates the ThingSpeak server and mobile application.

2.2 Components of the Robotic Smart IV Stand

Major components used in our system are under:

- Node MCU: Heart of our system
- Load Cell: 20-kg load cell we have used
- Battery: 12 V, 12 Ah Battery we have used
- Charge Controller: To control overcharging and undercharging of battery
- LED Indicators: To indicate level of bottle
- Buzzer: Sound alert to attendees.

3 Results on Server and Mobile

We have tested our system and found satisfactory results. Node MCU updates ThingSpeak server. We have created level field on server which shows the level of bottle. When bottle get empty than ThingSpeak server will send alert on ward

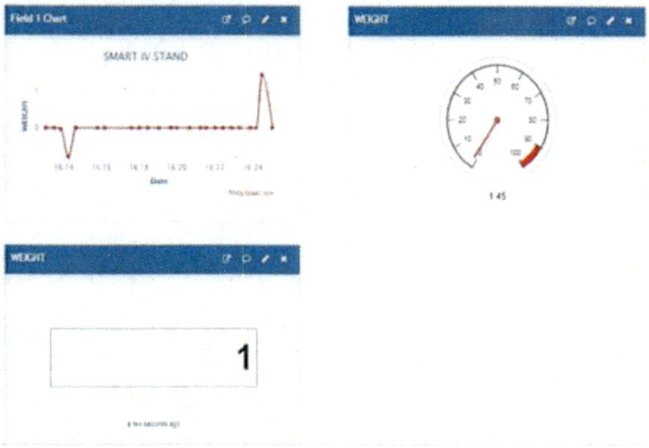

Fig. 4 Result on ThingSpeak server

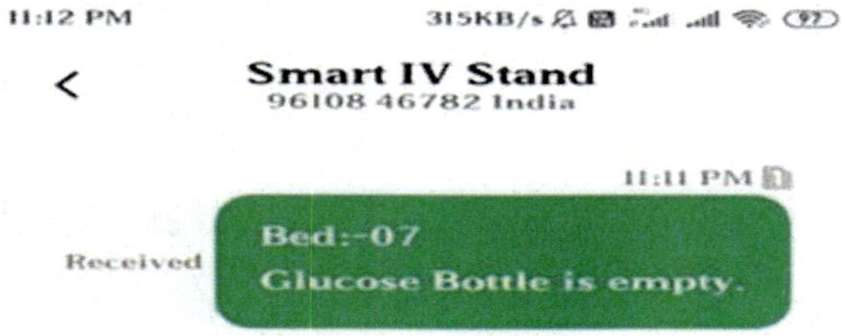

Fig. 5 Message alert on mobile

attendees mobile no and simultaneously update mobile application. Server results are shown below (Figs. 4, 5, 6).

4 Conclusion

Robotic Smart IV Stand was developed and successfully used in Geetanjali Medical Hospital, Udaipur, India. When the system started, it gives glow to the green LED, ensuring that the veins are being infused with the fluid. When the drip solution is about to end, it gives an alert message to the doctor/nurse on the phone through node MCU. We can also check the system update on Android mobile application and the red LED indicate that drip solution is critically low such that they can reach the patient timely to avoid the backflow of the blood. The capability of the novel robot to hold a bottle of fluid weighing up to 20 kg while maintained balance. The Smart Robotic IV Stand is very much practical to be used in the Hospitals, Army Medical Camps, Medical Dispensary, etc.

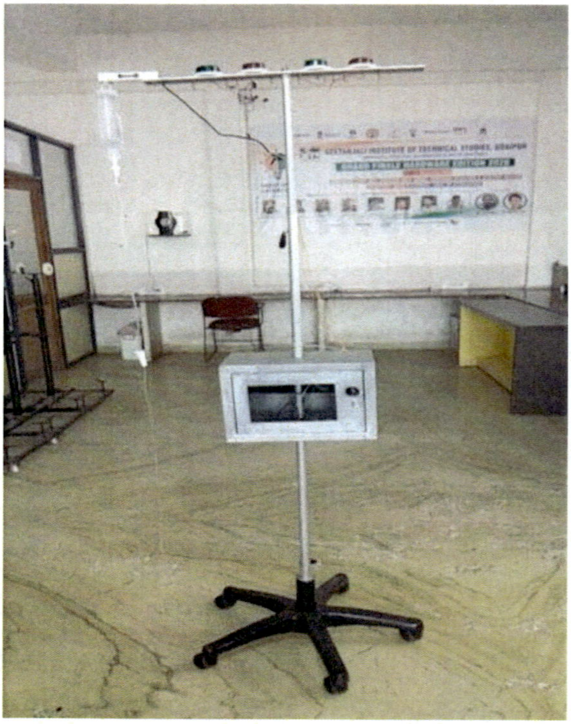

Fig. 6 Structure/photo of the robotic smart IV stand

References

1. Larson GR (1980) Larson Godfrey R, assignee. Handle for mobile intravenous stand. United States Patent US 4,225,104
2. Vignali M (2006) Intravenous stand design. Project partners from Huazhong University of Science and Technology, China
3. Schott JC, Schott MD, Fox JD (2000) Schott Jeffery C, Marc D, Fox Jeffrey D, assignee. Intravenous stand support assembly. United States Patent US 6,079,678
4. Adelman GZ (2006) Adelman Gregg Z, assignee. Transportable intravenous bag stand. United States Patent US 6,983,915
5. Wilt Jr CF (1985) St. Joseph's hospital, medical center, assignee. Apparatus for facilitating intravenous feeding during transportation of patient. United States Patent US 4,511,157
6. Binger M, Conway C, Goddard N, Jacobs N, Pysher C et al Project number: P15073 autonomous IV stand
7. Wiederhold BK, Riva G (2009) Affective robot for elderly assistance. Annual review of cyber therapy and telemedicine 2009: advanced technologies in the behavioral, social and neurosciences 144, 44
8. Michael M (2012) Arduino cookbook
9. Bancalari CG (1988) ICU medical, Inc., assignee. Collapsible pole and stand combination. United States Patent US 4,744,536

Chapter 24
Integrating Lean Green Technique Through Single Combined Strategy in Marble Industry

Saurabh Tege and Nilesh Diwakar

Abstract Today's scenario is technology-based and for huge growth, development with all resources all technology should be implemented at all stage s of manufacturing units. The quarrying of marble is not quite the same as the mining of different minerals. In marble quarrying, huge sizes of squares are exhumed. Prior extraction of marble was done physically by utilizing a drill and jib crane. In the marble mining industry, this issue arises as much that they willing not to implement latest trends because of a lack of knowledge and understanding. In this paper, the author should give a framework development through lean green technique, due to which wastages to be minimized in all aspects as well as environmental consideration also takes place.

Keywords Environmental design · Lean green technique · RFID · IoT · Sensor-based technology

1 Introduction

For quite a long while, fabricating rehearses have been for the most part centered on fulfilling or making requirements while retaining seriousness with regard to item consistency, marketing potential and growth. Specifically, lean assembling, first presented for the auto business in Japan, has to a great extent been viewed as quite possibly the most compelling assembling ideal models. Lean assembling furnishes associations with the devices to improve their intensity dependent on expanding an incentive to clients, regarding profitability, efficiency, quality, and costumers' fulfillment, by diminishing the assets utilization through waste end. This sort of assembly methods of logic focused on the interest of consumers, along with the increased demands of individuals for daily comforts, have spurred an increasing interest in

S. Tege (✉)
Geetanjali Institute of Technical Studies, Udaipur, Rajasthan, India

N. Diwakar
SSSUTMS, Sehore, India

R. Mathur et al. (eds.), *Emerging Trends in Data Driven Computing and Communications*, Studies in Autonomic, Data-driven and Industrial Computing,
https://doi.org/10.1007/978-981-16-3915-9_24

the object, fulfilled by a considerable measure of goods produced, wound up in an evolving era of pollution and squanders. In this particular case, organizations are urged to become more constructive with regard to their natural and social status, heading towards more feasible assembly rehearsals respect to the prominent conceptualization [1] of Triple Bottom Line (TBL) maintenance (appeared in Fig. 1), which recommends that an organization would have the option to accomplish manageable outcomes gave it is fit for improving ecological, social, and financial exhibitions all the while analysts accept lean assembly with a revived interest in producing greener arrangements capable of limiting squanders in this line, yet additionally of lessening, by expanding, altering adjusting and refreshing lean approaches, the natural and social negative effects of the generally utilized modern practices with sustainability [2, 3] through various technologies such as RFID, Sensor-based technology, IoT, etc.

Albeit some encouraging outcomes have been accounted for the composition by applying lean methods to achieve greener targets [4–6, 7]. Many analysts agree that lean techniques have not yet advanced to the stage of improvement needed to ensure manageability [5, 8]. In this way, they also suggested that, in order to produce fair performance, green production activities should also be considered [9]. Lean Green practices are centered on lessening unsafe discharges, disposing of the utilization of inefficient assets, reusing, and limiting well-being hazards all through the whole assembling measure, by limiting the natural impression during the entire item life cycle [10]. On one hand, there are scientists that concur that green

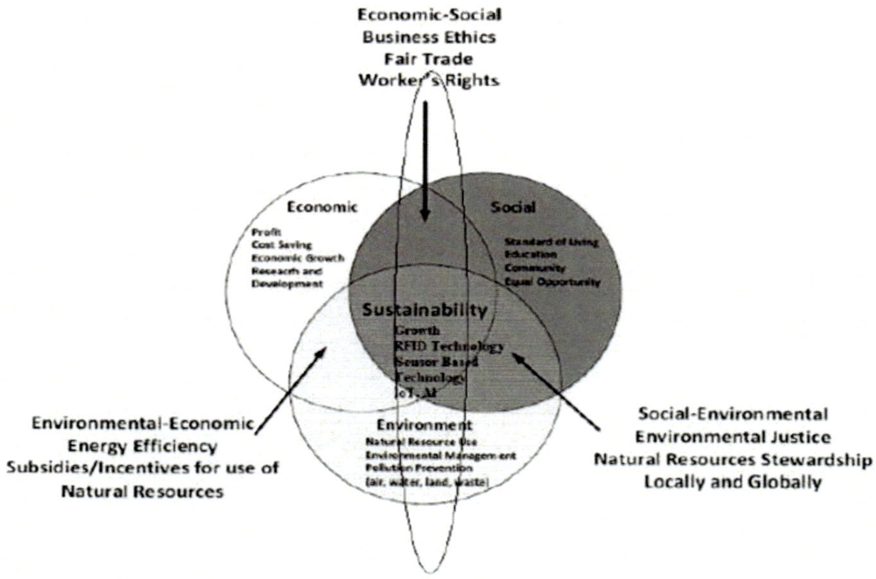

Fig. 1 Triple Bottom Line (TBL) model for attaining sustainability

techniques, for example, Environmental Design vs Green management of Supply Chain, improve the association of the monetary and the ecological frameworks, creating a practical turn of events and undertaking incorporation [11–14]. Then again, there are scientists that express that green practices are sufficiently not to guarantee manageability [15]. Besides, professionals regularly contend that green practices can be a weight for arriving at enhancements with respect to plan and creation measures. In this clashing setting, specialists and experts accept that when executed solely, neither lean nor green practices have shown the option of preserving the usual balance between economical, ecological, and social perspectives [8, 7, 16, 17]. At that point, the mix with lean green approach and also characteristics can be strengthened and their drawbacks can be draw, which has been suggested to meet the existing management requirements.

2 Previous Research Study

Lean A short writing audit of the connected investigations recently led in various pieces of the world has been examined here to comprehend the meaning of directing the current exploration. These were partitioned into subsections; Harmful effects of marble squanders, Life Cycle Assessment, Cleaner Production and uses of marble squanders in helpful items.

Gursel et al. (2014) audit the qualities and shortcomings of cement LCIs to date, and offers an exploration activity guide for development in the nature of future LCA concentrates in that area. This examination guide offers Life Cycle Assessment-based choice emotionally supportive network for makers. An exhaustive writing review was directed in such manner and their fundamental outcomes were made and looked at. It was uncovered that in endeavor any LCA study the gathering of sound information of life cycle stock (LCI) is a significant and basic advance. Based on which ensuing LCA and LCIA (life cycle sway evaluation) can be based.

Hanieh et al. (2014) zeroed in on the natural, financial, and social effect of stone and marble preparing industry. The goal of the investigation was to presents the life cycle of the interaction, ID of contamination area of interest and proposes alleviation measures for productive utilization of crude materials including common stone, water utilization, and energy use during handling. Recuperation, Reuse, and Recycle, 3Rs standards are utilized to boost the yield and limit the losses to improve measure proficiency. This will give a casing work to future activity in this area.

Kushwah et al. (2014) contemplated that the utilization of marble slurry in easing. Properties of marble slurry are Bulking is 42%, which is generally outrageous at 5% sogginess. Fineness modulus was found to be 0.93. According to these limits, Marbleslurry can be utilized in as curing help.

Ioannidou et al. (2014) detailed Life Cycle Assessment (LCA) to distinguish best waste treatment framework office having less natural effects. For that reason, information were gathered from two stone preparing offices in the investigation region and assessed. Furthermore, the utilization of other divider treatment frameworks like slender cladding and rock walling utilized were likewise assessed. The outcomes show that the stone item generally utilized in the design rather than slim cladding or rock divider framework. It was because of the accessibility of crude materials locally. Consequently, it was inferred that the utilization of stone items in divider framework has less natural area of interest as contrast with other cycle.

Li et al. (2014) completed Life Cycle Inventory examination (LCI) to make possible enhancements in concrete industry. The itemized life cycle stock (LCI) of concrete assembling plant included total info and yields. The useful units are 1 Tonne of Portland Ordinary concrete.

The information contain not just the conventional things, for example, crude materials incorporates limestone, dirt, gypsums, energy (coal and power), and admixtures (heater slag and fly debris), yet in addition new water which isn't focused in other writing. The yield information contain ozone-depleting substances as well as the perilous air contamination just as commotion and hefty metal outflows which are normally disregarded by different analysts. The information were estimated nearby and assessed for to fuse it into the utilizations of diminishing toxins and waste warmth recuperation advancements in concrete industry.

Rajni et al. (2014) use of stone waste as a marble powder in mortar and concrete, both pozzolanic and non-pozzolanic, were studied. The stone waste can be utilized for developing insignificant exertion building materials, for instance, block, square, tiles, etc.

Uygunoglu et al. (2014) explored the utilization of marble squander and reused total in the creation of Self-Compact Concrete. For that reason, arrangement of control examples of cement containing marble squander and reused totals were arranged and tried for droop test, J-Ring test, air content, compressive, and rigidity along modulus of flexibility were resolved. Results show that no huge contrast were seen in the mechanical properties of self-compacting solid utilizing marble squander and reused totals. It was reasoned that the marble squander and reused totals can be utilized in self-compacting concrete.

W. Rehman et al. (2014) found that the marble waste squares are lower in strength than common ended squares, yet their insignificant exertion, straightforwardness and speed of creation recommend them to be used in zone where higher strength isn't concern. They can be utilized in sanctuary for people in quake and flood impacted zones.

Singh et al. (2015) thought about that Finishing Material Marble Paste is better than divider mud. Which has more strength and more affordable. Marble stick is a mix of slurry powder with white concrete and relaxed or hydrated lime close by staying trained professional. Goodly influencing the environment. Put aside energy and money may be used somewhere else for the headway of the country.

Dharma et al. (2015) Studied that the marble waste and different kinds of waste can be used as filling material as 15% overriding with regular soil on thoroughfare advancement. Improve water conductivity, increases in the degree of coarser particles. It reduces quite far, raises beyond what many would consider possible and decreasing in the flexibility record of the earth in.

Nabil et al. (2015) investigated that the conceivable utilization of stone slurry powder in fake stone creation. The test results showed that the compressive strength extended by solid extent, the compressive strength of the fake stones is generally impacted by the solid to stone powder percent, compaction squeezing factor and reestablishing time.

Jehangir et al. (2015) researched the wellbeing hazard related with Marble effluents. They dealt with renal (kidney) stone development in the inhabitants of locale Rajsamand in Khyber Pakhtunkhwa region of Rajsamand. Kidney stone was seen in 1000 patients utilizing Marble wastewater polluted drinking water. In addition, the finding of this examination shows that marble preparing industry represents a potential human wellbeing hazard because of constant openness of marble effluents. This instigated persistent state of renal stone development requires appropriate consideration and alleviations to ensure the human wellbeing.

3 Lean and Green Technique

Finally, as shown by the investigation of selected manufacturing units, this review finds that typically generating units do not zero in on the principle of waste reduction, asset profitability, authoritative improvement, and source decrease, among the key similarities between the two practices; while differentiating their fundamental core, the concept of waste, the form of consumer, the assembly processes. In Fig. 2, relation between lean green has shown very well i.e. shown waste reduction in form of quality, cost, customer satisfaction, process waste etc.

4 Incorporating Green and Lean Activities Through a Single Combined Strategy

The standard and contrary concentrations of lean and green approaches are addressed at a high level. Consequently, it is then the problem of whether working together is genuinely equitable and appropriate. Furthermore, the composition is not inherently find exceedingly negligible observational proof of active instances in lean-green schemes, but the findings that can actually be found are often contradictory. On the one side, there are scholars who fight against the coordination between lean and green methods when they can synchronously make a collaborative effort to minimize waste. Energy, material and time usage, store network the load up and thing life cycle smoothing out. Of course, there are those that articulate that lean and green methods are not mostly possible, there are a few places where it can be especially difficult to go with them. For instance, in Fig. 3 stock levels are maybe the most essential points of view when merging lean and green approaches. For the present circumstance, while

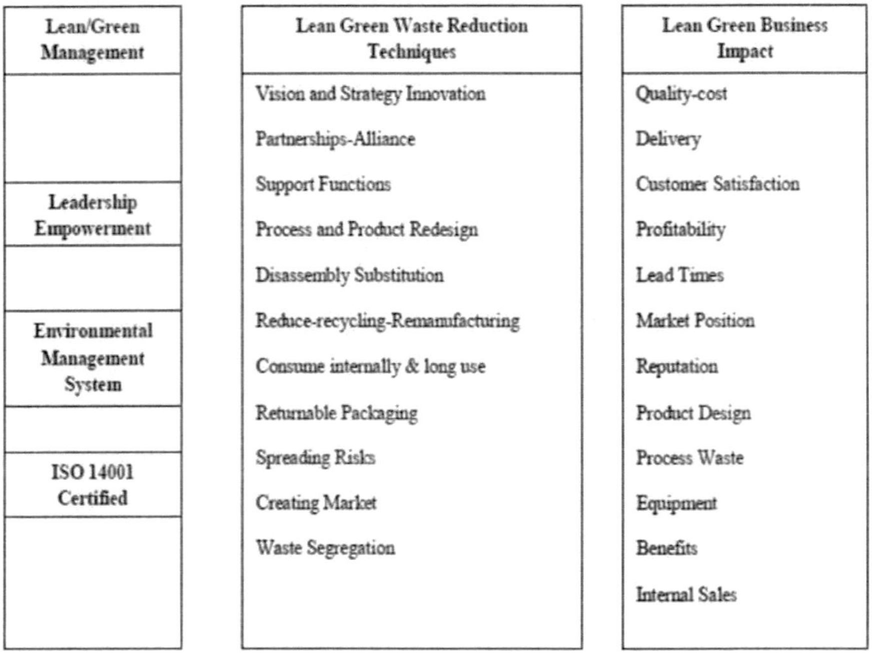

Fig. 2 Relation between lean green techniques

lean practices base on making, moving and squeezing little package expectation to fulfill customers' essentials.

At the point where everything is said in fact, experts who are reluctant to join the two programs are stressed whether lean practices based on waste decreasing from the extra-value viewpoint of the consumers, expense, efficiency and lead times would be beneficial, irrespective of whether green practices focusing on biological objections that are not by and wide according to lean objections are combined. In this line, manufacturers propose that the core guideline test involves completing how can environmental issues be coupled with lean principles without reducing the economic sustainability of the last-mentioned, but still maintaining the economical, ecological, and social agreements? Furthermore, it is reported that green techniques would not, for the most part, restrict the negative effects that going green can have on organizational perspectives in a combined context have, being similarly clear the converse route around. Finally, in, makers moreover express those lean-green philosophies ought to stand up to the very troubles when openly executed, lean and green techniques ought to go up against that. For example, in, it is included that, as it is generally the situation of green practices, experts may have to focus on changed techniques to complete the lean-green technique; whereas in, it was reported that one of the dimensional parameters where high equipment price is needed to implement the lean green proactive strategy.

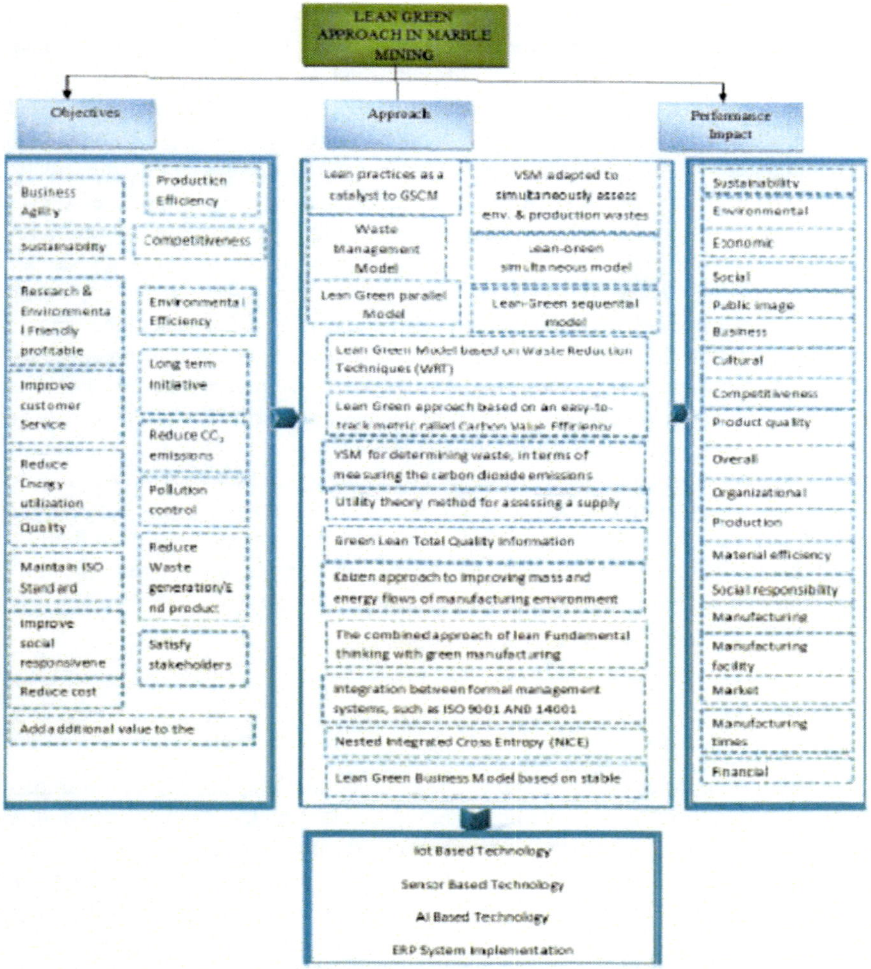

Fig. 3 Summary of the recommended lean green solution approach and their performance impact

In order to obtain green outcomes, they cannot substitute green activities, but they can provide a clear enhancement, agile and transparency culture for staff to create an effective and amazingly beneficial atmosphere for the use of green exercises. Makers agree with these observations, sharing that green chips will be a catalyst away from the lean community, empowering the collection of traditional activities. Similarly, a detailed definition of such a reactant effect can be found where it has been demonstrated that the influence of lean activities on the implementation of operating stock companies can be increased by hindering tainting and reuse. In order to obtain green outcomes, they cannot substitute green activities, but they can provide a clear enhancement, agile and transparency culture for staff to create an effective and amazingly beneficial atmosphere for the use of green exercises. Makers agree with these observations, sharing that green chips will be a catalyst away from the lean community, empowering the collection of traditional activities. Similarly, a detailed definition of such a reactant effect can be found where it has been demonstrated that the influence of lean activities on the implementation of operating stock companies can be increased by hindering tainting and reuse. Interdisciplinary gatherings aware of consolidating lean and green practices in two amassing associations are coordinated. There are too many technologies used such as IoT, RFID in transportation for monitoring and controlling and ERP system which incorporated all inventory stocks. In a similar manner, several researchers who have worked in the industry for a long time commonly studied about how green can be lean, for instance, the ones in and have contemplated that, since lean activities are not centered around green goals, In order to obtain green outcomes, they can not substitute green activities, but they can provide a clear enhancement, agile and transparency culture for staff to create an effective and amazingly beneficial atmosphere for the use of green exercises. Makers agree with these observations, sharing that green chips will be a catalyst away from the lean community, empowering the collection of traditional activities. Similarly, a detailed definition of such a reactant effect can be found where it has been demonstrated that the influence of lean activities on the implementation of operating stock companies can be increased by hindering tainting and reuse.

5 Conclusion

In marble mining industry, this issue arises as much that they willing not to implement latest trends because lack of knowledge and understanding. In this paper, author should give a conceptual framework development through lean green technique, due to which wastages to be minimized by using technologies such as sensor based technology, IoT, RFID technology in all aspects as well as environmental consideration also takes place. By using all these long term benefited taking by mining industries which will give benefit to industry as well as whole supply chain.

References

1. Bortolotti T, Boscari S, Danese P (2015) Successful lean implementation: organizational culture and soft leanpractices. Int J Prod Econ 160, 182–201. [CrossRef]
2. Fliedner G, Majeske K (2010) Sustainability: the new lean frontier. J Prod Invent Manag 46, 6–13
3. Peto O (2012) Lean in the aspect of sustainability. In: Club of economics in Miskolc TMP; TMP: Miskolc, Hungary, pp 54–58
4. Fliedner G (2008) Sustainability: a new lean principle. In: Proceedings of the 39th annual meeting of the decisionsciences institute, Baltimore, MD, USA. 22–25 Nov 2008

5. Dieste M, Panizzolo R (2018) On the relationship between lean practices and environmental performance. In: Proceedings of the IOP conference series: earth and environmental science; IOP Publishing: Bristol, UK, vol 151
6. Junior MJAP, Mendes JV (2017) Operational practices of lean manufacturing: potentiating environmental improvements. J Ind Eng Manag 10, 550–580
7. Maruthi GD, Rashmi R (2015) Green manufacturing: it's tools and techniques that can be implemented in manufacturing sectors. Proc Mater Today 2, 3350–3355. [CrossRef]
8. Marhani MA, Jaapar A, Bari NAA, Zawawi M (2013) Sustainability through lean construction approach: a literature review. Procedia Soc Behav Sci 101, 90–99. [CrossRef]
9. Garza-Reyes JA (2015) Green lean and the need for six sigma. Int J Lean Six Sigma 6, 226–248. [CrossRef]
10. Sun J, Han B, Ekwaro Osire S, Zhang HC (2003) Design for environment: methodologies, tools, and implementation. J Integr Des Process Sci 7, 59–75
11. Hibadullah SN, Fuzi NM, Desa AFNC, Zamri FIM (2013) Lean manufacturing practices and environmentalperformance in malaysian automotive industry. Asian J Financ Account 4, 366–374. [CrossRef]
12. Chin TA, Tat HH, Sulaiman Z (2015) Green supply chain management, environmental collaboration andsustainability performance. Procedia CIRP 26, 695–699. [CrossRef]
13. Chen Y (2008) The driver of green innovation and green image: green core competence. J Bus Eth 81, 531–543. [CrossRef]
14. Gehin A, Zwoli´nski P, Brissaud D (2008) A tool to implement sustainable end-of-life strategies in the productdevelopment phase. J Clean Prod 16, 566–576. [CrossRef]
15. Cherrafi A, ElFezazi S, Govindan K, Garza-Reyes JA, Benhida K, Mokhlis A (2016) A framework for theintegration of green and lean six sigma for superior sustainability performance. Int J Prod Res 55, 4481–4515. [CrossRef]
16. Choudhary S, Nayak R, Dora M, Mishra N, Ghadge A (2019) An integrated lean and green approach forimproving sustainability performance: a case study of a packaging manufacturing SME in the UK. Prod Plan Control 30, 353–368. [CrossRef]
17. Xu XS, Walker H, Nairn A, Johnsen T (2007) A network approach to understanding 'green buying': a literature review. In: Presented at the 23rd IMP conference, Manchester, UK, 30 Aug–1 Sept 2007

Chapter 25
Industrial Internet of Things: A Survey

Uma Sharma, Meenakshi Sharma, Ramendra Singh, Ajay Chauhan, and Pooja Pathak

Abstract Internet of Things (IoT) is a robust tool for the renewal of industrial systems and its applications. Industrial Internet of Things (IIoT) is the blend of industrial automation and IoT. It provides automatic sensing, smart processing with correlated transmission and autoimmune systems. With IIoT data shielding is of big concern as cyber-attacks are increasing day by day. IIOT demands a lot because of operational, safety, security, and regulatory issues of industries. This paper presents the current status of IOT in industries and also gives the area where more work is to be done.

Keywords IoT · Industrial IoT · Sensors · Big data · Cloud computing · Cyber physical systems

1 Introduction

The Internet of Things (IoT) is an emerging technology that is trending nowadays and brought the development in various fields in recent years. IoT is taken as a third information technology revolution. It has changed the traditional concepts of human–computer interaction, machine-to-machine interaction to large extent and

U. Sharma · M. Sharma · R. Singh · A. Chauhan (✉) · P. Pathak
Department of Electronics and Communication Engineering, Inderprastha Engineering College, Ghaziabad, Uttar Pradesh, India
e-mail: ajay.chauhan@ipec.org.in

U. Sharma
e-mail: uma.randev@ipec.org.in

M. Sharma
e-mail: meenakshi.sharma@ipec.org.in

R. Singh
e-mail: ramendra.singh@ipec.org

P. Pathak
e-mail: pooja.pathak@ipec.org.in

R. Mathur et al. (eds.), *Emerging Trends in Data Driven Computing and Communications*, Studies in Autonomic, Data-driven and Industrial Computing,
https://doi.org/10.1007/978-981-16-3915-9_25

greatly improved the efficiency of the entire system and changed the people's life. The one sector where it has shown a large impact is the industrial sector and, with it, the industrial transformation has begun and we come up with Industrial IoT (IIoT). IIoT is attracting the attention of the industrial sector as it is expected to be able to detect and correct the problems associated in the different processes taking place within the industry with to large extent thereby improving the production efficiency.

IIoT takes information from the connected equipment in the plant. After this, the data is processed using new analytical tools and latest software systems. Data taken from the connected devices can be used with greater flexibility for better performance as compared to the data taken from individual devices. In IIOT, combination of analytical software, hardware, communications, and networking technologies work together.

The outline of the paper is structured as follows: Sect. 1 gives introduction of Industrial IoT. Section 2 compares the IIoT with IoT. Section 3 provides the architecture of IIoT. Section 4 reviews the technologies that enable IIoT. Section 5 describes applications of IIoT in industries. Research challenges and future prospective are discussed in Sects. 6 and 7 gives the conclusion.

2 IoT and IIoT

IoT enables us to connect things to the internet and to the network that uses internet technology. This is done by embedded sensors, electronics, and software. In IOT, data is captured, shared, and depending on the recent information, the action can be planned. IoT is changing our life style, way of working, business, and also giving the basis of new industrial transformation, that is, Industry 5.0. IOT can be studied in two broader categories, one is Public IOT and other is Industrial IOT.

In Public IOT, the electronic equipment are interconnected to increase the awareness and to save the time and money. So, public IOT is the interconnection of machine and user. It is human-centered and data volume handled in this is not very high. Connectivity in IOT is flexible which leads to ad hoc or mobile networking.

On the other hand, IIOT is the interconnection of industrial machines, control systems, communication systems, and the business processes. IIOT is the basis of arrival of digital and smart production. Here, the large volume of data is collected and it is given to analytical software which leads to optimized industrial operations. IIOT is the need of smart production according to the dynamically changing demands. IIOT is machine-oriented which monitors the process in production and apply new techniques to have self-driven systems. In IIOT, data volume handled is very high. The comparison of Public IoT and Industrial IoT has been summarized in Table 1 [1].

Table 1 Comparison between public IoT and industrial IoT

Parameters	Public IOT	Industrial IOT
Working model	Human-oriented	Machine-oriented
Present status	New devices interconnections	Existing devices interconnection
Communication	Ad hoc network	Centralized network
Complexity	Critical in medical applications	Critical in respect of timings, reliability, security, and privacy
Capacity	Compact data	Bulky data

3 IIoT Architecture

IIoT is the key technology that integrates large number of components, heterogenous devices, and systems to a common platform to improve the efficiency of industrial sector. Many reference architecture has been proposed by the researchers in the past for different applications. Figure 1 gives the basic architectural approach which could be the part of any IIoT framework. IIoT architecture is a layered architecture which consists of four basic layers as shown in Figure 1 [2].

1. **Data sensing layer**: In the architecture of IIOT, the data sensing layer connects the physical devices through sensors and actuators. With the sensing devices, control equipment is also used like PLCs or other controllers. The combination of field equipment and control equipment is responsible for the collection of data related to various processes from industrial site [3].

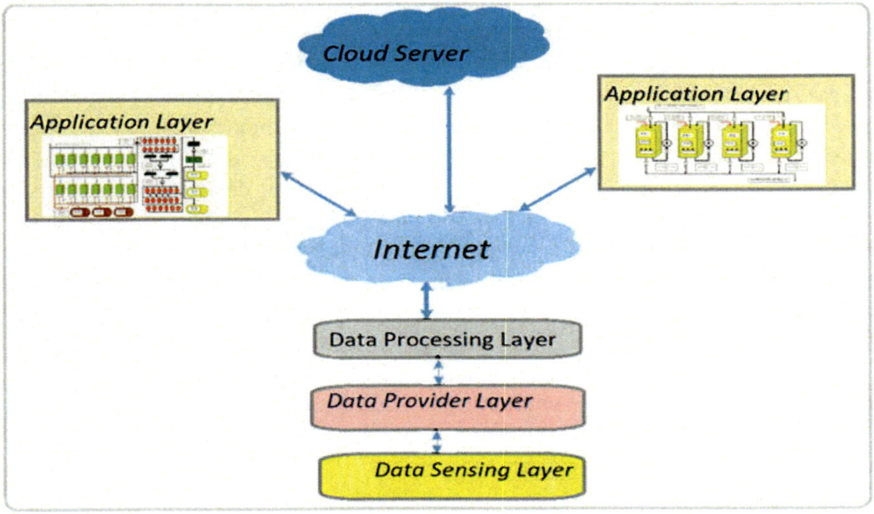

Fig. 1 Architecture of industrial IOT system

2. **Data Provider layer**: This layer is used either to take the data from sensors through field buses and store it in a memory so that the data can be passed on to the upper layers or it is used to take the data from memory when it is received from the other layers. So, the buses used in this layer are bidirectional. Memory is allocated according to Extensible Markup Language data files (XML). These files are used to give the structure to the data with tags and text.
3. **Data Processing Layer**: At this layer, temporary processing of data is done. Here, output of processing of physical and virtual data is given to the next layer through the transmission network. A middleware software such as Data Distribution Service (DDS) is used for real-time systems. Many other middleware systems are also used according to the requirement and development of IIOT systems.
4. **Application or Service Layer**: For the remote monitoring and controlling, the particular application of industry, application, or service layer is used. For the interaction of the things, specific application can be developed like SCADA (Supervisory Control and Data Acquisition). At this level, developed software is used for monitoring the industrial environmental things and for sending the commands as per the requirement [2].

4 Enabling Technologies in IIoT

The potential of any technology depends on other enabling technologies which helps in understanding and gaining the knowledge of it. In the same way, IIIoT is supported by various enabling technologies without which it has no existence. These technologies include IoT, cloud computing, big data analytics, artificial intelligence, industrial sensing technology, augmented and virtual reality cyber-physical system.

Internet of Things: If we consider the associated plant situation, IoT gadgets aid real-time information assortment and activation. Being the primary segment in IIoT, these gadgets trail the plant resources all around. Without human intervention, the entire cycle that begins with crude stuff and culminates into finished products is checked by employing IoT gadgets to accomplish huge decrease in the overall cost. The IoT gadgets incorporated in completely associated IIoT framework are implemented throughout the manufacturing plant offices going from stockrooms to creation offices and distribution centers. The setup, observation, and support of these gadgets is a difficult undertaking and require profoundly qualified specialized staff [4].

Big Data Analytics: Huge advancement in big data analytics prompted advanced knowledge in IIoT-based frameworks [5]. The data collecting technologies allows to huge amount of data from different sources such as sensors, smart devices web-enabled data sources, and many more. At different IIOT layers, this data is processed by using different data processing tools and algorithms [4].

Cloud Computing: The prominent development of information in IIoT needs profoundly disseminated elite registering frameworks to oversee, measure, dissect, and reserve the information. Cloud computing innovations offer network and capacity administrations across all the offices in an IIoT framework. Every associated gadget and implementation are straightforwardly annexed with clouds at the backend [4].

Cyber-Physical System: Cyber-Physical Manufacturing Systems (CPMS) play an indispensable role, checking actual cycles, and making a virtual copy of the actual world also called as advanced twin, which is accessible and open for people to control and work the whole production process. In context to smart manufacturing, CPMS empowers actual objects like machines, which are furnished with the IIoT, to speak with people through web network [6].

Augmented and Virtual Reality: IIOT and Augmented Reality (AR) have opened another area of its own, and the applications are far and wide all through the industries. AR advances empower to screen the laborers and machines during activities and quickly produce modified or notifies so that error can be reduced. The Virtual Reality innovations encourage in envisioning the setups and re-arrangements of industrial capacities prior to genuine executions in IIoT frameworks [4].

Industrial sensing and Networking Technology: This technology is used for monitoring real-time production. Industrial networking technologies include the classical fieldbus technologies and rising communication standards, such as LoRa, NB-IoT, 5G, etc. [7].

5 Applications

Applications of IOT are in its growing stage. As IoT is evolving itself, it has got its application in many areas. Our discussion is about the applications of IOT in various industries. Here, the few industrial IOT applications are given in brief.

1. **Management of logistics and Transportations**: Logistics and transportation industries apply IOT to observe the real-time movement of things from one place to final destination in the distribution and shipping processes. Sensing, networking, and data processing capabilities of vehicles are growing through IOT. Drive system is developed by some vehicles to provide present location and driving directions [8, 9].
2. **Safety in hazardous industries**: Human safety is increased by IOT in the industries with hazardous environment. Based on the area of application, IIoT devices themselves will take quick action required for the overall safety of the manpower. For example, if the temperature goes above the specific limit in case of boiler or a furnace IIoT devices give alarm to stop detonation or explosion [8, 9]. In order to avoid and decrease injuries in mining sector, there is an urgent need of IIoT to prevent the mining disaster.
3. **Smart factory**: Equipment allowed by IIoT can sense the environment and relay information to managers, enabling them to control their factory units remotely. Remote manager can maintain the machinery incase of deviation from specified parameters according to the information provided by the IIoT sensor.

Similarly, production process and inventory process can be managed as per the data provided by IIoT sensors [9].

4. **Quality Control**: When IIoT is employed in industry while manufacturing any product, it enhances the quality of the product produced. The reason being that various important parameters are maintained in the range that ensures quality production [9].

6 Future Prospective and Challenges

IIoT showed a lot of potential and development in industry sector. But due to its diverse and complex structure numerous important confrontations require to be resolved before its widespread deployment. These challenges include coexistence of other wireless technologies and heterogeneity, inter-operability, energy efficiency, mixed criticality, fault tolerance, scalability, functional safety, security challenges, and legacy long-lived industrial systems [4, 7]. Few of challenges which provide open areas where the research is to be conducted are discussed here briefly.

Energy Efficiency: Numerous IIoT applications require batteries for their operation. This requires the plan of low-power sensors which don't need battery substitution over their lifetimes. This makes an interest for energy-productive plans. To supplement such plans, upper layer approaches can assume significant parts through energy-effective activity. Numerous energy productive plans for remote sensor organization (WSN) have been proposed, however, those methodologies are not promptly applicable to IIoT. IIoT applications need a concentrated implementation of various mechanisms. Sensed information can be sent in questioned structure or in a constant structure which when send in dense deployment can devour a lot of energy. Green systems administration is consequently essential in IIoT to diminish power utilization and operational expenses [1].

Security: IIoT is a combination of industrial automation frameworks and IoT frameworks. It highlights extensive detecting, interconnected transmission, wise handling, self-association, and self-support. Its applications range intelligent transportation, shrewd industrial facilities, and knowledge. Numerous territories, for example, power grid and smart environment detection. With the broad use of IIoT innovation, the network safety dangers to IIoT frameworks are expanding step by step, and data security issues have become a significant challenge in its growth [3].

Coexistence of wireless technologies: As the IIoT connectivity is growing rapidly, there could be the deployment of many devices in same spectrum, i.e., the devices will lie in close proximity of each other. Because of the dense deployment of devices, it is a major challenge to combat with the interference between the devices and keep it to the minimum level. Moreover, coexistence of wireless technologies also made it difficult to choose the appropriate technology for a particular application as it is not possible to have all the features from one technology [1, 5].

Heterogeneity: The setup of IIoT framework incorporates an amalgamation of various diversified technologies, for example, robotics, IoT devices, sensors, actuators, gateways, edge nodes, cloud data servers different wired/wireless communication, and cellular networks. Integration and collaboration between these heterogenous devices and technologies become more challenging when the factors like synchronization, data privacy, sharing of resources come into the picture. Much more research and work are to be done in this area as more flexible and efficient techniques are require to bring all this together [5].

Fault Tolerance: With hundreds of devices working together in the industrial sector management of faults in the industrial internet sector becomes the major issue to work upon. The faults can occur due to unreliable communication, malfunctioning of local machines, cloud computing defects, and malicious attacks that severely degrade the performance of the overall system. Therefore, developing a fault-tolerant device is important. Mechanisms for recovery to respond to unexpected failures are required; otherwise, it could conduct to degradation or unavailability of service [7, 10].

Scalability: There are many issues of scalability as IIoT connect number of components machine and factories for its deployment. There are three major issue related with the scalability.

a Scalability of data: As the number of devices increases the large amount of sensing data is generated continuously and it require to pre-process before its uploading can be done. This large amount of frequently generated data offers a challenge to scalability of system

b Scalable collaboration: In customary modern frameworks, the control frameworks are generally freely designed and don't scale. Consequently, enhancing the interoperation turns into a confrontation to empower heterogeneous gadgets and frameworks to team up and communicate.

c Scalable Management: The integration of different modern segments and frameworks places the challenge of management and maintenance. Modern management technology, in order to attain scalability, needs to be incorporated into the framework [7].

7 Conclusion

IIoT is the creating innovation that will change the working of enterprises totally. Since IIoT will possess each industry in future, it is essential to comprehend. IIoT innovation and its related applications in the business. IIoT applications are getting progressively incredible, more affordable, and more modest. In spite of the fact that this innovation is popular, the threats it poses are protection and security, which should be amended. As IIoT is multi-viewpoint, this paper attempts to give vivid presentation regarding clear vision about IIoT innovation, its details, and utilizations in ventures by investigating different examination papers, white papers, and online data set.

References

1. Sisinni E et al (2018) Industrial internet of things: challenges, opportunities, and directions. IEEE Trans Ind Inform X(X)
2. Ungurean I et al (2020) Software architecture for the industrial internet of things—a conceptual model. J Sens 20(19). https://doi.org/10.3390/s20195603. 30 Sept 2020
3. Chen H et al (2019) Research on industrial internet of things security architecture and protection strategy. In: International conference on virtual reality and intelligent systems (ICVRIS)
4. Khan WZ et al (2020) Industrial internet of things: recent advances, enabling technologies and open challenges. J Comput Electr Eng 81, 106522
5. Rehman MH et al (2018) Big data analytics in industrial IoT using a concentric computing model. IEEE Commun Mag
6. Yu W et al (2019) Implementation of industrial cyber physical system: challenges and solutions. In: IEEE international conference on industrial cyber physical systems (ICPS). 6–9 May 2019
7. Li JQ et al (2016) Industrial internet: a survey on the enabling technologies, applications, and challenges. IEEE Commun Surv Tutor
8. Da Xu L, He W, Li S (2014) Internet of things in industries: a survey. IEEE Trans Ind Inform 10(4)
9. Bhati MS (2018) Industrial internet of things (IIoT): a literature review. Int J Res Eng Appl Manag (IJREAM) 04(03). ISSN: 2454-9150
10. Breivold HP, Sandström K (2015) Internet of things for industrial automation—challenges and technical solutions. In: IEEE international conference on data science and data intensive systems

Chapter 26
Analysis and Measurement of BER and SNR for Different Block Length in AWGN and Rayleigh Channel

Sonam Gour, Gaurav Kumar Soni, and Abha Sharma

Abstract Rapid change in the technology for seamless transmission of data at very high rate poses a great challenge for system designers. Signal is highly affected by the Noise and Error, reliability of the signal in the communication system is directly affected by the noise. Different types of Error control and correction codes are used for analyzing and controlling the error in the signal. Signal is directly affected by the Noise or error; it will change the reliability of the communication system. The occurrence of the errors is controlled by the Error control coding procedure. Shannon coding is also used for the error correction in the signals. The amount of redundancy added in any signal is increased as per the increment in the delay. In this paper, the author proposes a new interleaver for tolerating the latency situation and increasing the performance. It becomes more important to design and analyze the performance of interleaver for turbo code. It presents the comparative analysis of turbo code performance over an interleaver for various block lengths for two different types of channel which is AWGN and Rayleigh channel. It also provides the comparative analysis for the different fading channels.

Keywords Shannon · Fading · Interleaver · Noise

1 Introduction

In telecommunication, the channel coding becomes an important aspect for the design constraints. It is defined by the power and bandwidth constraints in the modern communication system. The error in the digital communication is controlled by the channel encoder circuit in the transmitter section and decoder circuit in the receiver.

Every communication system has three parts: a transmitter for transmitting the signal, one receiver for reception, and the channel for connecting to them. This concept is applied in both the systems, either analog or digital. Shannon provides a block range with long signals for supporting the capacity limit. The main drawback

S. Gour (✉) · G. K. Soni · A. Sharma
Arya College of Engineering & Research Centre, Kookas, Jaipur, RJ, India

R. Mathur et al. (eds.), *Emerging Trends in Data Driven Computing and Communications*, Studies in Autonomic, Data-driven and Industrial Computing,
https://doi.org/10.1007/978-981-16-3915-9_26

of the Shannon coding is that he cannot propose a practical implementation of the coding scheme. In the encoder circuit, if redundancy is increased, also the delay has increased, and the tolerance limit has not been defined by Shannon. In this paper, author compares the different types of interleaving technique calculating the latency of the encoder. For correcting the error in the signal, turbo codes encoded the signal twice, and checks or corrects the error. Turbo code uses a Soft output decision algorithm for measuring probability of error in particular bits in the process of decoding. The "hard" decision only determines the corresponding value as per the 0 or 1 decision, but the "soft" decision of the encoder checks the value and passes to the previous encoding circuit [1]. Maximum Likelihood Algorithm (MAP) and Soft Output Viterbi Algorithm (SOVA) lie in this category. MAP algorithm minimizes the probability of symbol error, and SOVA minimizes the probability of sequence error [2]. Multipath fading is used for the fast transmission rate and transmitting the data in the mainstream for more capacity. If we reduce the bit error rate (BER), the downlink speed of the system is also reduced, and it provides a great scope in the 5G technologies [3]. The Rician channel fading is not directly affected by the Bit Error Rate. It only affects the channel when utilized modulation type has been changed [4].

2 Interleavers and Channel Modeling

The performance of any Turbo code is analyzed by the interleaver. It lowers the error in the transmission and shows the good performance. The technique developed for the forward error correction is more accurate and correct as compared to the burst error. Author proposed one new technique of the interleaving for making forward error correction. It is more robust with respect to respective burst errors [5]. A TurboAE-MOD shows the moderate modulation in the block length modulation scheme and it performs canonical modulations with the code stacked performs comparable to modern codes stacked. It represents the Gaussian-like received values at the decoder [6]. The main purpose for using the interleavers is to reduce the error floor condition in the turbo codes. It occurs due to the small minimum distance captured by the encoder. The following figure shows the water fall region and the error floor region (Fig. 1).

The delay was also reduced by the 16-bit Turbo code model. Turbo Encoder with 16-bit input shows the more precise delay and latency as compared with the 8-bit input delay model [7].

3 Turbo Encoder and Decoder

Turbo encoder and decoder are used for encoding and decoding the signal. Here, we are representing the encoder and decoder used in the Turbo Code.

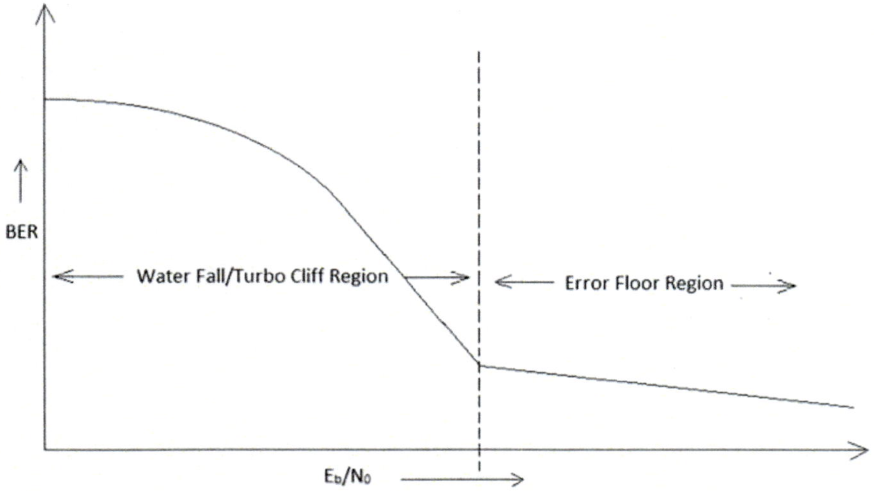

Fig. 1 BER curve showing the waterfall region and the error floor

3.1 Encoder

Turbo encoders generally used two convolution codes and single interleaver. Interleaver is used between the two encoders. The encoder in the turbo system is designed by the Recursive Systemic Convolution Codes. The structure provides a good result. The following Fig. 2 shows the turbo encoder scheme.

Interleaver in the turbo code is designed for the two purposes: first, if the input of the second encoder design interleaved, output of both the encoders is quitely different. Output is usually quite different from the output of the first constituent encoder. Second is that the code is the parallel combination of two codes so that it can gain knowledge by exchanging the knowledge of two codes [8].

3.2 Decoder

The structure of turbo decoder with the iterative technique is shown in Fig. 3. Interleaver provides the link of the two components in decoders as well as encoders. Turbo decoder has three inputs: first are the encoded channel output bits, parity bits transmission from the connected component encoder circuit, and the information from the other decoder for analyzing the concerning bits. Prior information is generated by the other decoder and provides the soft and the hard output. The decoder circuit also provides the correct probabilities of each bit for checking the correction in the bits. The soft outputs are represented in the form of Log Likelihood Ratios (LLRs).

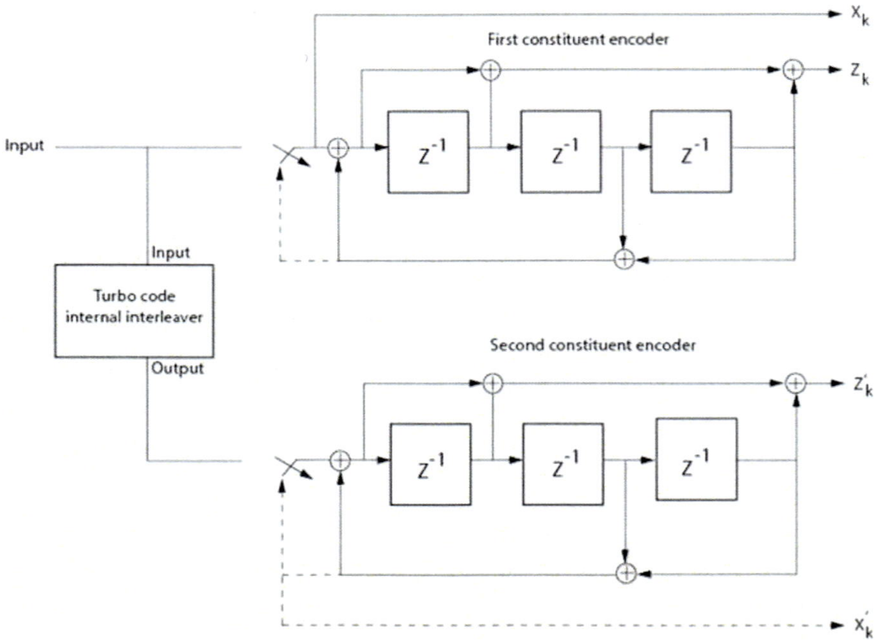

Fig. 2 Turbo encoder schematic diagram [8]

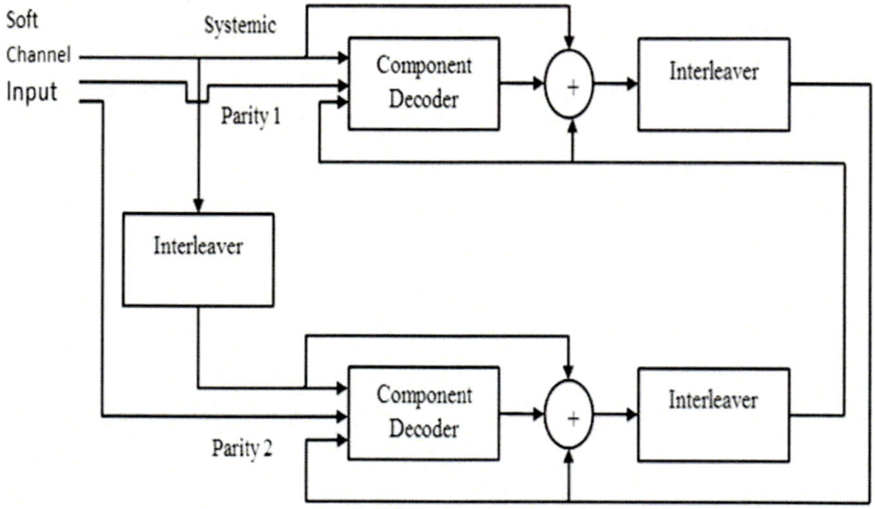

Fig. 3 Turbo decoder schematic diagram [9]

The sign of the bit is determined by the polarity of the LLR, and amplitude explains correct decision probability.

4 Proposed Methodology

In this paper, the performance of the turbo code is examined by an interleaver. It is evaluated by simulating the AWGN channel and Rayleigh channel. With this technique, the wide variety of channel distribution is achieved. The simulation process is conducted by the MATLAB.

Single-path Rayleigh fading and dual-path Rayleigh fading are used in the AWGN channels with the 10 us delay spreading. Error correcting schemes are gained by the two recursive non-systematic parallel connections [10]. Denoising Turbo-based transmission is used for separating the background and foreground noise to the transmitted signal [11]. In any encoder and decoder circuit, the trade-off is applied in the receiver complexity, running time, and performance [12]. The bit error rate is getting smaller by Eb/N0. The interleaving technique in the turbo code is used for reducing the burst error. The proposed technique is shown as follows (Fig. 4).

5 Result and Discussion

As per the graph shown between the SNR and BER region of the Turbo Code, it represents the different AWGN noise at the separate level. It shows the comparison between a random interleaver and proposed interleaver for calculating the noise in AWGN channel and Rayleigh channel. The result shows that the proposed technique interleaver is better than the AWGN channel interleaver. There is comparison of BER performance in random interleaver and a new proposed interleaver for AWGN channel and Rayleigh channel. The comparison between the proposed and random interleaver for AWGN is shown in the following Fig. 5. It is the combination of random interleaver and matrix interleaver. The RSC ½ encoder are used at the receiver side by using the Log-MAP algorithm. The large block length is achieved by this interleaver technique with the low BER.

After comparing the previous graph, the next graph shows the comparison in the performance of the interleaver for the Rayleigh Channel. The below Fig. 6 shows the average BER (bit error rate) is better than the random interleaver. The performance is analyzed by using the block length of 512.

The following Figs. 7 and 8 show the comparison between the AWGN and the Rayleigh techniques for the maximum block length of 1024 bits. It shows that the proposed interleaver has a better BER than the random interleaver.

The below Fig. 9 represents the BER performance for the one to third iteration for the proposed interleaver for AWGN channel. The block length for the coding

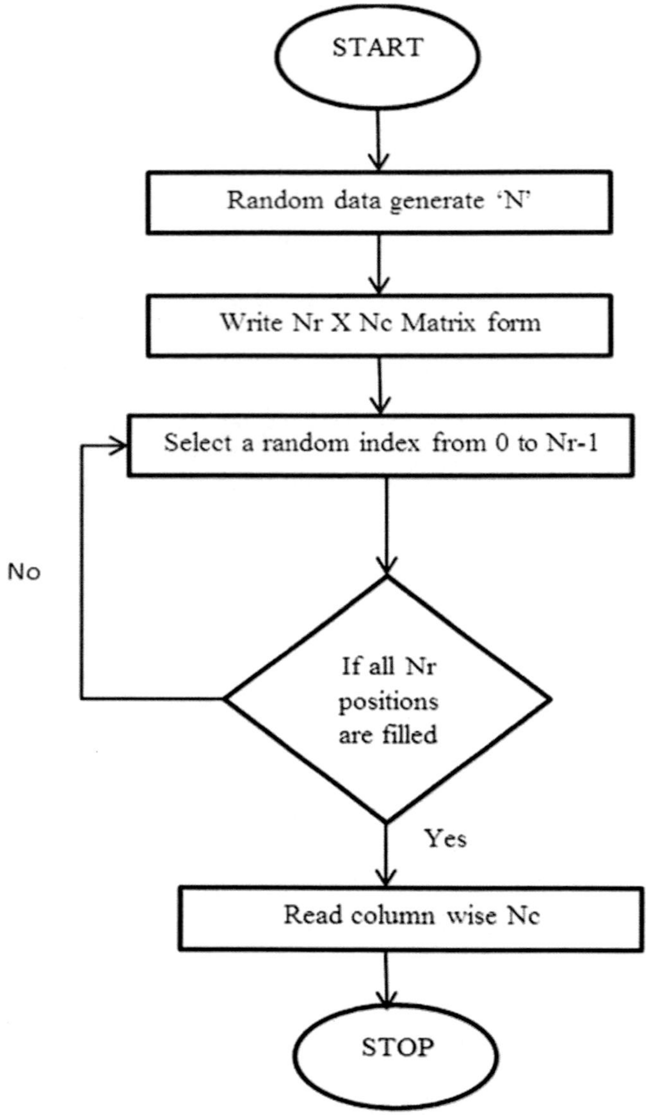

Fig. 4 Flow chart of proposed work

technique is 10^4. The graph shows that the number of iterations increases as per decrement in the BER.

The effect of the BER and SNR Rayleigh channel in the proposed interleaver and one to third interleaver is compared as per Fig. 10. As the BER decreases as per the iteration, the SNR increases.

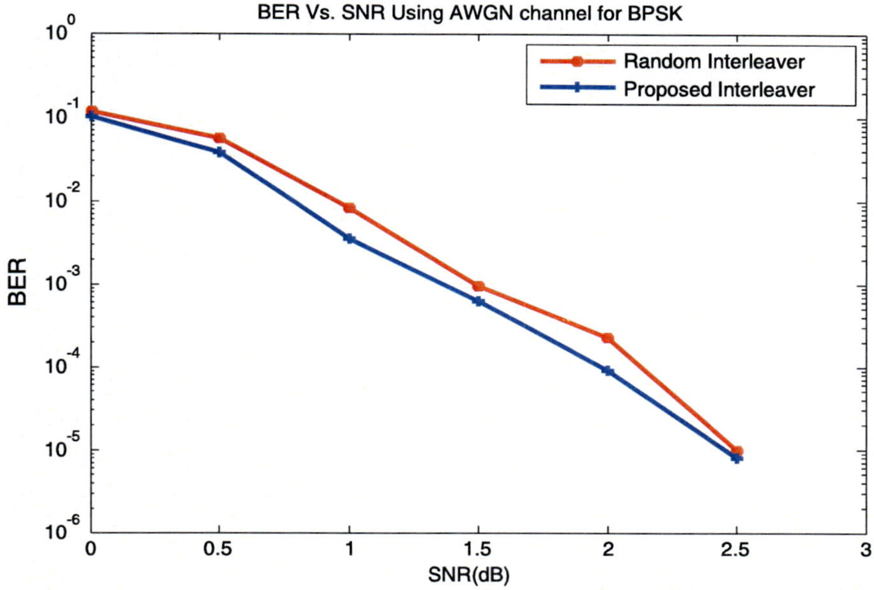

Fig. 5 Comparison between random interleaver and proposed interleaver for AWGN channel for N = 512

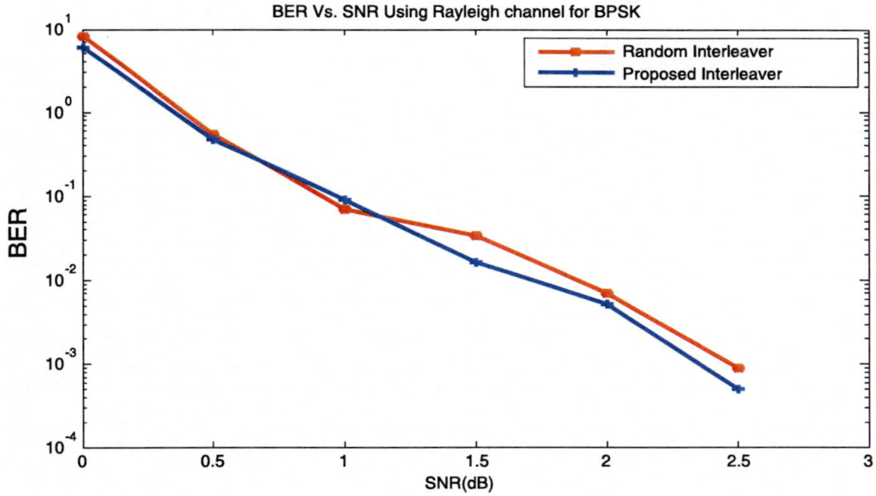

Fig. 6 Comparison between random interleaver and proposed interleaver for Rayleigh channel for N = 512

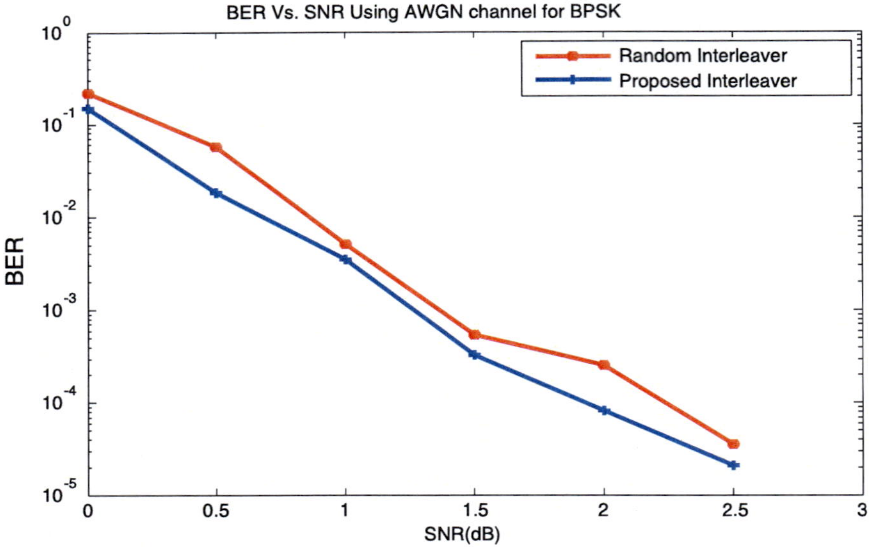

Fig. 7 Comparison between random interleaver and proposed interleaver for AWGN channel for N = 1024

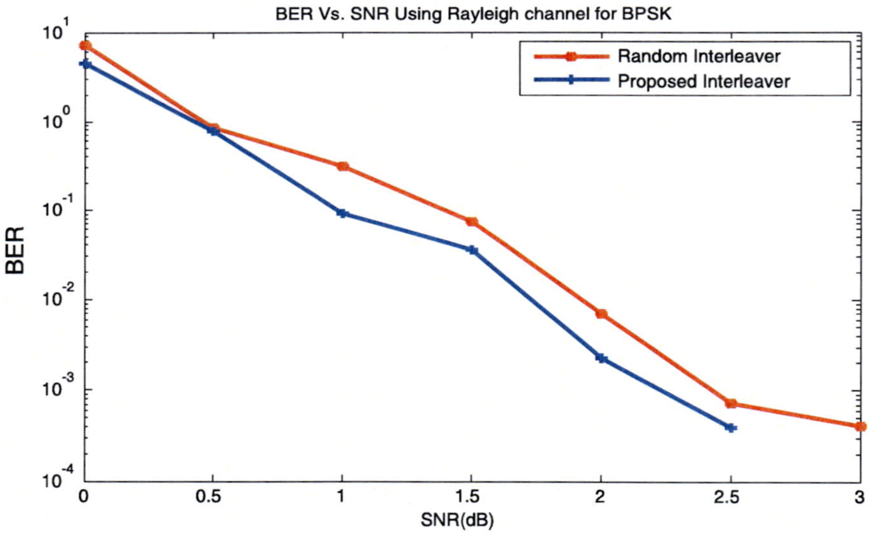

Fig. 8 Comparison between random interleaver and proposed interleaver for Rayleigh channel for N = 1024

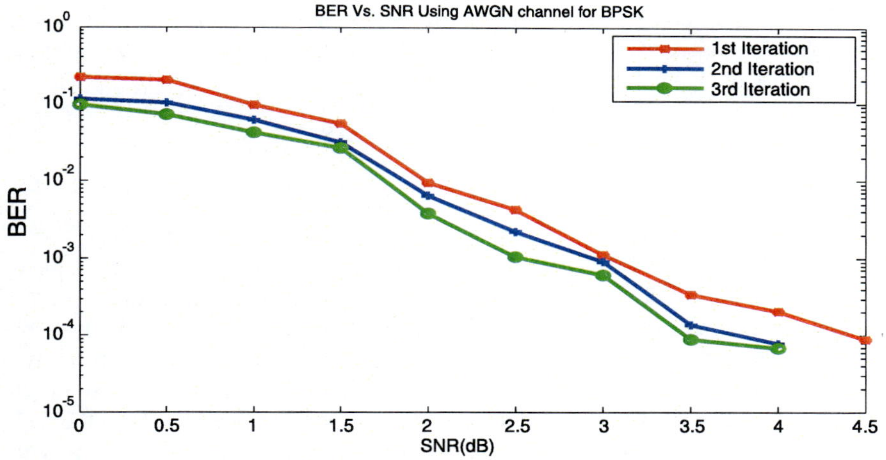

Fig. 9 Three iteration of proposed interleaver for AWGN channel $N = 10^4$

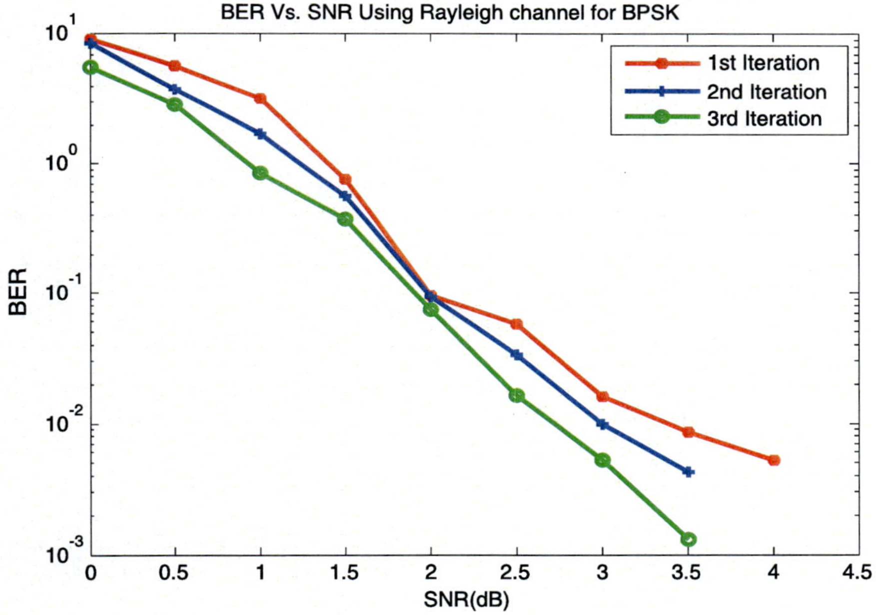

Fig. 10 Three iteration of proposed interleaver for Rayleigh channel $N = 10^4$

The coding is conducted by the interleaver. It has been simulated by using MATLAB for the AWGN and Rayleigh channel. By this channel distribution, a wider variety of channel distribution is achieved.

6 Conclusion

The paper discusses the Turbo code with new interleaver techniques. The experiment is carried out for the different block lengths for conducting the experiment for the AWGN channel and Rayleigh channel. It represents the performance analysis of turbo encoder and decoder. With the use of parallel concatenation of two recursive non-systematic convolution code, the BER rate and the SNR ratio of the system are reduced and presents better performance. The interleaver is designed for the 512 and 1024 length. The design is proposed for the Log-map decoder algorithm. The simulation result shows that the proposed techniques have the capacity for reducing the low bit error rates with the use of Eb/N0. The interleaver reduces the burst error rate in the simulation circuit.

References

1. Berrou C, Glavieux A, Thitimajshima P (1993) Near Shannon Limit error-correcting coding and decoding: Turbo codes. In: Proceedings of the international conference on communications, (Geneva, Switzerland), pp 1064–1070
2. Hagenauer J, Hoher P (1989) A Viterbi algorithm with soft-decision outputs and its applications. IEEE Globecom, pp 1680–1686
3. Jangir D, Shankar G, Jain BB, Soni GK (2020) Performance analysis of LTE system for 2 × 2 Rayleigh and Rician fading channel. In: Proceedings of the international conference on smart electronics and communication (ICOSEC), pp 961–966
4. Singh H, Soni GK, Suryawanshi P, Shankar G (2020) Performance analysis and BER comparison of OFDM system for 4 × 4 MIMO fading channel in different modulation scheme. In: Proceedings of the third international conference on smart systems and inventive technology (ICSSIT), pp 255–259
5. Feng W, Yuan J, Vucetic BS (2002) A code matched interleaver design for turbo codes. IEEE Trans Commun 50(6):926–937
6. Jiang Y, Kim H, Asnani H, Kannan S, Oh S, Viswanath P (2020) Joint channel coding and modulation via deep learning. In: 21st international workshop on signal processing advances in wireless communications (SPAWC)
7. Kumar PP, Gowreesrinivas KV, Samundiswary P (2016) Design and analysis of turbo encoder using Xilinx ISE, 2016. In: International conference on control, instrumentation, communication and computational technologies (ICCICCT), pp 328–332
8. Barbulescu A, Pietrobon SS (1994) Interleaver design for turbo codes. Electron Lett IEEE 30(25):2107–2108
9. Takeshita O, Costello D (2000) New deterministic interleaver designs for turbo codes. IEEE Trans Inf Theory 46(6):1988–2006
10. Dehdashtian S, Hashemi M, Salehkaleybar S (2021):Deep-Learning based blind recognition of channel code parameters over candidate sets under AWGN and multi-path fading conditions. IEEE Wirel Commun Lett

11. Xue Z, Yuan X, Yang (2020) Denoising-based Turbo message passing for compressed video background subtraction
12. Kocian A, Badiu M, Fleury BH (2017) Francesca Martelli3 and Paolo Santi: a unified message-passing algorithm for MIMO-SDMA in software-defined radio. J Wirel Commun Netw

Chapter 27
Software Development Life Cycle Performance Analysis

Shilpa Sen, Mayank Patel, and Ajay Kumar Sharma

Abstract In 1960s SDLC (Software development life cycle) was initiated to develop high quality systems in large scale. Software development life cycle is one of the most liked techniques nowadays to develop a powerful and high quality software product. Here various types of methods present in software development life cycle like waterfall, iterative, prototype, v-shaped, spiral and etc. each method has drawback and benefit. According to any one of these models, the software will be developed. This method helps a lot to developer's team to develop a product in easy and faster way. This technique is referred to the decrement of problems and increment of value of the software project. The SDLC methodology is very hard without a sequence order process to develop a good quality product (Manzoor Ahmad Rather: "A Comparative Study of Software Development Life Cycle Models", IJACSA (2015).).

Keywords SDLC model · V-shaped · Prototyping · Difference · Disadvantage

1 Introduction

In the world of technology it is important to build a good quality product which meets user requirement. SDLC is the system development life cycle which helps us to manage and meet the product approach in the symmetric disciplinary way. It is usual procedure followed by software companies for understand requirement, design, implement, checking and evolution of product. Software engineering is referred to the quality priority, procedure, technique and devices that are used in process of developing a product. This process includes various kinds of task and activities and also this process gives the method to implementing a technique that covers requirement, design, implement, checking and evolution of product. SDLC is separated into different parts that allow any software company to handle the project in easy way. It is very famous and important technique develops software [1].

S. Sen (✉) · M. Patel · A. K. Sharma
Geetanjali Institute of Technical Studies, Udaipur, India

R. Mathur et al. (eds.), *Emerging Trends in Data Driven Computing and Communications*, Studies in Autonomic, Data-driven and Industrial Computing,
https://doi.org/10.1007/978-981-16-3915-9_27

1.1 Phases of SDLC

There are various phases in SDLC given below (Fig. 1).

Requirement Analysis.

In this phase project developer's team make sure that they will gather all the requirements to client about project like user interface, functional requirement, framework selection, system technical architecture, and project planning [2, 3].

Design.

This phase is about finalizing design, user interface, requirement detail specification, and system technical architecture [4].

Implementation.

In this phase developers implement the code according to finalized design and before sending for the main testing of project they test the code that is called unit testing [5].

Testing.

This phase is the most important part of the software development life cycle because they check code quality and test according to requirements. After this phase project will be sent for deployment [6, 7].

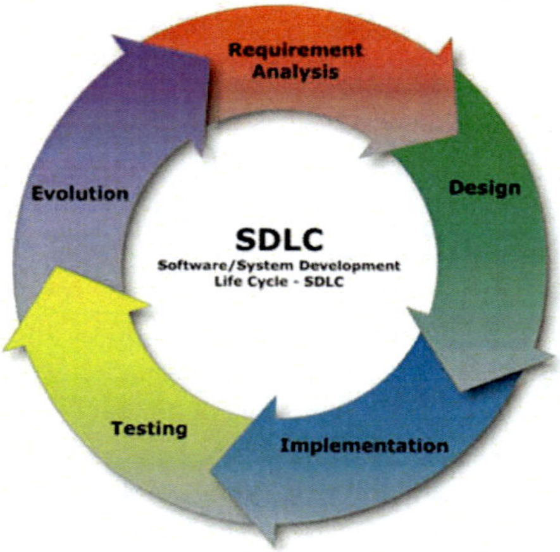

Fig. 1 Phases of SDLC

2 Types of Models

2.1 Waterfall Model

In 1970 Royce proposed this model named as "waterfall model". The waterfall model is a popular form of the systems development life cycle model for software engineering [8]. The waterfall model describes a classic approach of development method as sequential and linear. Waterfall model has different approaches for each interpretation of development. It works like the output of each step in this model is the input of next step [9]. Once a development of interpretation is finished, development proceeds to the next interpretation and no overlapping happens. The interpretation includes Requirements definition, Design, Implement, Checking, and Maintenance [2, 10] (Fig. 2).

Requirement Analysis. All possible requirements are captured in requirement gathering documents.

Design. Based on requirement analysis software architecture design was finalized.

Implementation. Coding of the project in the little unit with its functional testing.

Checking. To make sure that the project is meeting customer requirements.

Maintenance. Fixing issues and release new version with the issue patches as required.

Advantages. It is understandable, simple and easy to use and easy to manage as each interpretation has its outputs and review process. It is well for smaller products where requirements are very well cleared.

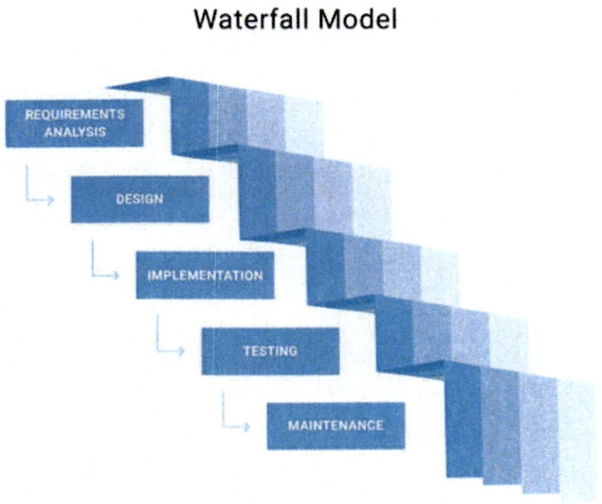

Fig. 2 Waterfall model

Disadvantages. When the project is checking interpretation, it is not possible to turn behind and update something which is not well. It is highly risky and uncertain method.

2.2 *The Iterative Model*

When some problems were seen in the first version of the model, then a new version of the model was created which was named as iterative model [2, 11]. It is the combination of both prototype model and waterfall model, and it is upgraded version of waterfall model. Iterative model comes when waterfall model is come up with a lot of drawbacks [12]. It splits the task into different fragments so that the product developing group can do their task faster and in an easy way. This model will work in the loop of Design, Implementation, Testing, and Review until software is deployed. This model is more flexible and it includes more features then waterfall model (Fig. 3).

When To Use Iterative Model

It uses when the product is big; the product's requirement is completely defined and the large requirement is clearly defined.

Advantages

It is faster, simpler, and easier to manage the classical model.

Disadvantages

This method is not well for smaller projects. It is not understandable and we cannot change requirements on multiple times.

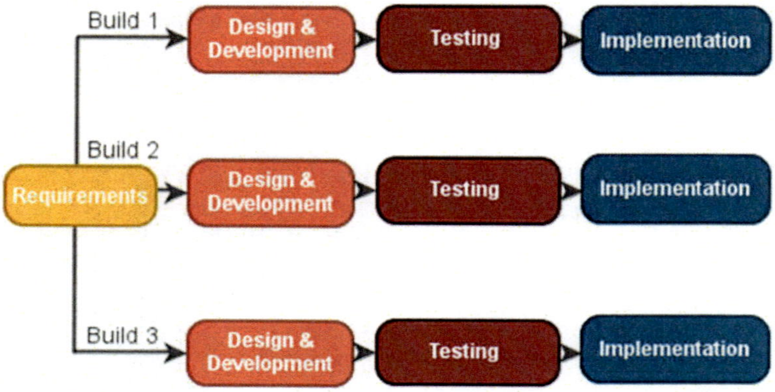

Fig. 3 Phases of SDLC

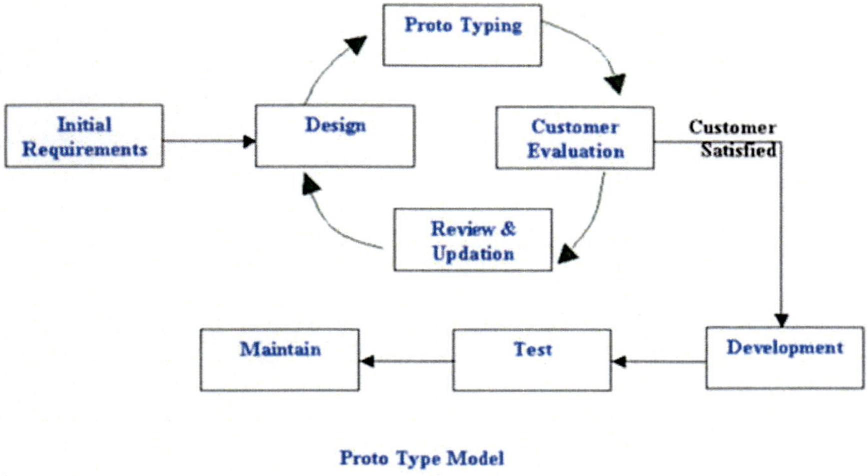

Fig. 4 Prototype model

2.3 *The Prototype Model*

Prototype means a kind of dummy also you can say a toy model of data which created according to client's requirement to develop a product [13]. This model is created only when the customer's ideas are not clear. This model is also called as "Throwaway model" because if the customer is not clear with their requirements then the prototype model will be created again. Then it will process to iterative model for design and developing a product (Fig. 4).

Advantages

This method is faster in design, its outcome is good, and it is simple to find bugs, it is understandable to developers.

Disadvantages

It is costly than other methods; it is not flexible and not good for big projects.

2.4 *The Spiral Model*

In 1986s Barry Boehm referred to this method. It is the most important model among all models in SDLC [14]. This model discards the mistakes of all models. Spiral model can analyze higher risk and also it can find an alternative solution to make better product without any errors. This model has four steps named as planning, Analyses of Risk, Design, Implement, Checking, and Maintenance. And its diagram

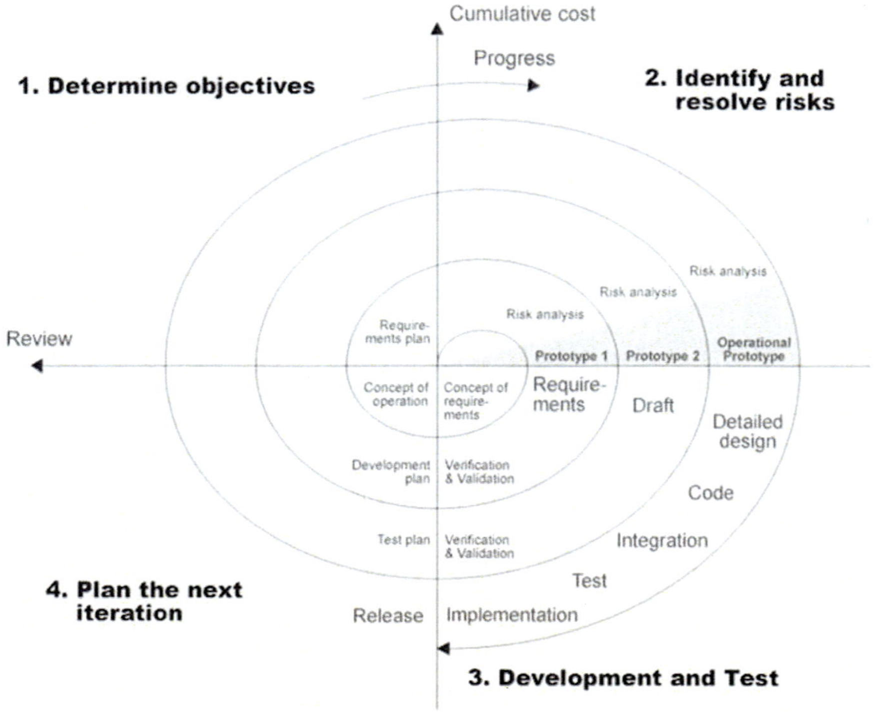

Fig. 5 Spiral model

(Fig. 5) looks like a spiral. This model is used when the product is large and has more time duration to complete the product.

Advantages

It can easily handle risk and it is better for big products; it is flexible according to requirement; it takes less time to deploy the product.

Disadvantages

It is more complicated than all models of SDLC; it is not good for small products and it is not easy in managing time during development.

2.5 V–Model

V-model is much-disciplined systems development life cycle model for software engineering which also can be considered an upgrade version of the waterfall model [15]. V-model describes relationships between the testing stage which is parallel to

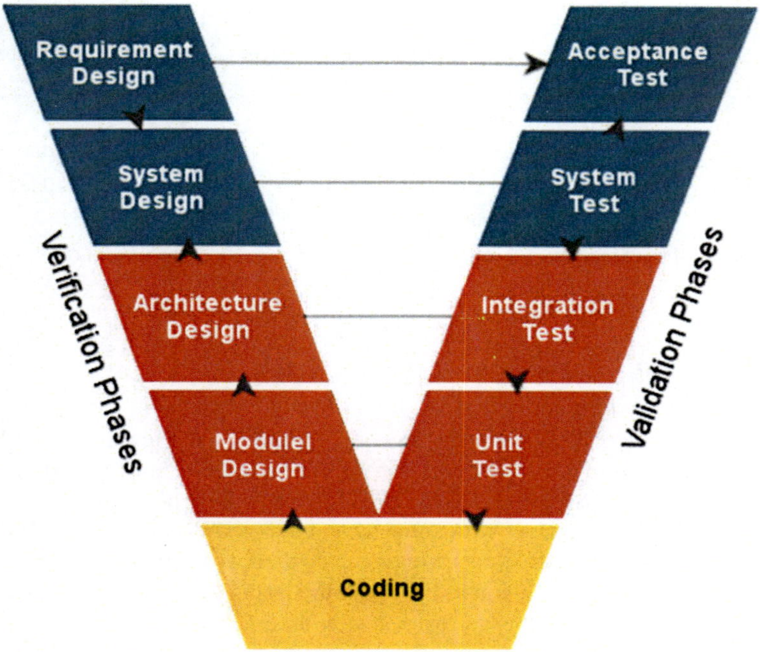

Fig. 6 V–Model

each development stage, i.e., In V-model testing is done on each interpretation that is parallel with development in a steady way. It is known as the Verification Model or Validation Model (Fig. 6).

Advantages

It is easy, understandable, simple and better for big products and it is easy errors.

Disadvantages

It is not flexible, more complex, not frequent, and highly risk; it is very high in cost and not good for small products.

3 Difference Between Models

See Table 1.

Table 1 Difference between all models

Model	The waterfall model	The iterative model	The prototype model	V-Shape model	Spiral model
Payment	Not high	Not high	High	Very high	High
Time	Long	Less	Less	Less	Flexible
Elasticity	Low	Low	High	Low	Flexible
Safety	Low	Limited	Low	Limited	High
Evaluation of risks	Only initially	Low	High	Low	Low

4 Conclusion

Since 1960 still today SDLC has made the work of project handling team easy to build software. All models of software development life cycle have their own advantages and disadvantages [1]. Software development life cycle brought new forms at every level and updates at each level with some advantages and disadvantages. By these different types of models a developer's team can do their work easier and faster. Main conclusion of software development life cycle models is to simplify the process of developing a software project easy, faster, and with less risk.

References

1. Rather MA (2015) A comparative study of software development life cycle models. IJACSA
2. Navita (2017) A study on software development life cycle & its model. IJERCSE
3. Boehm B (2000) In: Hansen WJ (ed) Spiral development: experience, principles, and refinements
4. Pressman R, Software engineering: a practitioner's approach, 6th edn. McGraw-Hill Publication
5. Aggarwal KK, Yogesh Singh Software Engineering, 3rd edn
6. Karlm (2006) Software lifecycle models, KTH
7. Rastogi V (2015) Software development life cycle models-comparison, consequences. Int J Comput Sci Inf Technol 6(1). ISNN: 0975–9646
8. Maheshwari S, Jain DC (2012) Comperative analysis of different type of models in software development life cycle. Int J Adv Res Comput Sci Soft Eng 2(5):285–290
9. Munassar NMA, Govardhan A (2010) A comparison between five models of software engineering. Int J Comput Sci 7(5):94–101
10. Raval RR, Rathod HM (2013) Comparative study of various process model in software development. Int J Comput Appl 82(18):16–19
11. Ben-Zahia MA, Jaluta I (2014) Criteria for selecting software development models. In: 2014 global summit on computer & information technology (GSCIT). IEEE
12. Velmourougan S et al (2014) Software development life cycle model to build software applications with usability. IEEE
13. Velmourougan S et al (2014) Software development life cycle model to improve maintainability of software applications. IEEE

14. Thitisathienkul P, Prompoon N (2014) Quality assessment method for software development process document based on software document characteristics metric. In: 2014 Ninth international conference on digital information management (ICDIM). IEEE
15. Davis A.M, Bersoff E. Hm Comer E.R, A strategy for comparing alternative oftware development life cycle models. In: IEEE transactions on Software Engineering, vol 14, Issue 10. Publication Year: 1988

Chapter 28
Scrum Framework Based on Agile Methodology in Software Development and Management

Upasana Ameta, Mayank Patel, and Ajay Kumar Sharma

Abstract With the advancement in technology, the demands for software has also increased. Companies are applying different methods to increase performance and business-related profits. The methodology employed in software development is software development life cycle (SDLC), but with the developing technology and growing market, the software-related requirements changes rapidly. So, another methodology called "Agile" came into existence. This is an improved version of SDLC which increases speed, flexibility, and responsiveness in software development. Scrum is a framework, that is based on Agile methodology, that follows a repetitive and incremental procedure. This paper starts with the introduction of Agile methodology and scrum, followed by procedure and description of Scrum values.

Keywords Agile methodology · SDLC · Scrum · Sprint · Product backlog

1 Introduction

To meet the need of growing market demands, companies are required to deliver reliable, economic, and quick-delivering software. Also, companies attempt to limit the development time and maximize their business profit. Software development can be understood as creating a new software to take place of an old one or rectify the existing one.

For developing faster and accurate software, there is a need for some kind of methodology [1]. A software development life cycle (SDLC) is used to create high-quality software in minimum time [3]. SDLC is used for software development, as well as maintenance [1].

Agile method is also one method that can be employed in software development. It is a kind of small-term software development methodology that mainly focuses on how quickly a system can adapt changes. Agile methodology focuses more on communication than the procedure, a basic working software is more important than

U. Ameta (✉) · M. Patel · A. K. Sharma
Geetanjali Institute of Technical Studies, Udaipur, India

R. Mathur et al. (eds.), *Emerging Trends in Data Driven Computing and Communications*, Studies in Autonomic, Data-driven and Industrial Computing,
https://doi.org/10.1007/978-981-16-3915-9_28

a full-fledged system. The most important aspect is—'being responsive to change [1]. It breaks the project into small modules and these modules are executed iteratively. The main difference between traditional SDLC and Agile methodology is the client involvement and responsiveness [4].

Scrum is a framework under the Agile methodology that works on the concept of repetition and increment [5]. Scrum aims of delivering high-quality software in the shortest time, which are flexible enough to incorporate changes. Scrum breaks the project into small modules that are mini-projects in themselves, thus opposing the traditional sequential approach [4].

2 Literature Review

2.1 *Software*

A computer program that is responsible for interaction between the user and a variety of hardware is called software. Software translates the commands of the user into an understandable format for the hardware. So, it basically serves as a "translator" between the two [1]. Software is made up of a number of programs, functions. and procedures including the data. The demand for software is increasing and the quality of the software is very important for developers as well as customers [2].

2.2 *SDLC (Software Development Life Cycle)*

SDLC is a life cycle for developing a software by following several steps [1]. For developing software in a systematic and sequential way, the SDLC is very useful [6]. This life cycle enables us to develop quality software in given timeline according to the requirements and also for maintaining it [1]. Following are several stages of SDLC:

1. Survey and feasibility study
2. Analysis
3. Determine user requirements
4. Finding best possible solutions
5. Requirements gathering (hardware and software)
6. Designing
7. Coding
8. Implementation
9. Maintenance

From the several models of SDLC, the most-used model is the waterfall model. Other than this, fountain, spiral, rapid, incremental, prototyping, etc. are the SDLC models [1].

2.3 Agile

An enhanced version of SDLC called the "Agile methodology" was first introduced in 2001 [2]. The word Agile means quick, active, brisk, and lightweight [1]. Agile methodology is a way to develop software by breaking the project into small, time-boxed iterations. Each iteration undertakes the pieces of project development in a specific duration [4]. This type of development process promises rapid delivery of software, inculcate changing requirements, enhances communication, measures progress from time to time, and ensures client interactions [2]. An important aspect of Agile is that it reduces risks because projects are not carried out in one go, in fact, they are built in pieces, which decreases the chances of entire system failures. Rapid applications Development (RAD), Dynamic system Development method (DSDM), and SCRUM are the frameworks that comes under Agile Software Development [4].

2.4 Traditional Versus Agile

Agile methodology is just a variation in SDLC with the agenda to develop flexible software in a given time. Agile does not follow the hierarchical procedure of conventional SDLC, rather it tends to work in an interactive and iterative format. The most important difference between the two is that agile methodology is more into collaborating with clients and is responsive to change [4].

Agile software development is the latest trend in the industry and 65% of the industry have moved from traditional SDLC to Agile development [4]. However, this development process also has some limitations. Agile methodology works best for small and medium scale projects, but for large scale projects, companies still prefer to use traditional approaches [7].

Both approaches have their merits and demerits which needs to be considered. Refer to Table 1. For a detailed comparison between Traditional Software Development approach and Agile Software Development approach.

2.5 Scrum

Jeff Sutherland developed the concept of Scrum in 1993. Scrum is a framework used in software development and management. It uses values and principles of agile methodology to develop responsive and efficient software [1]. The area of focus for

Table 1 Comparison between traditional SDLC and agile methodology

	Traditional SDLC	Agile
User requirements	All the requirements mentioned by the customer are well-studied before implementation starts	Also welcomes late changing requirements by clients
Responsiveness	Does not incorporate changes	Is flexible and adapt changes easily
Testing	One time testing is done after coding of the entire project is completed	Testing is done iteratively after each module completion
Client involvement	Not much involvement is seen	Clients are invited to see progress from time to time
Focus	Complete documentation by sequentially following the procedure	Developing a working software with maximum interactions

the Scrum framework is to develop flexible, high-quality software, deliver it in the minimum possible time and increase the business value [3]. It does not follow the conventional sequential approach, rather the entire team works together in achieving a common objective [1].

The key principle of Scrum is that it welcomes late requirement changes. This means that if the customer changes his mind and asks for adding some new features during the project development, it is readily acceptable by the developers [2]. Scrum encourages self-organizing teams by maximum casual and verbal interactions between team members and maintaining discipline throughout the project. It follows an empirical approach to meet the responsive requirements [4].

It is a kind of "Inspection and Adaption" framework. This means that the software development process is inspected on a timely basis and if any changes are needed to be incorporated into the system, they are adapted quickly [5]. Scrum framework is a repetitive life cycle for developing high-quality and responsive software. It can be defined as a combination of a set of activities or iterations that are undertaken using a variety of tools and techniques.

3 Methodology

3.1 Scrum Roles

For carrying out the process, some specific roles are required as mentioned below

Product Owner. The role of the product owner is to identify and illustrate the characteristics and specifications of the software that has to be developed [1]. The product owner has the authority to decide the priority of the features that need to be built by the team members [3]. The product owner puts all the basic initial requirements of the project into the product backlog [1]. This person directly communicates with the customer to understand the software requirements and demonstrate these details to the SCRUM team. It can be said that the product owner is the mediator between customer and developer [2].

Scrum Team. The product backlog developed by the Product Owner is executed by the Scrum team. The user stories from the backlog that need to be worked on, in the following, sprint may be allotted to different team members according to their skills, knowledge, and understanding [2]. A Scrum team may consist of 4–9 people, who are responsible for running the project [4].

Usually, a Scrum team comprises of following people:

Project Manager. A person who is responsible for monitoring and controlling the entire project.

Business Analyst. A person who is responsible for analyzing the business.

System Analyst. A person who conducts feasibility studies and analyses for the software.

Programmer. A person who designs the system, writes the code, and undertake internal testing.

Tester. A person who runs and check the program, whether it is appropriate against the specifications or not [1].

Scrum team members should be cooperative so that there is a state of well-communication among the team and it encourages teamwork [5].

Scrum Master. Scrum Master is responsible for introducing the Scrum process to the team members. He is the one who looks after the implementation of SCRUM process within the team during the entire project development [1]. Scrum Master is a role in the Scrum framework that coaches the team in developing the software using SCRUM values. Scrum Master also solves the issues that are encountered during daily Scrum meetings. This person is more into training than controlling [2]. Scrum Master encourages teamwork and builds healthy relations between team members. A major task of this person is to analyze and monitor the progress of work as well as the team [4].

Refer to Fig. 1 that depicts the scrum roles and their responsibilities.

Every individual role holds its own importance in the Scrum process. If any of them is not present during the Scrum flow, it becomes difficult to ensure a smooth working flow. This will eventually hamper the development process, slow down its progress, and may lead to inappropriate results.

Fig. 1 Scrum roles

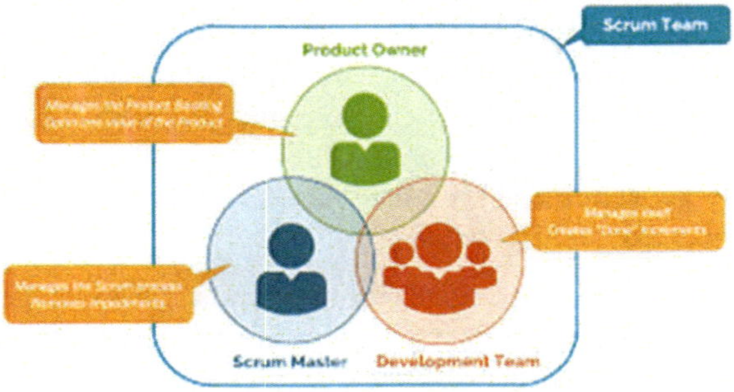

3.2 Sprint

A Sprint is an iteration in which a piece of work is undertaken in a time period of about two to four weeks. This time period is fixed for all the iterations that are going to be executed throughout the project [2]. The user stories from the product backlog that needs to be executed, are divided into a number of sets. These sets are implemented in their respective allotted sprints [4]. Before the beginning of every sprint, there is a meeting which covers the identification of work to be done in the following sprint and estimated goal prediction. The sprint ends with a review meeting to inspect sprint work [4]. At the end, the result of sprints is combined to form a project [5]. To cancel or re-schedule the sprint, the product owner needs to be approached, as only he has the authority to do it [2].

3.3 Scrum Process

The scrum process is a combination of several stages performed by different roles.

Determining product backlog. The project developed using the SCRUM process begins with an illustration of the system that needs to be prepared. The product owner illustrates the plan into a product backlog [1].

A product backlog is a list of features called user stories. These features are nothing but the requirements of the system [3]. The product owner lists down all the features that need to be included in the system. Then he prioritizes this list of features, according to his immediate and later requirements. Now, this list of features is used to build a product backlog [1]. All the user stories and functionalities are registered into the product backlog. Which backlog items need to be developed in the followings sprint are filtered by the product owner on the basis of priority and feasibility [2]. Then the product owner informs team members about the features to be included in the system. All the user stories in the product backlog are executed by the team. After sorting the list of features based on end-user's perspective, the development of the system proceeds [4].

There is a term "sprint" used in scrum, which means the goals that we need to achieve in the next 30 days. It is an iteration of about 2–4 weeks. The authority to cancel a sprint is only given to the Product Owner [1].

Purpose. The stories that need to be developed are identified according to priority which makes the depiction easy for the team and leads to efficient development.

Sprint Planning. After determining the product backlog, a meeting is summoned at beginning of each sprint, called "Sprint Planning meeting". Product owner and all the team members attend this meeting [1]. Following are the activities that are undertaken in this meaning:

1. The goals and agenda of the sprint are identified.
2. The importance and purpose of each user story is discussed.
3. An approximate timeline to acquire the goals is made [4].

4. Team asks a number of questions about the user stories and functionalities to the product owner, so that they can make a sprint backlog. A sprint backlog is just a detailed version of a product backlog [2].

Also, each team member determines that how many hours he will spend to execute each user story. After deciding the working hours, the team can start carrying out the first story is the product backlog. Multiple user stories from the product backlog can be worked on together by the respective team members [1]. The first sprint is generally of 2 weeks because the users-stories are still just in a documentation format. Then a consistent time period of sprints set the entire project (say 30-days/sprint).

This meeting can last up to 4 h and it is dedicated to making plans for what will be done in the following sprint [5].

Purpose. This kind of open sprint planning meeting provided a summarized view to the team and clarifies the roles and responsibilities of every individual [4].

Daily Stand-up meeting. To monitor performance, a meeting should be held daily to report what has each team member done, the previous day [1]. A short meeting of about 15 min is daily held, usually at the start of the day [3]. This meeting is called "Daily stand-up meeting".

This meeting is chaired by Scrum Master, who is responsible for monitoring performance and solving the problems that the team members encounter. All the team members need not the present in this meeting. The team member, that are executing the specific features, needs to attend the meeting [2].

All the team members discuss and coordinate their daily work. Each team member discusses the following 3 questions-

- What you did yesterday?
- What will you do today?
- What problem is he facing [5]?

The basic agenda of this meeting is to monitor the progress of the project against the sprint goals, exchange daily progress reports, solve the issues, and optimize time for scrum [3].

At the end of each meeting, the renewed time of completion, i.e., the remaining time to complete the project is calculated [4].

A *burn-down chart* is a graphical representation that shows the collective time remaining for the project. It shows how much more time is required to complete the project. It basically shows the progress of the project, efforts of the team, and the remaining effort. The progress of the project is good if the average pending effort is less than estimated effort [1].

Purpose. Daily meetings lead to transparency, timely monitoring and shared understanding. There is a development of informal communication between team members that makes the work process easy and fast (Fig. 2).

Sprint Review. After completion of every sprint, a 4 h meeting is held, referred to as sprint review [5]. It does not involve any presentation of slides or documentations.

A demo of the running features of the software is presented by the whole team. These are the features that have been developed during the last sprint period. Product

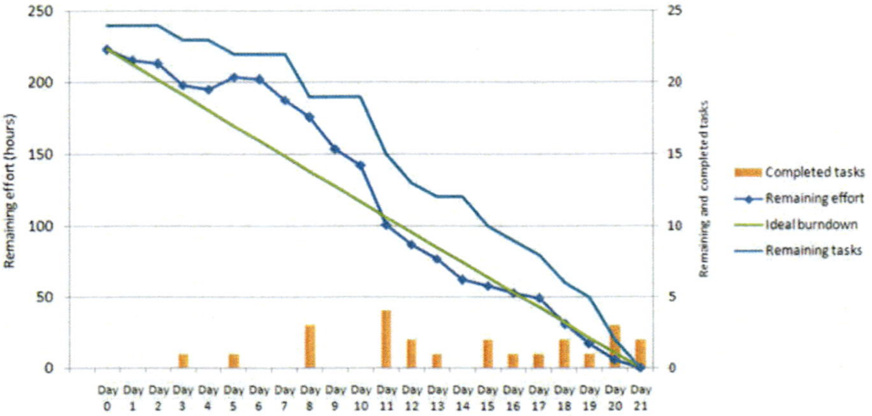

Fig. 2 A burn-down chart for monitoring

Owner attends this review meeting to see the state and progress of the project [4]. Sometimes, the customer wants to see that what the progress of the development of the software is. So, the product owner invites the customer to attend the sprint reviews meeting and present the online demo of features developed so far [5].

Purpose. The purpose of this meeting is to discuss the state of the project at this very point and review all the work done in the sprint. The plan for the next sprint is also discussed.

Sprint Retrospective. Sprint Retrospective is a time-boxed meeting, where following discussions take place-

- The task that is encountering issues [1].
- The tasks that can/cannot be continued for the next upcoming sprints.
- Ways to improve the developments process [3].
- Lesson for upcoming sprints [4].

Purpose. Such kind of meetings also monitors teamwork. What went good in the last sprint and what improvisation can be made in the next sprint is determined [3].

Scrum process is easy to carry if performed properly (Fig. 3).

4 Scrum Values

4.1 Courage

Scrum teams are well-introduced to the complexity of project development. They have the courage to take the best decision in difficult situations. The scrum master

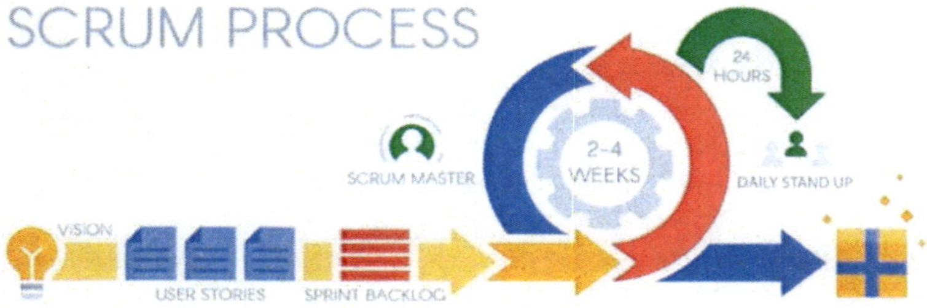

Fig. 3 Scrum flow

has to manage the entire team, encourage communication, and moral values between team members for good teamwork [5].

4.2 Focus

From product owner to all the team members including scrum master, all should be focused on working together and achieving the sprint goals. Scrum masters should focus on smooth scrum flow and solving the encountered issues [3].

4.3 Commitment

The entire scrum process needs commitment for work. Product owner should be committed to choosing the correct priority for backlog. Team should execute that backlog and commit to do whatever it takes to acquire the goal. Scrum master should be committed towards solving problems and assisting the team throughout [1].

4.4 Respect

Since scrum framework encourages teamwork, it is necessary for team-mates to have respect for each other. Product owner is in charge of the entire project, but he is only responsible to check whether work is done or not. He is not questionable to the team, that is how work is being done. Scrum master should not control the team, rather he should facilitate it. Each team member should be treated with equal importance [5].

Table 2 Benefits of scrum framework

Case	Description
Flexibility	Scrum is the most flexible framework as it is responsive to change. Traditional SDLC models like waterfall does not support this
Quality	Scrum is designed in such a way that it develops high-quality and efficient software in a specific time
Client involvement	Constant feedback from end-user is taken and they are also invited to see demos after every sprint
Monitoring	Daily monitoring helps to keep an eye on the progress of the project and also measures individual productivity
Live application view	At the end of each sprint, the client can see a demo of running features that have been developed Since testing is done after each sprint, issues are encountered and solved early
Quick testing	Since testing is done after each sprint, issues are encountered and solved early. It is efficient in comparison with waterfall where testing is done after the whole project is implemented
Estimations	Work and time estimations are done early to avoid the missing deadlines
Quick delivery	Since a working software is delivered initially to the client, rather than a full-documented software, the delivery time is quite short
Utility	Scrum framework makes sure that the usage of time and money in an efficient manner, which is essential for business profit

4.5 Openness

Product owner is the head of the team, but he should be open to welcome ideas from team members as well. Team members should be transparent and open to each other and also to scrum master. They should openly discuss ideas and solutions among themselves [5]. A sense of open-casual interaction should be there between product owner, team and scrum master [2] (Table 2).

5 Limitations

Just the way every coin has two sides, there are some limitations of scrum framework as well. It is no exception.

Lack of commitment. If there is a lack of commitment in the product owner, team or scrum master, then it will hamper the scrum flow [1].

Scalability. It is difficult to apply this methodology to large scale projects; whose development is complex. Traditional SDLC models are preferred in these cases [7].

Team Management. Sometimes, it becomes difficult for the scrum master to manage the team together as there may be a lack of cooperation between them. Teamwork is highly necessary.

Challenging. Daily meeting between large teams is difficult to manage. Also, when a team member leaves a project in the middle, it is difficult to train a new one in his place [5].

Need Skill and Training. Proper knowledge about the values and principles of this methodology is required. Lack of enough knowledge will raise issues while following the Scrum flow between large teams.

6 Conclusion and Future Scope

This review paper concludes that integrating Scrum and Agile is a good approach towards software development. Taking constant feedback from the customers would help, the software development cycle adapts changes, therefore, overcoming the limitation of the traditional SDLC waterfall model. On the basis of the analysis made, the following conclusions can be pinned:

- Scrum offers a high degree of responsiveness
- Open work atmosphere for self-organizing teams is provided
- Encourages sharing of knowledge and informal communication
- The live demo of working software is displayed to the customer after each iteration.
- Real time speed and progress measurement.

In future, enhancement can be done in scrum framework with respect to business change. Also, scrum could be implemented on other domains as well.

References

1. Permana PAG (2015) Scrum method implementation in a software development management. IJACSA
2. Singhal A, Divya G (2014) Scrum: an agile method. IJETMAS
3. Ng GC (2018) A study of an agile methodology with scrum approach to filipino company-sponsored IT capstone program. Int J Comput Sci Res
4. Jeldi NP, Chavali VKM (2013) Software development using agile methodology using scrum framework. Int J Sci Res Publ
5. Hossain EM, Babar MA, Paik HY (2009) Using scrum in global software development: a systematic literature review. In: Fourth IEEE conference on global software engineering
6. Modi HS, Singh NK, Chauhan HP (2017) Comprehensive analysis of software development life cycle models. IRJET 4
7. Leau YB, Loo WK, Tham WY (2012) Software development life cycle agile versus traditional approaches. IPCSIT, vol 37. IACSIT Press, Singapore
8. Dingsoyr T et al (2010) Agile software development: current research and future directions. Springe Berlin
9. Sutherland J (2001) Agile can scale: inventing and reinventing scrum in five companies. Cutter IT J
10. Ambler S (2008) Agile practice and principles survey: Feb 2008. http://ambysoft.com/surveys/agileFeburary2008.html. Accessed 11 Apr 2009

11. Stober T, Hansmann U (2010) Agile software development: best practices for large software development projects. Springer, New York
12. Langer AM (2012) Guide to software development designing and managing the life cycle. Springer, London, New York
13. Elbert C (2005) SpringerLink (Online Service): best practices in software measurement how to use metrics to improve project and process performance. Springer, Berlin
14. Maher P (2009) Weaving agile software development techniques into a traditional computer science curriculum. In: Proceedings of 6th IEEE international conference on information technology: new generation
15. Misra SC, Kumar U, Kumar V, Grant G (2006) The Organizational changes required and the challenges involved in adopting agile methodologies in traditional software development organizations. IEEE, pp 25–28

Chapter 29
A Comparative Study and Tool to Early Predict Diabetes Using Various Machine and Deep Learning Based Techniques

Shrusti Porwal, Chintal Kumar Patel, Deepa Tilwani, and Shri Krishn Bansal

Abstract Diabetes or polydipsia is a very usual disease in humans. The reason for Diabetes is a set of metabolic disorders. In this disorder, blood sugar levels of the human body become extremely excessive for a long interval of time. This disease has a detrimental effect on various organs of humans, especially the nerves and arteries. By early prediction of this disease, we can control it easily and as a result we can successfully save human lifes. This research work was mainly accomplished for saving human lifes from such type of disease. In this work, we have traversed and compared various possible threads related to this disease using machine and deep learning techniques. All the prediction techniques that are provided in this paper give very coherent outcomes to extricate knowledge. We have also proposed a prediction tool that can predict the positive and negative results for polydipsia on the basis of the input provided by the patient. Dataset can be collected from some of the diabetic and nondiabetic persons. We have taken the dataset from UCI machine learning Repository. Extracting understanding from such data can be useful for predicting diabetics in patients at an early stage. In this work, we have taken some well-liked ML-based algorithms, namely, SVM, Logistic Regression, XGBOOST, Voting Classifier, Naive Bayes, KNN, Random Decision Forests, Decision Tree, and ANN. All algorithms were compared and evaluated using the UCI machine learning diabetes dataset.

Keywords Machine learning · ANN · Polydipsia · Logistic regression · XGBOOST · Random forest · Naive Bayes · KNN · SVM · Decision tree · Deep learning

---

S. Porwal (✉) · D. Tilwani · S. K. Bansal
The LNM Institute of Information Technology, Jaipur, India

C. K. Patel
Geetanjali Institute of Technical Studies, Udaipur, India

R. Mathur et al. (eds.), *Emerging Trends in Data Driven Computing and Communications*, Studies in Autonomic, Data-driven and Industrial Computing,
https://doi.org/10.1007/978-981-16-3915-9_29

1 Introduction

Diabetes or polydipsia is a very common disease in the human body that attacks hormonal insulin. The consequence of this disease is the irregular metabolism of carbohydrates in the human body. And this irregular metabolism increases the blood sugar level in the human body. Excess sugar in the blood reduces organ functioning in the human body. High blood sugar mainly affects the arteries, nerves, and blood veins in the human body. The body of a diabetic patient is inadequate or unable to extract energy from the food that human body take in the form of fruit, vegetable, water, etc. As Conveyed by the World Health Organization (WHO), in 2016, approx. 1.6 million people died due to diabetes or polydipsia. And also, stated by the WHO prediction that by 2030, polydipsia will be the 7th main reason for loss of life. It is stated by IDF that moreover 150 million cases of polydipsia are still undiagnosed.

1.1 Categories of Diabetes

1. Type 1: This is the type of diabetes is called an autoimmune disease. In this category, the immune system strikes and destroys the cells in the organ of the digestive system where the insulin is made. Till now it is not clear that what is the causes this type 1 diabetics.
2. Type 2: This type of diabetes occurs when the human body becomes resistant to insulin, and the human body automatically builds up sugar in the human blood.
3. Prediabetes: This is the type of diabetes that occurs in the human body when a person's blood sugar is higher than a normal human being. This type of diabetes generally occurs when females are getting to be pregnant.
4. Gestational: This type of diabetes occurs during pregnancy. In this case, the human body's blood sugar increases because of some Insulin-blocking hormones produced by the placenta in the human body. This type of disease often spots it in middle or late pregnancy.

Hence, the foremost motive of our research paper is an early prediction about whether a person is having diabetes or not using predefined machine learning and deep learning [1, 2] models. And we are also establishing some comparison between some of the techniques.

2 Related Work

Researchers have administered, explored, and researched the reason for diabetes using machine and deep learning techniques to bring out knowledge from accessible health datasets. For example, Aljumah et al. [3] proposed in the research paper to develop a predictive analysis model. This model uses support vector machine algorithm to predict diabetes in young and old patients. Sonar et al. [4] proposed a comparative study of different machine learning algorithms to predict diabetics in person. The proposed model development is based on the comparison methods such as Decision Tree, Support Vector Machine, ANN, Naive Baye, and SVM algorithms. According to the given classification algorithm, Decision Tree model gives the precision value of 85%, Support Vector Machine gives the precision value of 77.3%, and Naive Bayes model gives the precision value of 77%. Alic et al. [5] visited again the dataset of San Antonio Heart Study. With that dataset, the author developed a machine learning algorithm-based model that predicts the future chance of type-2 diabetes in the human body. For building that prediction model, the author used the SVM machine learning algorithm. The described model takes 10 features as an input which will be useful for making strong predictors of future diabetes. Islam et al. [6] Compared the health dataset with Random Forest (RF) Algorithm, Logistic Regression (LR) Algorithm, and Naive Bayes Algorithm. They applied Percentage Split evaluation and tenfold Cross-Validation techniques on the given health dataset, and by these measures, they generalize that Random forest is the best and accurate among the other health datasets. They also develop an easily approachable, convenient application for the human being to check the probability of having diabetes in an early stage. Khan et al. [7] developed a smart mobile health application tool. This tool is based on machine learning. This tool is having capabilities to check whether a person is having the chance of being non-diabetic, prediabetic or diabetic.

By this tool, without consulting any doctor or any medical health test, one can diagnose about the same. Sisodia et al. [8] compared three machine learning algorithms named Decision Tree, Support vector machine (SVM), and Naive Bayes. The author performed an experiment to discover whether a person is suffering from diabetes mellitus in the very initial stage. All the experiments were done on Pima Indians Diabetes Dataset (PIDD). This required dataset is available at the UCI machine learning repository.

3 Methodology

To accomplish our aim, we have taken a health dataset to train our tool. In the proposed work, we have compared various ML and Deep learning-based algorithms. We have established this comparison for evaluation purpose, so that we can generalize which of the algorithm is best suited for our prediction. In the methodology part, we will discuss few points that we have used in our proposed method.

3.1 Model Diagram

For the purpose of prediction, we have created one simple, easy, and user-friendly tool. The step-by-step diagram or flow chart of that is given in Fig. 1. As an input, we have taken a dataset of the patient in excel format that includes features and label information. Then, we have done the selection process of the features. For the same, Pearson coefficient and chi-square test is used. Then, we have applied various prediction algorithms to the dataset. According to the input, the tool predicts positive or negative results for polydipsia [9–11].

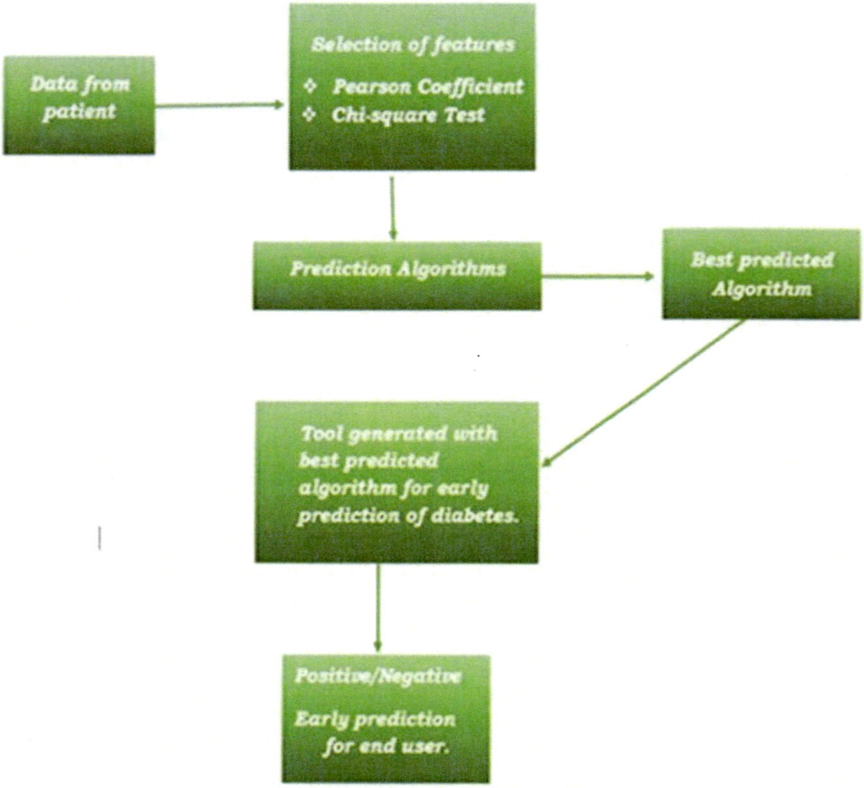

Fig. 1 The Flow chart of the proposed tool

3.2 Tool Description

People need an easy, user-friendly, and globally accessible diabetics prediction system that provides immediate help to the mass of people who are having the risk of diabetes. So, to help the mass of people, we have created one tool that takes some of the input from the users related to their lifestyle, age, gender, and medical history. And on basis of that input, it generates positive or negative results for disease. A representation of our proposed tool is shown in Fig. 1 (Figs. 2 and 3).

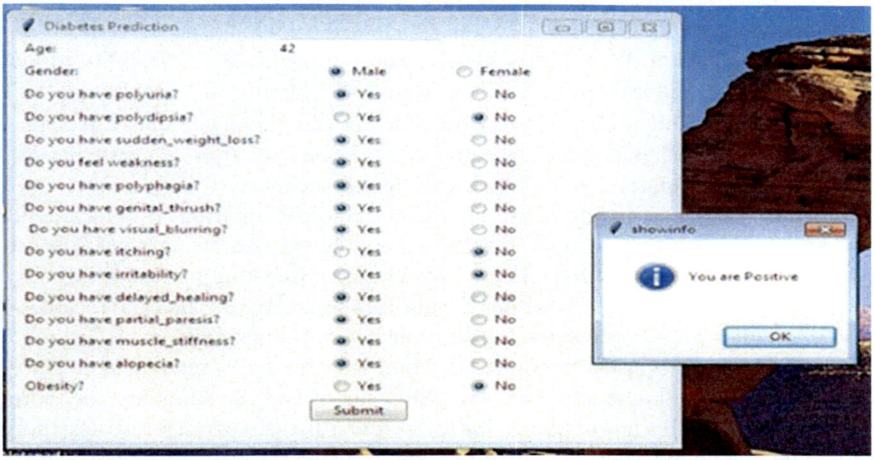

Fig. 2 Tool snapshot for person result positive for diabetes

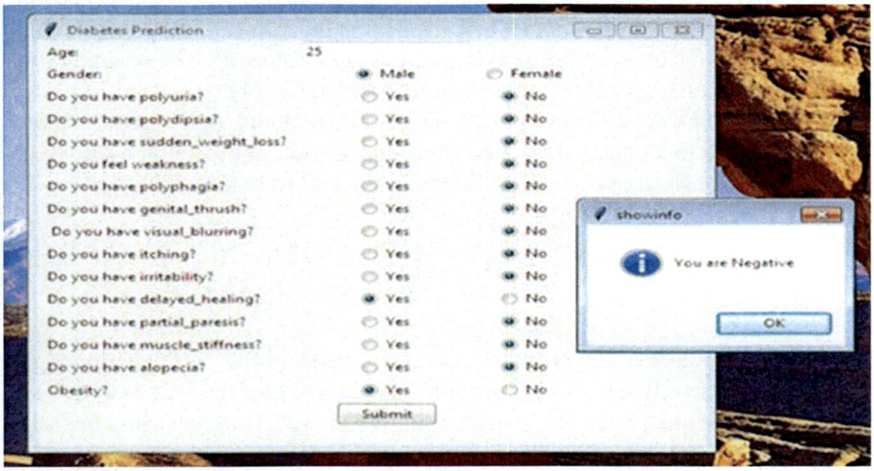

Fig. 3 Tool snapshot for person result Negative for diabetes

3.3 Quick Description of About the Algorithms Used

- Support vector machine (SVM): It is a standard supervised ML-based model. It is useful for classification related problems. In the support vector machine, two classes of sample training are taken and checked for the highest margin hyperplane to distinguish the two categories [12].
- Naive Bayes: Naïve Bayes machine learning algorithm is a classification technique that is used to define features. Naïve Bayes classification techniques are based on the method of conditional probability. This algorithm is considered to be a very robust algorithm for classification related problems.
- Logical Regression: It is also based on supervised machine learning-based classification algorithm. We can use this algorithm to predict the probability of the target variable. Logical regression is of two types: binomial and multinomial.
- Decision Tree: This is also a supervised machine learning technique. It is applicable for solving classification related problems. Decision tree technique is used to predict some futuristic thing like it will rain tomorrow.
- XGBoost: XGBoost is one of the distributed gradient libraries. This technique of prediction is highly flexible, portable, and efficient. We can apply XGBoost ML-based algorithms with the help of the Gradient Boosting framework. It also provides simultaneous tree boosting. By this technique, we can solve data science-based problems very efficiently. It also gives good accuracy.
- Random forest: It is also a supervised machine learning-based classification algorithm. This algorithm makes the forest with a no. of trees. In summary, the more trees in the forest the more efficient the forest looks. In the random forest classifier, the more the no. of trees in the forest will higher the accuracy.
- KNN: The KNN algorithm is an effortless, easy-to-implement, and very simple supervised ML-based algorithm. We can use this algorithm to resolve both regression as well as classification problems. KNN algorithm takes some labels and that points are used to train the model. KNN works on the nearest label point method. The new point is always compared with the nearest value (k). The meaning of k is the number of neighbours it checks in K-Nearest Neighbors.
- ANN: Artificial Neural Networks is a computational model. ANN model is influenced by the human brain. This algorithm can be used in the field of AI, which includes Image Recognition, Voice Recognition, and Robotics (Table 1).

4 Proposed Work

The Skeleton structure of the proposed method is given in Fig. 1. Prediction algorithms such as Naive Bayes (NB), Decision Trees (DT), Logistic Regression (LR), XGBOOST, Random Forest (RF) Algorithm, SVM, KNN, voting classifier, and ANN will be fed to the health dataset that contains information about the patient's prodrome and also provides label data to train the model. The consistency of the algorithms was checked using Pearson coefficient and chi-square test. And on the

Table 1 Some accuracy estimates

Measures	Definitions	Formula
Accuracy (A)	It is used to determine the accuracy of any of the algorithms in predicting instances	A = (TP + TN)/(sum of no. of samples)
Precision (P)	Is show Classifier's preciseness/accuracy is measured by Precision	P = TP/(TP + FP)
Recall (C)	It is used to estimate the classifier's entirety or sensitivity	R = TP/(TP + FN)
F1-score (F1)	It is used to measure the model's accuracy on a dataset	F1 = TP/TP + ½(FP + FN)

basis on these parameters, we have concluded that which algorithm is best suited for prediction purpose. The proposed tool takes the symptom of the user as an input, and on the basis of that input, the tool predicts negative or positive results.

This health dataset includes the report of 520 people who are having symptoms related to diabetes. We have also provided label data to train our tool. This dataset that we are using to train our tool was created by a direct questionnaire to patients at Sylhet Diabetes Hospital. This hospital is in Sylhet, Bangladesh. The questionnaires were asked to the people who are just now diagnosed as diabetic or some people who are still nondiabetic but they are having some symptoms for diabetes. There are a total of 16 features available in the dataset. Two classes—positive or negative are used to identify whether a person is at the risk for diabetes or not. We have checked the performance of different machine learnings on our health dataset and also compared detailed accuracy information that is shown in the given tables in the paper. We have made an effort to compare NB, LR, Decision Tree, Random Forest, SVM, KNN, XGBOOST, and ANN. Our Approach is to train models by feeding data from many patients, and then to use a trained data as a classification model for other patients. Before training the classifiers, feature selection is done using Pearson's coefficient [13] and chi-square test [14], and compared both after training. The top ten features are selected for Pearson's coefficient and chi-square test and these features are used to train classifiers.

Table 2 Pearson's coefficient features results

Performance parameters	Class	NB	KNN	Decision Tree	SVM	LR	RF	XGboost	ANN
F-score measure	Positive	0.81	0.93	0.97	0.96	0.92	0.97	0.97	0.971
	Negative	0.88	0.89	0.95	0.93	0.86	0.95	0.95	
	Weighted	0.86	0.91	0.96	0.95	0.89	0.96	0.96	
Recall	Positive	0.86	0.91	0.97	0.98	0.91	0.97	0.97	0.971
	Negative	0.84	0.92	0.95	0.89	0.87	0.95	0.95	
	Weighted	0.84	0.91	0.96	0.95	0.89	0.96	0.96	
Precision	Positive	0.90	0.95	0.97	0.94	0.92	0.97	0.97	0.971
	Negative	0.78	0.85	0.95	0.97	0.85	0.95	0.95	
	Weighted	0.86	0.92	0.96	0.95	0.89	0.96	0.96	
Accuracy		0.86	0.91	0.96	0.95	0.89	0.96	0.96	0.961

5 Result and Conclusion

The algorithms are very effective in distinguishing positive and negative classes from the data. Although the NB (Naïve Bayes) classifier is a very popular classifier and is highly accepted by many of the data prediction algorithms, the accuracy of the health dataset was the lowest for both the Pearson's coefficient and chi-square test features. However, the best results have been achieved using the ANN algorithm, where 97.11% of instances were correctly classified using chi-square test features. All the classifiers are compared on the basis of features selection and accuracy metrics. Comparing Chi-square test features combined with the ANN model gives the best result.

5.1 A Detailed Description of Results is in the Following Tables

See Tables 2 and 3.

5.2 Graphical Representation of Accuracy on Both the Parameters

See Figs. 4 and 5.

Table 3 Chi-Square features results

Performance parameters	Class	NB	KNN	Decision Tree	SVM	LR	RF	XGboost	ANN
F- score measure	Positive	0.88	0.91	0.97	0.95	0.91	0.98	0.96	0.979
	Negative	0.81	0.86	0.95	0.91	0.85	0.96	0.92	
	Weighted	0.86	0.90	0.96	0.93	0.89	0.97	0.94	
Recall	Positive	0.86	0.88	0.98	0.95	0.89	0.98	0.97	1.0
	Negative	0.84	0.92	0.92	0.89	0.87	0.95	0.89	
	Weighted	0.86	0.89	0.96	0.93	0.88	0.97	0.94	
Precision	Positive	0.90	0.95	0.96	0.94	0.92	0.97	0.94	0.959
	Negative	0.78	0.81	0.97	0.92	0.82	0.97	0.94	
	Weighted	0.86	0.90	0.96	0.93	0.89	0.97	0.94	
Accuracy		0.86	0.89	0.96	0.93	0.88	0.97	0.94	0.971

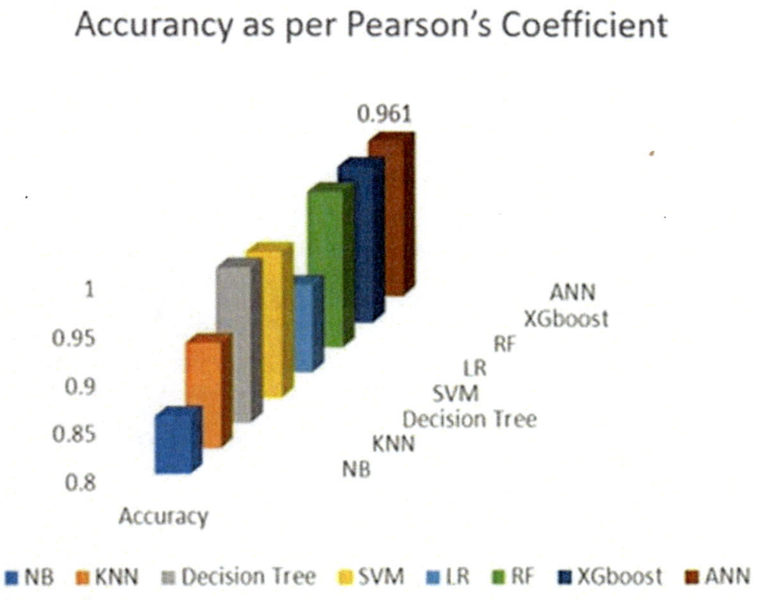

Fig. 4 Accuracy of algorithms according to Person's Coefficient

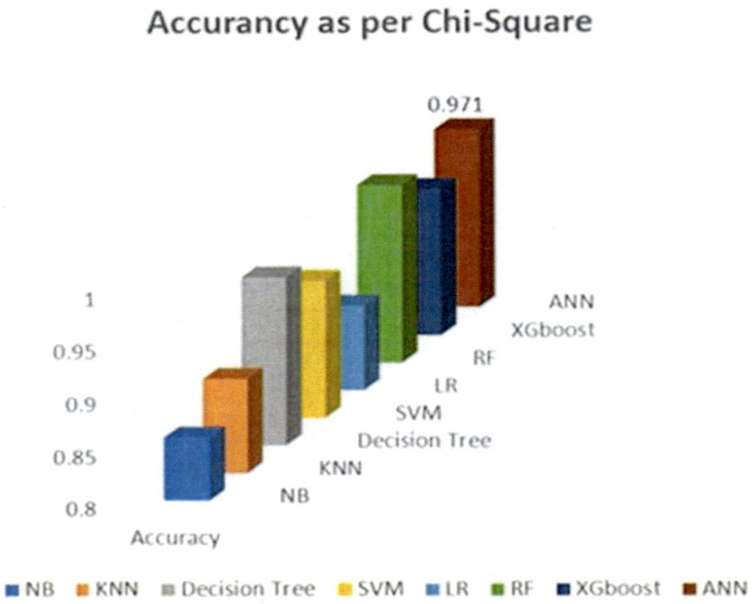

Fig. 5 Accuracy of algorithms according to Chi-Square Test

References

1. Vasileios T, Konstantinos K, Vasileios C, Athanasios K (2019) A comparative assessment of machine learning algorithms for events detection. https://doi.org/10.1109/SEEDA-CECNSM.2019.8908366 Piraeus, ©2019 IEEE
2. Susmita R (2019) A quick review of machine learning algorithms. ISBN: 978–1–7281–0212–2, https://doi.org/10.1109/COMITCon.2019.8862451, Faridabad india, ©2019 IEEE
3. Aljumah MGA, Mohammad KS (2013) Application of data mining: diabetes health care in young and old patients.J King Saud Univ—Comput Inf Sci25(2):127–136
4. Priyanka S, JayaMalini K (2019) Diabetes prediction using different machine learning approaches. IEEE Xplore Part Number: CFP19K25-ART; ISBN: 978–1–5386–7808–4
5. Lejla A, Hasan TA, Marelyn R, Muhammad AG, Khalid Q (2019) Predicting diabetes in healthy population through machine learning. 2372–9198/19/$31.00 ©2019 IEEE
6. Faniqul Islam MM, Rahatara F, Sadikur R, Humayra YB (2020) Likelihood prediction of diabetes at early stage using data mining techniques. Springer, Singapore. https://doi.org/10.1007/978-981-13-8798-2_12
7. Nabila SK, Mehedi HM, Anusha K, Muhammad NI (2017) Diabetes predicting mHealth application using machine learning. WIECON-ECE. 978–1–5386–2621–4/17/$31.00 ©2017 IEEE
8. Deepti Sisodia and Dilip Singh Sisodia (2018) Prediction of diabetes using classification algorithms. ICCIDS. https://doi.org/10.1016/j.procs.2018.05.122,132(2018)1578-1585
9. Choubey DK, Paul S, Kumar S, Kumar S (2017) Classification of Pima Indian diabetes dataset using naive bayes with genetic algorithm as an attribute selection. In: Communication and computing systems: proceedings of the international conference on communication and computing system (ICCCS 2016), pp 451–455

10. Orabi KM, Kamal YM, Rabah TM (2016) Early predictive system for diabetes mellitus disease. In: Industrial conference on data mining, Springer. Springer, pp 420–427
11. Tiebin L, Rodolfo V, Marta G, Muin JK (2010) Application of support vector machine modeling for prediction of common diseases: the case of diabetes and pre-diabetes. BMC Med Inf Decis Mak 10. https://doi.org/10.1186/1472-6947-10-16, arXiv:arXiv:1011.1669v3
12. Kavakiotis I, Tsave O, Salifoglou A, Maglaveras N, Vlahavas I, Chouvarda I (2017) Machine learning and data mining methods in diabetes research. Comput Struct Biotechnol J 15:104–116. https://doi.org/10.1016/j.csbj.2016.12.005
13. Mohammed CA, Fatima ZEM, Loubna L, Najib N, Noureddine EM, Mohamed O (2020) The Pearson correlation coefficient applied to compare multi-criteria methods: case the ranking problematic. DOI: https://doi.org/10.1109/IRASET48871.2020.9092242©2020IEEE
14. Rao CR (2002) Karl Pearson Chi-Square test the dawn of statistical inference. https://doi.org/10.1007/978-1-4612-0103-8_2, ISBN978–1–4612–6613–6

Chapter 30
Single-End Current-Based Scheme for Detection and Classification of Distant-End and Adjacent-End Faults in Transmission Lines

Asifa, Shoyab Ali, and Gaurav Kapoor

Abstract This paper introduces a single-ended current-based scheme using FT (Fourier transform) and WT (wavelet transform) for the detection of distant-end and adjacent-end faults in the transmission line. By simulation investigation using MATLAB, the validity of the suggested scheme is represented. Extensive simulation studies verify that the presented scheme generates acceptable performance for all simulated faults.

Keywords Transmission lines · Fourier transform · Fault detection · Wavelet transform

1 Introduction

The two momentous components of the digital relaying scheme are fault detection and fault classification. The overall performance of the digital relaying scheme is decided by the speed and accuracy of these components. Many tools have been introduced up to now for the digital relaying of transmission lines. Several of those tools are: fuzzy inference and microcontroller [1], discrete wavelet transform (DWT) and Clarke's transformation [2], estimated voltage drop (EVD) [3], and empirical mode decomposition (EMD) with random forest classifier [4]. Moreover, extra schemes are: smoothed pseudo wigner-ville distribution [5], S-transform [6], deep learning [7], positive sequence components [8], wavelet transformation [9], and random forest [10].

In this paper, a scheme is proposed for fault detection in transmission line based on the wavelet index values (WIV).

Asifa · S. Ali
Department of Electrical Engineering, Vedant College of Engineering and Technology, Bundi, India

G. Kapoor (✉)
Modi Institute of Technology, Kota, India

R. Mathur et al. (eds.), *Emerging Trends in Data Driven Computing and Communications*, Studies in Autonomic, Data-driven and Industrial Computing,
https://doi.org/10.1007/978-981-16-3915-9_30

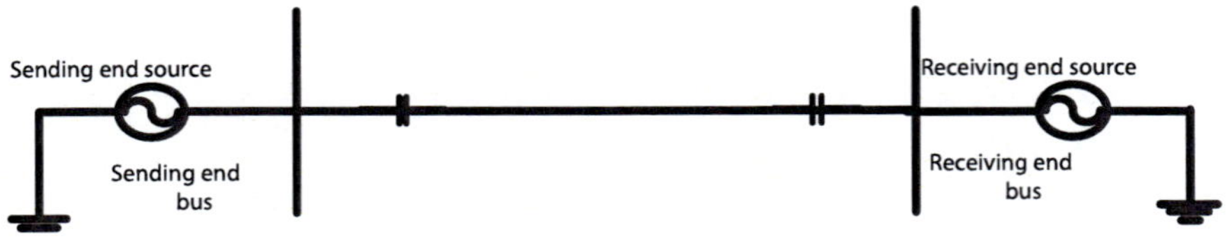

Fig. 1 Simulated model

The application of the scheme is presented in Sect. 2. Presented fault detection scheme is provided in Sect. 3. The effects of simulations are illustrated in Sect. 4, while Sect. 5 concludes the paper.

2 Application of the Scheme

The presented scheme is applied to the system of 500 kV illustrated in Fig. 1. The model has two 50 Hz line sections of 150 km each. Numerous faults are incepted at different locations through MATLAB simulations in SIMULINK.

3 Presented Fault Detection (FD) Scheme

The block diagram of FD scheme is presented in Fig. 2.

4 Simulation Effects

The effects of simulations in view of numerous fault scenarios to bear out the implementation of the scheme are provided in this section.

4.1 Performance Under Fault Application Time Variation

The scheme performance during ACG fault with fault application time (FAT) of 0.05 s at 2% length (6 km) with FI (fault impedance) of 25 Ω and GI (ground impedance) of 20 Ω is revealed in Fig. 3. It is demonstrated that ACG fault is detected at 0.065 s, which specifies that the scheme took 15 ms time for ACG fault detection. Likewise, the performance of the scheme during BCG fault applied at 0.3 s at 95% length (285 km) with the same values of FI and GI is presented in Fig. 4. The BCG fault is detected at 0.31 s accurately. The FDT in this case is 10 ms. Results of some faults

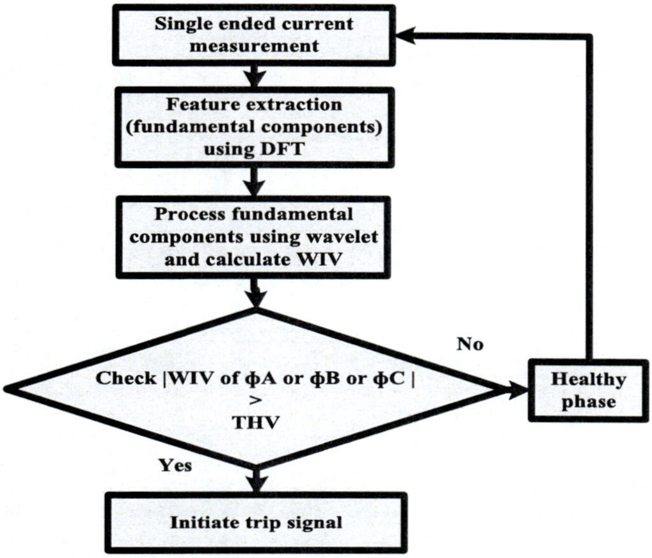

Fig. 2 Block diagram of FD scheme

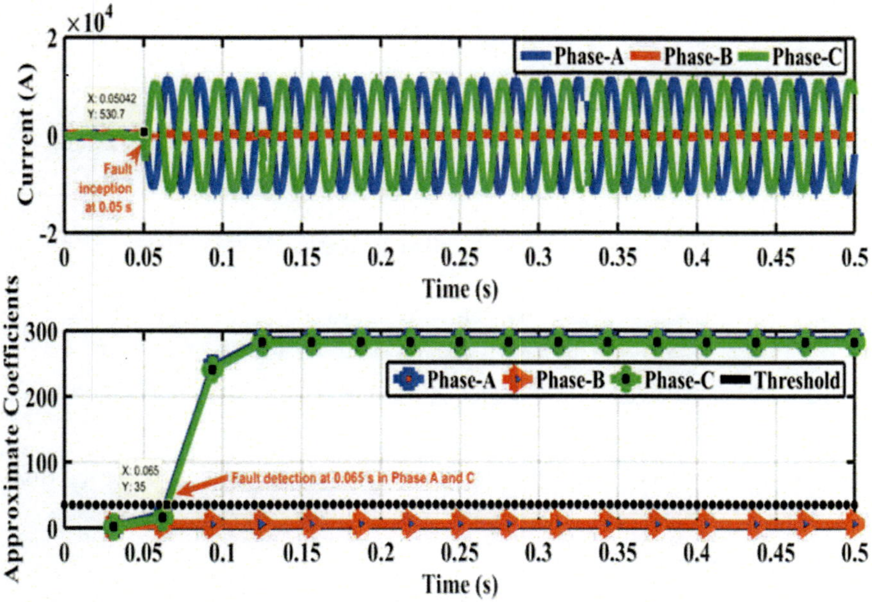

Fig. 3 Scheme performance during ACG fault applied at 0.05 s and detected at 0.065 s at 2% length (6 km)

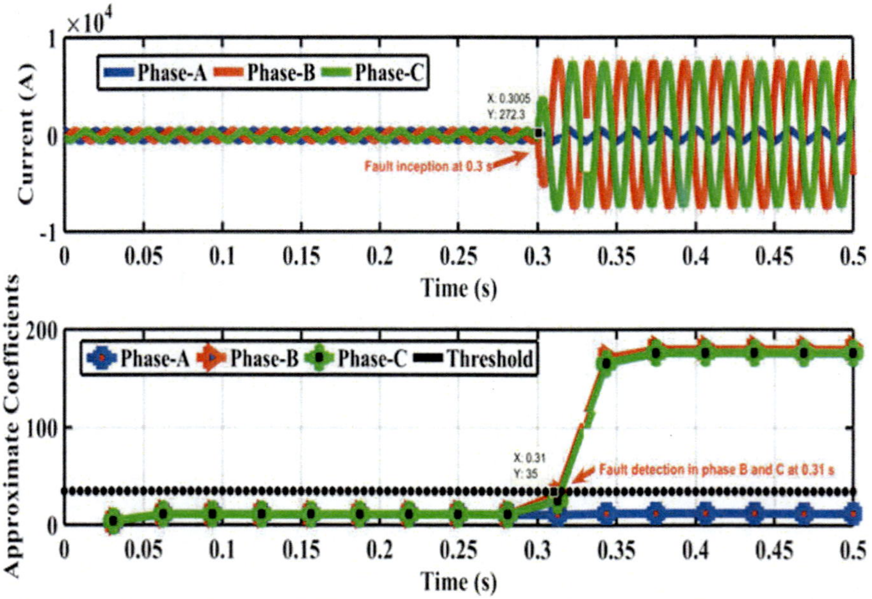

Fig. 4 Scheme performance during BCG fault applied at 0.3 s and detected at 0.31 s at 95% length (285 km)

Table 1 WIVs for faults under variation in fault application time

Fault case	FAT (s)	FL (%)	Wavelet index values			Fault detection time (ms)		
			WIV-A	WIV-B	WIV-C	FDT-A	FDT-B	FDT-C
ACG	0.05	2	285.3219	6.0144	282.3420	15	–	15
AB	0.1	4	278.1968	271.9104	6.2599	5	5	–
ABC	0.15	5	316.0670	316.1513	316.1867	10	10	10
ABCG	0.2	85	210.4119	210.4103	210.3975	5	5	5
AG	0.25	90	88.5174	12.1769	11.3488	15	–	–
BCG	0.3	95	12.5820	179.5997	176.1639	–	10	10

are given in Table 1. The scheme is found resistant to FAT variation as viewed from Table 1.

4.2 *Performance Under Fault Impedance Variation*

The performance during AG fault with fault application time (FAT) of 0.1 s at 6% length (18 km) with FI of 7 Ω and GI of 15 Ω is validated in Fig. 5. It is shown that

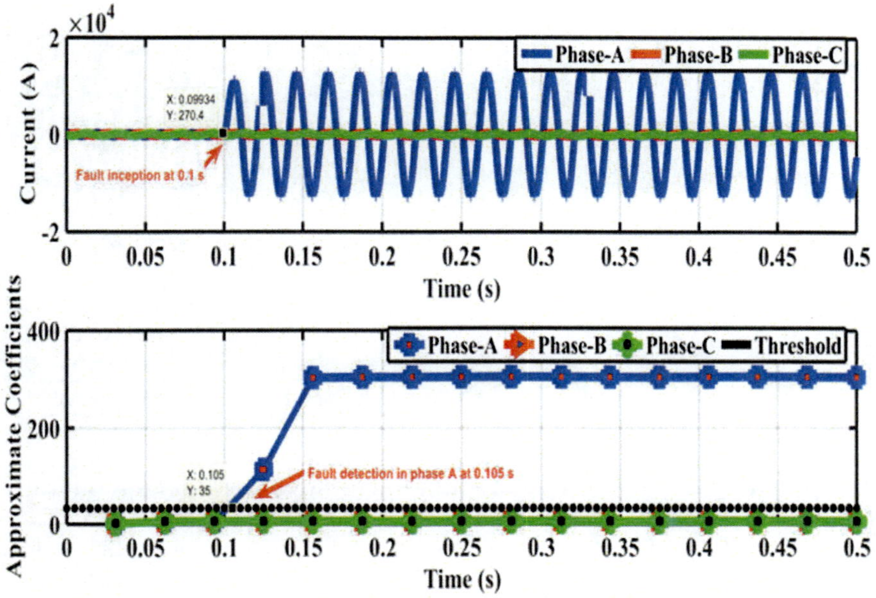

Fig. 5 Scheme performance during AG fault applied at 0.1 s and detected at 0.105 s with FI = 7 Ω

the fault is detected in A-phase at 0.105 s, i.e., in 5 ms after the application of fault. Similarly, the performance during ACG fault created at 0.1 s at 96% length (288 km) with FI of 32 Ω and GI of 15 Ω is illustrated in Fig. 6. The ACG fault is detected at 0.11 s correctly. The fault detection time, in this case, is 10 ms. Effects of some further faults are given in Table 2. The scheme is found resilient to fault impedance variation as observed from Table 2.

4.3 Performance Under Ground Impedance Variation

The working of the scheme during ABCG fault with fault application time (FAT) of 0.2 s at 9% length (27 km) with GI of 6 Ω and FI of 22 Ω is authenticated in Fig. 7. It is shown that the fault is detected in ABC-phases at 0.205 s, i.e., in 5 ms after the inception of fault. Correspondingly, the operation during BCG fault incepted at 0.2 s at 87% length (261 km) with GI of 33 Ω and FI of 22 Ω is pointed up in Fig. 8. The BCG fault is suitably detected at 0.205 s. The fault detection time, in this case, is 5 ms. Results of some further faults are given in Table 3. As followed from Table 3, the scheme is found impervious to ground impedance variation.

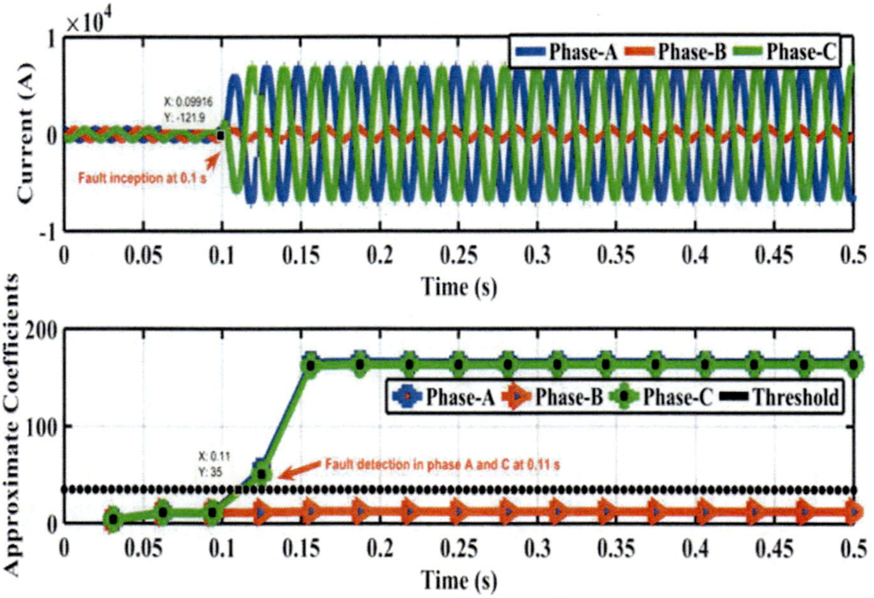

Fig. 6 Scheme performance during ACG fault applied at 0.1 s and detected at 0.11 s with FI = 32 Ω

Table 2 WIVs for faults under variation in fault impedance

Fault case	FI (Ω)	FL (%)	Wavelet index values			Fault detection time (ms)		
			WIV-A	WIV-B	WIV-C	FDT-A	FDT-B	FDT-C
AG	7	6	305.4713	7.1197	6.4888	5	–	–
BC	12	7	6.6136	450.4522	444.3757	–	5	5
ACG	17	8	365.8750	6.9977	360.0597	5	–	5
CG	22	82	11.0560	10.3486	102.7876	–	–	20
AB	27	92	183.3885	176.1194	10.8400	5	5	–
ACG	32	96	165.6488	12.7652	163.3652	10	–	10

5 Conclusion

This paper showed a new WT-based FT approach for detection/classification of distant-end and adjacent-end transmission line faults. Only currents of relaying-end are adequate to implement the scheme. The presented scheme is verified for various distant-end and adjacent-end faults under variation in fault application time, ground impedance, and fault impedance. It was monitored that the scheme detected the faults with reasonable accuracy. The minimum FDT is computed as 5 ms in most of the cases and the maximum FDT is computed as 25 ms only in one case.

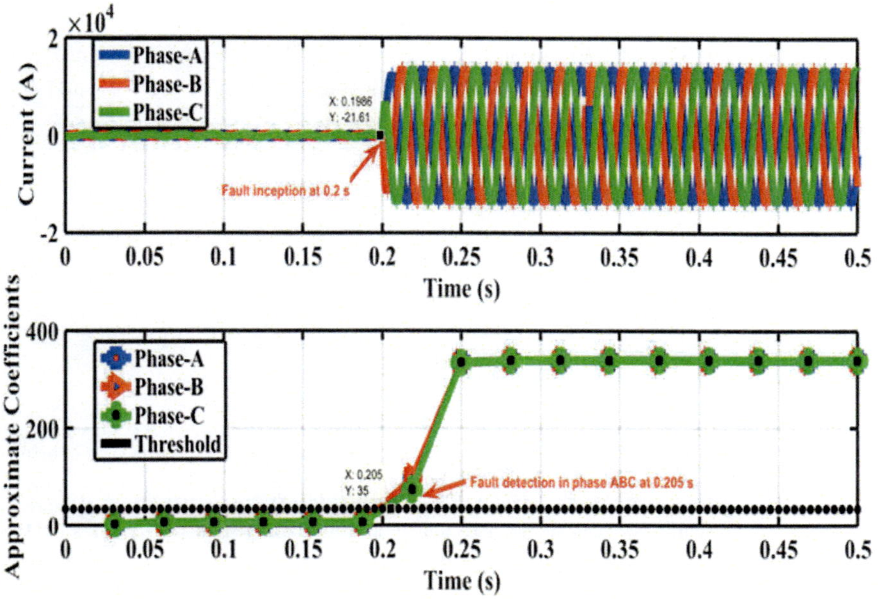

Fig. 7 Scheme performance during ABCG fault applied at 0.2 s and detected at 0.205 s with GI = 6 Ω

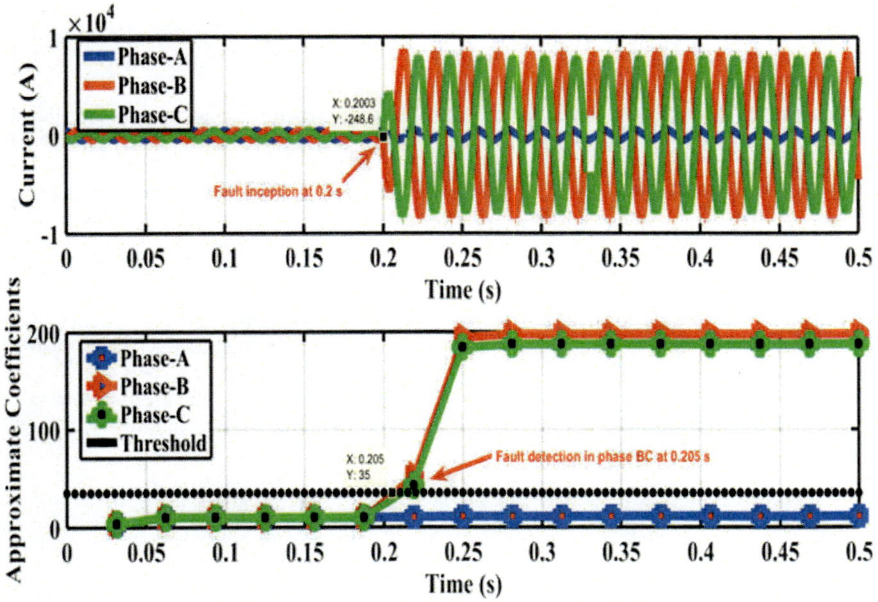

Fig. 8 Scheme performance during BCG fault applied at 0.2 s and detected at 0.205 s with GI = 33 Ω

Table 3 WIVs for faults under variation in ground impedance

Fault case	GI (Ω)	FL (%)	Wavelet index values			Fault detection time (ms)		
			WIV-A	WIV-B	WIV-C	FDT-A	FDT-B	FDT-C
ABCG	6	9	338.8122	338.6849	338.7286	5	5	5
AG	13	10	208.4626	7.7834	6.9622	5	–	–
ABG	18	11	291.7839	300.0248	7.4249	5	5	–
ACG	23	75	207.2915	9.6786	211.4336	5	–	5
BG	28	78	10.4772	9.5426	96.7651	–	25	–
BCG	33	87	11.2796	197.9949	187.8722	–	5	5

References

1. Yadav A, Ashok V, Pazoki M (2020) Protection scheme for hybrid transmission system using fuzzy inference system and microcontroller. Evol Intel 1–15 (2020)
2. Abd-Elhamed Rashad B, Ibrahim DK, Gilany MI, El'Gharably, Aboul'F (2020) Adaptive single-end transient-based scheme for detection and location of open conductor faults in HV transmission lines. Electr Power Syst Res 182:106252
3. Kang S-H, Hong H, Son Y-I (2020) A fault location algorithm using estimated voltage drop across TCSC bank in a series compensated transmission line. In: 15th international conference on developments in power system protection (DPSP 2020). The Institution of Engineering and Technology, Liverpool (UK), pp 1–6
4. Biswas S, Nayak PK (2020) A new approach for protecting TCSC compensated transmission lines connected to DFIG-Based wind farm. IEEE Trans Ind Inf
5. Na S, Kishor N, Uhlen K, Mohanty SR (2019) Detecting instant of multiple faults on the transmission line and its types using time–frequency analysis. IET Gener Transm Distrib 13(22):5248–5256
6. Hosseini HS, Koochaki A, Hosseinian SH (2019) A novel scheme for current only directional overcurrent protection based on post-fault current phasor estimation. J Electr Eng Technol 14:1517–1527
7. Mirzaei M, Vahidi B, Hosseinian SH (2019) Accurate fault location and faulted section determination based on deep learning for a parallel-compensated three-terminal transmission line. IET Gener Transm Distrib 13(13):2770–2778
8. Chatterjee B, Debnath S (2021) A new protection scheme for transmission lines utilizing positive sequence fault components. Electr Power Syst Res 190:106847
9. Wang Y, Ravishankar J, Phung T (2019) Wavelet transform-based feature extraction for detection and classification of disturbances in an islanded micro-grid. IET Gener Transm Distrib 13(11):2077–2087
10. Okumus H, Nuroglu FM (2020) A random forest-based approach for fault location detection in distribution systems. Electr Eng

Author Index

A
Agarwal, Muskan, 67
Ali, Shoyab, 345
Ameta, Upasana, 321
Asifa, Shoyab Ali, 345

B
Bansal, Shri Krishn, 333
Bhati, Tina, 195

C
Chaplot, Siddhi, 267
Charhate, Shrikant, 121
Chauhan, Ajay, 291
Choudhary, Naveen, 15, 153

D
Dabi, Ashmit, 267
Dehariya, Ashish Kumar, 237
Dhaked, Devender Kumar, 169
Diwakar, Nilesh, 281

G
Gahlot, Alok Singh, 83
Gamundani, Attlee M., 1
Ghosh, Soumalya, 67
Gour, Sonam, 203, 299

H
Hassan, Fardin, 67

J
Jain, Arpit, 111
Jain, Harshita, 247
Jain, Kalpana, 15, 153
Jain, Konika, 137
Jain, Milind D., 247
Jain, Ritesh Kumar, 91
Jain, Shruti, 137
Jat, Dharm Singh, 1, 111

K
Kapoor, Gaurav, 345
Khan, Latif, 195, 267
Khan Mr., Latif, 255, 275
Kommina, Lavanya, 53
Kothari, Rakshit, 15

L
Lalwani, Soniya, 99

M
Makuvaza, Auther, 1
Mandot, Manju, 213
Mathur, Rajeev, 255, 267, 275
Maurya, Vijendra K., 195
Mundra, Kavita, 213

P
Pal, Saurabh, 225
Pandey, Akhil, 37, 179
Pandey, Garima, 67

R. Mathur et al. (eds.), *Emerging Trends in Data Driven Computing and Communications*, Studies in Autonomic, Data-driven and Industrial Computing,
https://doi.org/10.1007/978-981-16-3915-9

Parasar, Deepa, 121
Patel, Chintal Kumar, 333
Patel, Mayank, 137, 247, 311, 321
Pathak, Pooja, 291
Patil, Pankaj Rambhau, 121
Payyavula, Yogeshwari, 53
Porwal, Shrusti, 333

R
Rathi, S. K., 91
Reddy, Gurram Deekshith, 53

S
Sabir, Mohammad, 195
Sahni, Gunjan, 99
Samma, Mohammad Ateek, 255, 275
Sen, Shilpa, 311
Sepo, Kamwayamunzi Fortune, 111
Seth, Darshan, 267
Sharma, Abha, 203, 299
Sharma, Ajay Kumar, 311, 321
Sharma, Khushi, 267
Sharma, Lokesh, 255, 275
Sharma, Meenakshi, 291
Sharma, Sheetal, 255, 275
Sharma, Uma, 291
Shrivastav, Vishal, 37, 179
Shubh, 67
Shukla, Pragya, 237
Singh, Ramendra, 291
Singh, Surbhi, 37
Soni, Divya, 267
Soni, Gaurav Kumar, 203, 299
Srivastava, Saurabh, 247

T
Tege, Saurabh, 281
Theerthagiri, Prasannavenkatesan, 53
Tilwani, Deepa, 333

V
Vagrecha, Shubham, 255, 275
Vemula, Prudvi Sai, 53
Vyas, Ruchi, 83

W
Wadhwani, Roshni, 179

Y
Yadav, Dhyan Chandra, 225